# lectures on classical electrodynamics

# lectures on classical electrodynamics

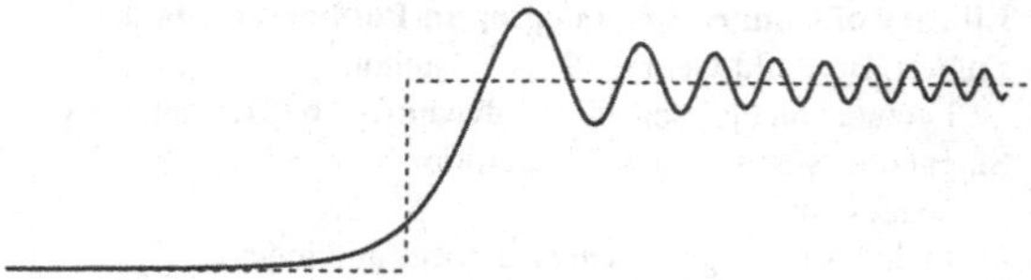

berthold-georg englert
National University of Singapore, Singapore

World Scientific

NEW JERSEY · LONDON · SINGAPORE · BEIJING · SHANGHAI · HONG KONG · TAIPEI · CHENNAI

*Published by*

World Scientific Publishing Co. Pte. Ltd.
5 Toh Tuck Link, Singapore 596224
*USA office:* 27 Warren Street, Suite 401-402, Hackensack, NJ 07601
*UK office:* 57 Shelton Street, Covent Garden, London WC2H 9HE

**Library of Congress Cataloging-in-Publication Data**
Englert, Berthold-Georg, 1953– author.
Lectures on classical electrodynamics / by Berthold-Georg Englert (National University of Singapore, Singapore). -- First edition.
pages cm
Includes bibliographical references and index.
ISBN 978-9814596923 (hardcover : alk. paper) -- ISBN 978-9814596930 (pbk. : alk. paper)
1. Electrodynamics. I. Title.
QC631.E54 2014
537.6--dc23
2014019553

**British Library Cataloguing-in-Publication Data**
A catalogue record for this book is available from the British Library.



Printed in Singapore

To my teachers, colleagues, and students

# Preface

This book on Classical Electrodynamics grew out of a set of lecture notes for a fourth-year undergraduate course that I taught at the National University of Singapore (NUS) in recent years. The presentation is rather detailed and does not skip intermediate steps that — as experience shows — are not so obvious for the learning student.

As a rule, students would have taken two earlier courses on Electricity and Magnetism in their second and third years, so that they arrive with a solid knowledge of the basic material. The fourth-year course, then, deals with advanced topics, many of which could well be part of a post-graduate course on advanced electrodynamics.

When preparing these lectures, I did not follow any of the standard textbooks on the subject, but I frequently consulted the notes I had from the lectures that Julian Schwinger gave at the University of California, Los Angeles (UCLA) in the early 1980s. Any similarity between this book and the published version of Schwinger's lectures* is, therefore, not accidental. But, of course, Schwinger covered much more ground in two quarters at UCLA than I could in one semester at NUS.

The material of this book is my personal selection for that one-semester fourth-year course, presented in full during twenty two-hour lectures. There is a strong emphasis on the properties of electromagnetic radiation as they follow directly from Maxwell's equations. Other lecturers will surely omit some of the material of my choice in favor of topics that I did not choose to include, such as the propagation of light in media or wave guides. Accordingly, there is no ambition of, and no attempt at, treating each and every aspect of electromagnetism in these notes — they just represent what I could and would deal with in one semester.

*J. Schwinger, L.L. DeRaad, Jr., K.A. Milton, and W.-y. Tsai, *Classical Electrodynamics* (Perseus Books, Reading, 1998)

This book owes its existence to the outstanding teachers, colleagues, and students from whom I learned so much. I dedicate these lectures to them.

I am grateful for the professional help by the staff of World Scientific Publishing Co., which was crucial for the completion. I acknowledge the invaluable support of Miss Lai Fun Kwong with sincere gratitude. Dr. Yin Lu deserves a particular thank-you for turning the original handwritten notes into an electronic version that I could then edit, and I am much obliged to Dr. Paul Condylis who took time off to record the diffraction patterns on pages 181–186.

I wish to thank my dear wife Ola for her continuing understanding and patience by which she is giving me the peace of mind that is the source of all achievements.

Singapore, May 2014 *BG Englert*

# Contents

# Glossary

Here is a list of the symbols used in the text; the numbers in square brackets indicate the pages of first occurrence.

### Miscellanea

| | |
|---|---|
| $\boldsymbol{0}$, $\mathbf{0}$, $\mathbf{1}$ | null vector, null dyadic, unit dyadic |
| $\nu!$ | factorial of number $\nu$ |
| $z^*$ | complex conjugate of complex number $z$ |
| $\boldsymbol{a}^{\mathrm{T}}$ | transposed row version of column vector $\boldsymbol{a}$ [24] |
| $\boldsymbol{a}\cdot\boldsymbol{b}$, $\boldsymbol{a}\times\boldsymbol{b}$ | scalar, vector product of vectors $\boldsymbol{a}$ and $\boldsymbol{b}$ |
| $\boldsymbol{a}\,\boldsymbol{b}$ | dyadic product of vectors $\boldsymbol{a}$ and $\boldsymbol{b}$ |
| $\lambda(\boldsymbol{k},\omega)$ | position-and-time Fourier transform of field $\lambda(\boldsymbol{r},t)$ [76] |
| $\partial^\mu$, $\partial_\mu$ | contravariant, covariant components of the 4-gradient [51] |
| $\frac{\partial}{\partial t}\lambda$, $\frac{\mathrm{d}}{\mathrm{d}t}f=\dot{f}$ | time derivative of field $\lambda(\boldsymbol{r},t)$, of function $f(t)$ |
| $\boldsymbol{\nabla}\lambda$ | gradient of field $\lambda(\boldsymbol{r},t)$ |
| $\boldsymbol{\nabla}\cdot\boldsymbol{F}$ | divergence of vector field $\boldsymbol{F}(\boldsymbol{r},t)$ |
| $\boldsymbol{\nabla}\times\boldsymbol{F}$ | curl of vector field $\boldsymbol{F}(\boldsymbol{r},t)$ |
| $\mathrm{tr}\{\ \}$ | trace of a dyadic [10] |

### Latin alphabet

| | |
|---|---|
| $\boldsymbol{a}$ | acceleration [120] |
| A | ampere, basic SI unit of current |
| $\boldsymbol{A}(\boldsymbol{r},t)$ | vector potential [3] |
| $\boldsymbol{A}_\perp$, $\boldsymbol{A}_\parallel$ | its tranverse, longitudinal parts [71] |
| $\boldsymbol{A}_s(\boldsymbol{r},t)$ | vector potential associated with surface current [175] |
| $\mathrm{Ai}(x)$ | Airy function [214] |
| $\boldsymbol{B}(\boldsymbol{r},t)$ | magnetic field [1] |
| $\boldsymbol{B}_\perp$, $\boldsymbol{B}_\parallel$ | its parallel, perpendicular components [45] |

| | |
|---|---|
| $c$ | speed of light, $c = 2.99792 \times 10^{10}\,\mathrm{cm\,s^{-1}}$ [1] |
| C | coulomb, SI unit of charge, $1\,\mathrm{C} = 1\,\mathrm{A\,s}$ |
| cm | centimeter, basic cgs unit of length |
| $\cos, \sin, \ldots$ | trigonometric functions |
| $\cosh, \sinh, \ldots$ | hyperbolic functions |
| $\boldsymbol{d}(t)$ | electric dipole moment [92] |
| $(\mathrm{d}\boldsymbol{r})$ | volume element [2] |
| $\mathrm{d}\boldsymbol{S}$ | vectorial surface element [2] |
| $\boldsymbol{D}(\boldsymbol{r}, t)$ | displacement field [37] |
| dyn | dyne, cgs unit of force, $1\,\mathrm{dyn} = 1\,\mathrm{g\,cm\,s^{-2}}$ |
| e | Euler's number, $\mathrm{e} = 2.71828\ldots$ |
| eV | electron-volt, $1\,\mathrm{eV} = 1.60218 \times 10^{-12}\,\mathrm{erg}$ |
| erg | ergon, cgs unit of energy, $1\,\mathrm{erg} = 1\,\mathrm{g\,cm^2\,s^{-2}}$ |
| $e$ | electric charge [6] |
| $e_0$ | elementary charge, $e_0 = 4.80320 \times 10^{-10}\,\mathrm{Fr}$ [40] |
| $e_j$ | electric charge of the $j$th particle [6] |
| $\boldsymbol{e}_z$; $\boldsymbol{e}_\parallel$, $\boldsymbol{e}_\perp$ | unit vector in $z$ direction [82]; for parallel, perpendicular polarization [149] |
| $E$, $\dfrac{\mathrm{d}E(\omega)}{\mathrm{d}\Omega}$ | energy [15], spectral distribution of radiated energy [111] |
| $E(\omega)$ | spectral density of radiated energy [112] |
| $\dfrac{\mathrm{d}E}{\mathrm{d}T}$, $\dfrac{\mathrm{d}E}{\mathrm{d}s}$ | energy loss per unit time, per unit distance [127] |
| $\boldsymbol{E}(\boldsymbol{r}, t)$ | electric field [1] |
| $\boldsymbol{E}_\perp$, $\boldsymbol{E}_\parallel$ | its parallel, perpendicular components [45] |
| $\boldsymbol{E}_\perp$, $\boldsymbol{E}_\parallel$ | its tranverse, longitudinal parts [72] |
| $\boldsymbol{E}_{\mathrm{diff}}(\boldsymbol{r})$ | diffracted electric field [194] |
| $\boldsymbol{E}_{\mathrm{inc}}(\boldsymbol{r})$ | incident electric field [169] |
| $f_m$ | Fourier coefficient [141] |
| $\boldsymbol{f}(\boldsymbol{r}, t)$ | force density [9] |
| $\boldsymbol{F}$, $\boldsymbol{F}_{\mathrm{rad}}$ | force [6], radiation-reaction force [161] |
| $F(T)$ | Fresnel integral [184] |
| $F^{\mu}{}_{\nu}$ | contravariant-covariant component of the field dyadic [52] |
| Fr | franklin, cgs unit of charge, $1\,\mathrm{Fr} = 1\,\mathrm{g^{1/2}\,cm^{3/2}\,s^{-1}}$ |
| g | gram, basic cgs unit of mass |
| $g$, $g_0$ | magnetic charge [46], unit magnetic charge [48] |
| $g_{\mu\nu}$ | metric tensor [50] |
| G | gauss, cgs magnetic-field unit, $1\,\mathrm{G} = 1\,\mathrm{g^{1/2}\,cm^{-1/2}\,s^{-1}}$ |
| $G$ | generator for infinitesimal variations [62] |

| | |
|---|---|
| $G_{\pm}(\boldsymbol{r}, t)$ | Green's function [78] |
| $G(\boldsymbol{r}, \boldsymbol{r}')$ | Green's function [166] |
| $\boldsymbol{G}(\boldsymbol{r}, t)$ | momentum density [10] |
| $\hbar$ | Planck's constant divided by $2\pi$, $\hbar = 6.62607 \times 10^{-27}$ erg s $= 0.658212$ eV fs |
| $\boldsymbol{H}(\boldsymbol{r}, t)$ | magnetic induction field [37] |
| $H(\boldsymbol{r}, \boldsymbol{p}, t)$ | Hamilton function [63] |
| i | imaginary unit, $\mathrm{i}^2 = -1$ |
| $I$ | current [102] |
| $\boldsymbol{j}(\boldsymbol{r}, t)$ | electric current density [1] |
| $\boldsymbol{j}_s(\boldsymbol{r}, t)$ | electric surface-current density [175] |
| $j^{\mu}$ | contravariant component of the 4-current density [51] |
| J | joule, SI unit of energy, $1\,\mathrm{J} = 1\,\mathrm{kg\,m^2\,s^{-2}}$ |
| $\mathrm{J}_m(z)$ | $m$th Bessel function [138] |
| $\boldsymbol{J}$ | angular momentum [16] |
| $k, \boldsymbol{k}$ | wave number [76], wave vector [57] |
| $k^{\mu}$, $k_{\mu}$ | contra-, covariant components of the wave 4-vector [57] |
| $\boldsymbol{K}(\boldsymbol{r}_{\perp})$, $K(y)$ | surface current density [175], relevant component [189] |
| kg | kilogram, basic SI unit of mass |
| $L$ | length [102] |
| $L; L_{\mathrm{kin}}, L_{\mathrm{int}}$ | Lagrange function [61]; kinetic part, interaction part [67] |
| $L_{\mathrm{emf}}, L_{\mathrm{p}}, L_{\mathrm{f}}$ | for the electromagnetic field [69], particle part, field part [73] |
| $L(\boldsymbol{v})$ | 4-dyadic of a finite Lorentz transformation [41] |
| m | meter, basic SI unit of length |
| $m$; $m_{\mathrm{el}}$ | mass [17]; electron mass, $m_{\mathrm{el}}c^2 = 510.999\,\mathrm{keV}$ [118] |
| $m_{\mathrm{crit}}$ | critical order of emitted harmonics [145] |
| $n$ | index of refraction [125] |
| N | newton, SI unit of force, $1\,\mathrm{N} = 1\,\mathrm{kg\,m\,s^{-2}}$ |
| $\boldsymbol{n}$ | unit vector [47] |
| $\frac{\mathrm{d}N}{\mathrm{d}s}$ | photon emission per unit length [127] |
| $\boldsymbol{p}$, $\boldsymbol{P}$ | momentum [62], total momentum [15] |
| $P(t)$; $\frac{\mathrm{d}P(t)}{\mathrm{d}\Omega}$ | power radiated; into solid angle $\mathrm{d}\Omega$ [90] |
| $P_{\parallel,\perp}(T)$ | total power of parallel, perpendicular polarization [152] |
| $P_m(T)$ | total power radiated into the $m$th harmonic [141] |
| $\frac{\mathrm{d}P_m(T)}{\mathrm{d}\Omega}$ | power radiated into the $m$th harmonic per solid angle [140] |

| | |
|---|---|
| $\left(\dfrac{\mathrm{d}P_m(T)}{\mathrm{d}\Omega}\right)_{\parallel,\perp}$ | its parallel-, perpendicular-polarization contributions [140] |
| $\dfrac{\mathrm{d}P(\omega,T)}{\mathrm{d}\Omega}$ | time-dependent power spectrum [120] |
| $P(\omega,T)$ | spectral power density [133] |
| $P_{\mathrm{tot}}$ | total mechanical power [162] |
| $\mathcal{P}$ | principal value [115] |
| $\mathsf{Q}(t)$ | electric quadrupole moment [94] |
| $\boldsymbol{r};\ \boldsymbol{r}_{\parallel},\ \boldsymbol{r}_{\perp}$ | position vector [1]; parallel, perpendicular components [24] |
| $\boldsymbol{r}_j(t)$ | position of the $j$th particle [6] |
| $\langle \boldsymbol{r}(t)^2\rangle$ | ensemble-averaged squared distance [20] |
| $\langle \boldsymbol{r}\rangle_E(t)$ | energy-averaged position [17] |
| $\langle \boldsymbol{r}\rangle_P(t)$ | momentum-averaged position [17] |
| $r_{\mathrm{cl}}$ | classical electron radius [159] |
| $R$ | circle radius [129] |
| $\boldsymbol{R}(t),\ \boldsymbol{V}(t)$ | position, velocity along a trajectory [80] |
| s | second, basic cgs and SI unit of time |
| $s$ | chordal distance [133] |
| $S$ | surface [2] |
| $\boldsymbol{S}(\boldsymbol{r},t)$ | energy current density, Poynting vector [8] |
| statV | statvolt, cgs potential unit, $1\,\mathrm{statV} = 1\,\mathrm{g}^{1/2}\,\mathrm{cm}^{1/2}\,\mathrm{s}^{-1}$ |
| $t$ | time [1] |
| $t_{\mathrm{e}};\ t_{\mathrm{ret}},\ t_r$ | emission time [90]; retarded time [81], to leading order [87] |
| $t_{\mathrm{D}}$ | duration [122] |
| T | tesla, SI magnetic-field unit, $1\,\mathrm{T} = 1\,\mathrm{kg}\,\mathrm{s}^{-2}\,\mathrm{A}^{-1}$ |
| $T$ | epoch time [120] |
| $\mathsf{T}(\boldsymbol{r},t)$ | momentum current density, stress tensor [10] |
| $u^{\mu},\ u_{\mu}$ | contra-, covariant components of the 4-velocity [58] |
| $U(\boldsymbol{r},t)$ | energy density [8] |
| $v,\ \boldsymbol{v}$ | speed [20], velocity [6] |
| $\boldsymbol{v}_j(t)$ | velocity of the $j$th particle [6] |
| $\boldsymbol{v}_E$ | energy velocity [17] |
| $\boldsymbol{v}_P$ | momentum velocity [17] |
| V | volt, SI potential unit, $1\,\mathrm{V} = 1\,\mathrm{kg}\,\mathrm{m}^2\,\mathrm{s}^{-3}\,\mathrm{A}^{-1}$ |
| $V$ | volume [2] |
| $V(\boldsymbol{r},t)$ | potential energy [61] |
| W | watt, SI unit of power, $1\,\mathrm{W} = 1\,\mathrm{kg}\,\mathrm{m}^2\,\mathrm{s}^{-3}$ |
| $\dot{W}(t)$ | power of the work done by moving charges [98] |
| $W_{12},\ \delta W_{12}$ | action [61], its infinitesimal change [62] |

| | |
|---|---|
| $x$, $y$, $z$ | cartesian coordinates of $\boldsymbol{r}$ [23] |
| $x^\mu$, $x_\mu$ | contravariant, covariant event coordinates [49] |

**Greek alphabet and Greek-Latin combinations**

| | |
|---|---|
| $\alpha$ | fine structure constant, $\alpha = 1/137.036$ [48] |
| $\gamma$; $\gamma_u$ | Lorentz factor [23]; for speed $u$ [57] |
| $\gamma$; $\gamma_{\text{rad}}$, $\gamma_{\text{diss}}$ | friction coefficient [160]; radiation, dissipation contribution [162] |
| $\delta_{\text{coor}}$ | coordinate change of an infinitesimal Lorentz transformation [27] |
| $\delta L$ | 4-dyadic of an infinitesimal Lorentz transformation [34] |
| $\delta\boldsymbol{r}$, $\delta t$ | infinitesimal variations about the actual trajectory [61] |
| $\delta\boldsymbol{v}$ | velocity of an infinitesimal Lorentz transformation [25] |
| $\delta X$ | infinitesimal Lorentz transformation of quantity $X$ [25] |
| $\delta(x)$; $\delta(\boldsymbol{r})$ | one-dimensional [43]; three-dimensional delta function [43] |
| $\delta'(x)$, $\delta''(x)$ | first [134], second derivative of $\delta(x)$ [121] |
| $\epsilon$ | small positive quantity [76], electric permittivity [125] |
| $\epsilon_0$ | permittivity of the vacuum (SI units only) |
| $\eta(x)$ | unit step function [102] |
| $\theta$ | rapidity [25], polar angle [78], scattering angle [158], diffraction angle [169] |
| $\lambda$, $\overline{\lambda}$, $\lambda\!\!\!^{-}$ | wavelength [102], typical value [128], reduced wavelength [163] |
| $\lambda(\boldsymbol{r}, t)$ | scalar field [3] |
| $\mu$ | magnetic permeability [125] |
| $\mu_0$ | permeability of the vacuum (SI units only) |
| $\boldsymbol{\mu}(t)$ | magnetic dipole moment [93] |
| $\pi$ | Archimedes's constant, $\pi = 3.14159\ldots$ |
| $\rho(\boldsymbol{r}, t)$ | electric charge density [1] |
| $\sigma$, $\sigma_{\text{tot}}$, $\dfrac{\mathrm{d}\sigma}{\mathrm{d}\Omega}$ | cross section [159], total c.s. [162], differential c.s. [158] |
| $\sigma_{\text{Th}}$, $\sigma_{\text{Ray}}$ | for Thomson scattering [158], for Rayleigh scattering [161] |
| $\tau$ | relative time [120], characteristic time [209] |
| $\boldsymbol{\tau}$ | torque [11] |
| $\phi$, $\varphi$ | angle [154], rotation angle [40] |
| $\Phi(\boldsymbol{r}, t)$, $\Phi_s(\boldsymbol{r})$ | scalar potential [3], for surface current [175] |
| $\omega$, $\overline{\omega}$; $\omega_0$ | circular frequency [57], typical value [128]; dominant value [160] |
| $\omega_0$, $\boldsymbol{\omega}_0$ | Larmor frequency, its angular velocity vector [129] |
| $\mathrm{d}\Omega$ | solid-angle element [85] |

# Chapter 1

# Maxwell's Equations

## 1.1 Review of familiar basics

Throughout this course on electrodynamics, we shall use cgs units, in which the electric field $\boldsymbol{E}$ and the magnetic field $\boldsymbol{B}$, the two aspects of the one electromagnetic field, have the same unit, namely $\mathrm{g}^{1/2}\,\mathrm{cm}^{-1/2}\,\mathrm{s}^{-1} = (\mathrm{erg}/\mathrm{cm}^3)^{1/2}$. As discussed in the appendix on electromagnetic units, the conversion to SI units is straightforward.

The fundamental *Maxwell's equations** then have the form

$$\begin{aligned} \boldsymbol{\nabla}\cdot\boldsymbol{E} &= 4\pi\rho\,, & \boldsymbol{\nabla}\cdot\boldsymbol{B} &= 0\,, \\ \boldsymbol{\nabla}\times\boldsymbol{B} - \frac{1}{c}\frac{\partial}{\partial t}\boldsymbol{E} &= \frac{4\pi}{c}\boldsymbol{j}\,, & \boldsymbol{\nabla}\times\boldsymbol{E} + \frac{1}{c}\frac{\partial}{\partial t}\boldsymbol{B} &= 0\,, \end{aligned} \tag{1.1.1}$$

where $\rho$ is the electric charge density, and $\boldsymbol{j}$ is the electric current density. All quantities appearing in Maxwell's equations: the fields $\boldsymbol{E}$ and $\boldsymbol{B}$ as well as the densities $\rho$ and $\boldsymbol{j}$, are functions of the position vector $\boldsymbol{r}$ and time $t$. When necessary or helpful, we shall make that explicit by writing $\boldsymbol{E}(\boldsymbol{r},t)$, for example, but most of the time the $\boldsymbol{r}$, $t$ dependence is left implicit. The letter $c$, of course, denotes the speed of light, $c = 2.99792 \times 10^{10}\,\mathrm{cm\,s}^{-1}$.

The $2\times2$ array of Maxwell's equations corresponds to two ways of grouping them in pairs. The top pair are two constraints (no time derivatives); the bottom pair are two equations of motion (yes time derivatives); the left pair are two inhomogeneous equations; the right pair are two homogeneous equations. So, for instance, the bottom left equation could be referred to as the inhomogeneous equation of motion.

There are also historical names. The top left equation is *Gauss's Law*;† the bottom left equation is *Ampère's Circuital Law*;‡ the bottom right

*James Clerk MAXWELL (1831–1879) †Karl Friedrich GAUSS (1777–1855)
‡André-Marie AMPÈRE (1775–1836)

equation is *Faraday's Induction Law*;* and we call the top right equation *Gilbert's Law*.† It states an important empirical fact: There are no magnetic charges — and, therefore, also no magnetic currents, which would appear on the right-hand side of Faraday's law.

## 1.2 Continuity equation; conservation of charge

The electric charge density $\rho$ and the electric current density $\boldsymbol{j}$ must always obey the *continuity equation*

$$\frac{\partial}{\partial t}\rho + \boldsymbol{\nabla}\cdot\boldsymbol{j} = 0\,, \tag{1.2.1}$$

which states the local conservation of electric charge. We recall the standard argument: The time derivative of the total charge enclosed by a volume $V$,

$$\frac{\mathrm{d}}{\mathrm{d}t}\int\limits_V (\mathrm{d}\boldsymbol{r})\,\rho = \int\limits_V (\mathrm{d}\boldsymbol{r})\,\frac{\partial}{\partial t}\rho\,, \tag{1.2.2}$$

can be expressed as a surface integral over the boundary $S$ of volume $V$,

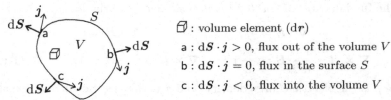

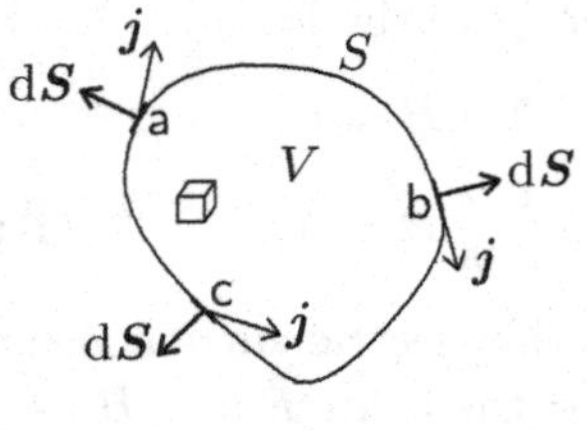


: volume element $(\mathrm{d}\boldsymbol{r})$

a : $\mathrm{d}\boldsymbol{S}\cdot\boldsymbol{j} > 0$, flux out of the volume $V$

b : $\mathrm{d}\boldsymbol{S}\cdot\boldsymbol{j} = 0$, flux in the surface $S$

c : $\mathrm{d}\boldsymbol{S}\cdot\boldsymbol{j} < 0$, flux into the volume $V$

$$\int\limits_V (\mathrm{d}\boldsymbol{r})\,\frac{\partial}{\partial t}\rho = -\int\limits_V (\mathrm{d}\boldsymbol{r})\,\boldsymbol{\nabla}\cdot\boldsymbol{j} = -\int\limits_S \mathrm{d}\boldsymbol{S}\cdot\boldsymbol{j}\,, \tag{1.2.3}$$

where the conversion of the volume integral to the surface integral is a familiar application of Gauss's Theorem. Therefore, any change of the charge inside $V$ requires a flux of current through the surface $S$,

$$\frac{\mathrm{d}}{\mathrm{d}t}\int\limits_V (\mathrm{d}\boldsymbol{r})\,\rho + \int\limits_S \mathrm{d}\boldsymbol{S}\cdot\boldsymbol{j} = 0\,. \tag{1.2.4}$$

Charges cannot just disappear; they must move away.

---

*Michael FARADAY (1791–1867) †William GILBERT (1544–1603)

We note that there is absolutely no possibility for $\rho$ and $\boldsymbol{j}$ to disobey the continuity equation, because then we would get a contradiction between Gauss's law and Ampère's law. The logic, however, is not that the continuity equation is implied by the left pair of Maxwell's equations; rather it is a pre-existing constraint.

## 1.3 Potentials, gauge invariance; radiation gauge, Lorentz gauge

Conservation laws in physics are usually accompanied by symmetries, and the conservation of electric charge is no exception. The relevant symmetry is the *gauge invariance.* We recall that we can parameterize the electric and magnetic fields in terms of a scalar potential $\Phi$ and a vector potential $\boldsymbol{A}$,

$$\boldsymbol{E} = -\frac{1}{c}\frac{\partial}{\partial t}\boldsymbol{A} - \boldsymbol{\nabla}\Phi\,, \qquad \boldsymbol{B} = \boldsymbol{\nabla}\times\boldsymbol{A}\,, \tag{1.3.1}$$

which is such that the right pair of homogeneous Maxwell's equations is automatically obeyed. Accordingly, the inhomogeneous Maxwell's equations will serve for the determination of $\Phi$ and $\boldsymbol{A}$.

Upon inserting this parameterization into the two left equations, we get

$$\left(\frac{1}{c^2}\frac{\partial^2}{\partial t^2} - \boldsymbol{\nabla}^2\right)\Phi - \frac{1}{c}\frac{\partial}{\partial t}\left(\boldsymbol{\nabla}\cdot\boldsymbol{A} + \frac{1}{c}\frac{\partial}{\partial t}\Phi\right) = 4\pi\rho \tag{1.3.2}$$

and

$$\left(\frac{1}{c^2}\frac{\partial^2}{\partial t^2} - \boldsymbol{\nabla}^2\right)\boldsymbol{A} + \boldsymbol{\nabla}\left(\boldsymbol{\nabla}\cdot\boldsymbol{A} + \frac{1}{c}\frac{\partial}{\partial t}\Phi\right) = \frac{4\pi}{c}\boldsymbol{j}\,. \tag{1.3.3}$$

This pair of coupled second-order equations looks worse than the original quartet of first-order equations, but this first impression is misleading. It is time to remember the freedom of gauge: The choice of $\Phi$ and $\boldsymbol{A}$ is not unique at all, we have the option of modifying the potentials in accordance with

$$\boldsymbol{A} \to \boldsymbol{A} + \boldsymbol{\nabla}\lambda\,, \qquad \Phi \to \Phi - \frac{1}{c}\frac{\partial}{\partial t}\lambda\,, \tag{1.3.4}$$

with any (not unreasonable) scalar field $\lambda$. This is called a *gauge transformation*, which is actually a historical misnomer with the blame on Weyl.*

*Claus Hugo Hermann WEYL (1885–1955)

Whereas a gauge transformation alters $\Phi$ and $\boldsymbol{A}$, there is no effect on $\boldsymbol{E}$ and $\boldsymbol{B}$, as we verify by considering a pure gauge field:

$$\boldsymbol{A} = \boldsymbol{\nabla}\lambda \quad \text{and} \quad \Phi = -\frac{1}{c}\frac{\partial}{\partial t}\lambda \tag{1.3.5}$$

imply

$$\boldsymbol{B} = \boldsymbol{\nabla} \times \boldsymbol{\nabla}\lambda = 0 \quad \text{and} \quad \boldsymbol{E} = -\frac{1}{c}\frac{\partial}{\partial t}\boldsymbol{\nabla}\lambda - \boldsymbol{\nabla}\left(-\frac{1}{c}\frac{\partial}{\partial t}\lambda\right) = 0\,. \tag{1.3.6}$$

As an immediate consequence we observe that, for given $\boldsymbol{E}$ and $\boldsymbol{B}$, there is a plethora of potentials $\Phi$ and $\boldsymbol{A}$.

We exploit this freedom of gauge by imposing additional conditions on $\Phi$ and $\boldsymbol{A}$. Two conventions are particularly useful and popular: the Lorentz gauge and the radiation gauge. The *Lorentz* gauge* is specified by requiring that

$$\boldsymbol{\nabla} \cdot \boldsymbol{A} + \frac{1}{c}\frac{\partial}{\partial t}\Phi = 0\,, \tag{1.3.7}$$

and the *radiation gauge* by

$$\boldsymbol{\nabla} \cdot \boldsymbol{A} = 0\,. \tag{1.3.8}$$

Let us assume that we have parameterized $\boldsymbol{E}$ and $\boldsymbol{B}$ by some potentials $\Phi_0$ and $\boldsymbol{A}_0$, and now we wish to find $\lambda$ such that

$$\boldsymbol{A} = \boldsymbol{A}_0 + \boldsymbol{\nabla}\lambda \quad \text{and} \quad \Phi = \Phi_0 - \frac{1}{c}\frac{\partial}{\partial t}\lambda \tag{1.3.9}$$

obey the selected gauge condition. In case of the Lorentz gauge, we need $\lambda$ such that

$$\left(\frac{1}{c^2}\frac{\partial^2}{\partial t^2} - \boldsymbol{\nabla}^2\right)\lambda = \boldsymbol{\nabla} \cdot \boldsymbol{A}_0 + \frac{1}{c}\frac{\partial}{\partial t}\Phi_0\,, \tag{1.3.10}$$

and in the case of the radiation gauge, $\lambda$ is a solution of

$$-\boldsymbol{\nabla}^2\lambda = \boldsymbol{\nabla} \cdot \boldsymbol{A}_0\,. \tag{1.3.11}$$

The latter is the Poisson† equation of electrostatics, with $\boldsymbol{\nabla} \cdot \boldsymbol{A}_0$ playing the role of $4\pi\rho$, and the former is the three-dimensional wave equation, in which the inhomogeneity term $\boldsymbol{\nabla} \cdot \boldsymbol{A}_0 + \frac{1}{c}\frac{\partial}{\partial t}\Phi_0$ does not vanish unless $\boldsymbol{A}_0$ and $\Phi_0$ already obey the Lorentz gauge condition. Since these equations

*Hendrik Antoon LORENTZ (1853–1929) †Siméon Denise POISSON (1781–1840)

have known solutions whatever functions of $\boldsymbol{r}$ and $t$ we have on the right-hand sides, we can surely enforce the Lorentz gauge (1.3.7) or the radiation gauge (1.3.8). Explicit solutions of the inhomogeneous wave equations that we meet in (1.3.10) as well as in (1.3.12) and (1.3.14) below are derived in Chapter 6; see (6.2.1) in particular.

It should be clear, however, that the gauge conditions alone do not enforce a unique choice of the potentials. In the Lorentz gauge, for example, we can still change $\boldsymbol{A}$ and $\Phi$ with the aid of a $\lambda$ that solves the homogeneous wave equation. There are many such $\lambda$s.

Now, in the Lorentz gauge (1.3.7), the equation pair (1.3.2) and (1.3.3) reads

$$\left(\frac{1}{c^2}\frac{\partial^2}{\partial t^2} - \boldsymbol{\nabla}^2\right)\begin{pmatrix}\Phi \\ \boldsymbol{A}\end{pmatrix} = \frac{4\pi}{c}\begin{pmatrix}c\rho \\ \boldsymbol{j}\end{pmatrix} \qquad (1.3.12)$$

that is: we have a decoupled set of inhomogeneous wave equations, from which $\Phi$ and $\boldsymbol{A}$ can be determined by integration. In the radiation gauge, by contrast, we get the instantaneous Poisson equation

$$-\boldsymbol{\nabla}^2\Phi = 4\pi\rho \qquad (1.3.13)$$

for the determination of the scalar potential $\Phi$, and another inhomogeneous wave equation,

$$\left(\frac{1}{c^2}\frac{\partial^2}{\partial t^2} - \boldsymbol{\nabla}^2\right)\boldsymbol{A} = \frac{4\pi}{c}\boldsymbol{j} - \frac{1}{c}\frac{\partial}{\partial t}\boldsymbol{\nabla}\Phi \qquad (1.3.14)$$

for the vector potential $\boldsymbol{A}$. Here, too, we can rely on familiar machinery for finding the solution.

A note on terminology: The term "Lorentz gauge" emphasizes that the gauge condition (1.3.7) is invariant under Lorentz transformations, as will be discussed in Chapter 3. With reference to L. V. Lorenz,* very likely the first to use this gauge, it is also known as the "Lorenz gauge." The radiation gauge is advantageous when studying electromagnetic radiation; in view of the instantaneous Coulomb† potential that solves (1.3.13) it is also known as the "Coulomb gauge" although Coulomb himself had nothing to do with it. History would give more justification to calling it the "Maxwell gauge" because it was Maxwell's preferred choice of gauge, but this terminology is not in use at all. The radiation-gauge condition (1.3.8) states that the vector potential is a transverse field (see Sections 5.4 and 7.4) and, therefore,

---

*Ludvik Valentin LORENZ (1829–1891) †Charles-Augustin de COULOMB (1736–1806)

the term "transverse gauge" is self-suggesting and, in fact, used by some authors.

## 1.4 Force, work, energy conservation

The left pair of inhomogeneous Maxwell's equations tells us how the charge and current distribution act as sources of the electromagnetic field. When there is action, there must be reaction, so that we have proper *inter*action between charges and fields. The *re*-action of the field on the charges is given by the mechanical effect of the Lorentz force. For a single charge $e$, at position $\boldsymbol{r}$ at time $t$, it is

$$\boldsymbol{F} = e\left(\boldsymbol{E}(\boldsymbol{r},t) + \frac{1}{c}\boldsymbol{v} \times \boldsymbol{B}(\boldsymbol{r},t)\right), \tag{1.4.1}$$

where $\boldsymbol{v}$ is the velocity of the charged particle. For a collection of charges, with strengths $e_j$, position vectors $\boldsymbol{r}_j(t)$, and velocity vectors $\boldsymbol{v}_j(t)$, the total force is

$$\boldsymbol{F} = \sum_j e_j \boldsymbol{E}\big(\boldsymbol{r}_j(t),t\big) + \frac{1}{c}\sum_j e_j \boldsymbol{v}_j(t) \times \boldsymbol{B}\big(\boldsymbol{r}_j(t),t\big) \tag{1.4.2}$$

or

$$\boldsymbol{F} = \int (\mathrm{d}\boldsymbol{r}) \left[\rho(\boldsymbol{r},t)\boldsymbol{E}(\boldsymbol{r},t) + \frac{1}{c}\boldsymbol{j}(\boldsymbol{r},t) \times \boldsymbol{B}(\boldsymbol{r},t)\right], \tag{1.4.3}$$

where

$$\rho(\boldsymbol{r},t) = \sum_j e_j \delta\big(\boldsymbol{r} - \boldsymbol{r}_j(t)\big) \tag{1.4.4}$$

and

$$\boldsymbol{j}(\boldsymbol{r},t) = \sum_j e_j \boldsymbol{v}_j(t) \delta\big(\boldsymbol{r} - \boldsymbol{r}_j(t)\big) \tag{1.4.5}$$

are the charge density and current density, respectively. Note that the continuity equation (1.2.1) is automatically obeyed by this $\rho$ and $\boldsymbol{j}$ because

$$\boldsymbol{v}_j(t) = \frac{\mathrm{d}}{\mathrm{d}t}\boldsymbol{r}_j(t) \tag{1.4.6}$$

for the trajectory of the $j$th charge.

The work done by the Lorentz force on the particle changes the mechanical energy of the particles, with the power (work per unit time) given by Joule's* expression

$$\int (\mathrm{d}\boldsymbol{r})\, \boldsymbol{j} \cdot \boldsymbol{E} = \int (\mathrm{d}\boldsymbol{r}) \sum_j e_j \boldsymbol{v}_j(t) \delta\big(\boldsymbol{r} - \boldsymbol{r}_j(t)\big) \cdot \boldsymbol{E}(\boldsymbol{r}, t)$$
$$= \sum_j e_j \boldsymbol{v}_j(t) \cdot \boldsymbol{E}\big(\boldsymbol{r}_j(t), t\big)\,, \tag{1.4.7}$$

where the contribution from the $j$th charged particle has the familiar form of "velocity times force,"

$$e_j \boldsymbol{v}_j \cdot \boldsymbol{E}(\boldsymbol{r}_j, t) = \boldsymbol{v}_j \cdot \left[ e_j \boldsymbol{E}(\boldsymbol{r}_j, t) + e_j \frac{1}{c} \boldsymbol{v}_j \times \boldsymbol{B}(\boldsymbol{r}_j, t) \right], \tag{1.4.8}$$

as it should be. We thus recognize that $\boldsymbol{j} \cdot \boldsymbol{E}$ is the power density, meaning that $(\mathrm{d}\boldsymbol{r})\, \boldsymbol{j} \cdot \boldsymbol{E}$ is the power for the transfer of mechanical energy to the charged particles inside the volume element $(\mathrm{d}\boldsymbol{r})$. We expect that this transfer of energy from the fields to the particles is accompanied by a matching change of the energy stored in the electromagnetic field.

To establish the statement about energy balance, we need to express the Joule's power density $\boldsymbol{j} \cdot \boldsymbol{E}$ solely in field terms, without reference to the particles. The first step replaces $\boldsymbol{j}$ by the curl of $\boldsymbol{B}$ and the time derivative of $\boldsymbol{E}$ with the aid of Ampère's law,

$$\boldsymbol{j} \cdot \boldsymbol{E} = \frac{c}{4\pi} \boldsymbol{E} \cdot \left( \boldsymbol{\nabla} \times B - \frac{1}{c} \frac{\partial}{\partial t} \boldsymbol{E} \right). \tag{1.4.9}$$

Now, keeping in mind that we are on the search for a statement of energy conservation, which we expect to be analogous to the continuity equation that states the conservation of electric charge, we would like to write the right-hand side as a combination of a time derivative (of an energy density) and a divergence (of an energy current density). Thus, we first note that

$$\boldsymbol{E} \cdot (\boldsymbol{\nabla} \times \boldsymbol{B}) = -\boldsymbol{\nabla} \cdot \big( \boldsymbol{E} \times \boldsymbol{B} \big) + \boldsymbol{B} \cdot \big( \boldsymbol{\nabla} \times \boldsymbol{E} \big)$$
$$= -\boldsymbol{\nabla} \cdot (\boldsymbol{E} \times \boldsymbol{B}) - \boldsymbol{B} \cdot \frac{1}{c} \frac{\partial}{\partial t} \boldsymbol{B}\,, \tag{1.4.10}$$

where the first step is a vector identity and the second step uses Faraday's law, and then combine the two time-derivative terms,

*James Prescott JOULE (1818–1889)

$$\boldsymbol{E}\cdot\left(-\frac{1}{c}\frac{\partial}{\partial t}\boldsymbol{E}\right)-\boldsymbol{B}\cdot\left(\frac{1}{c}\frac{\partial}{\partial t}\boldsymbol{B}\right)=-\frac{1}{c}\frac{\partial}{\partial t}\frac{1}{2}\left(\boldsymbol{E}^2+\boldsymbol{B}^2\right), \tag{1.4.11}$$

and we arrive at

$$\boldsymbol{j}\cdot\boldsymbol{E}=-\frac{\partial}{\partial t}\left[\frac{1}{8\pi}\left(\boldsymbol{E}^2+\boldsymbol{B}^2\right)\right]-\boldsymbol{\nabla}\cdot\left(\frac{c}{4\pi}\boldsymbol{E}\times\boldsymbol{B}\right), \tag{1.4.12}$$

or

$$\frac{\partial}{\partial t}U+\boldsymbol{\nabla}\cdot\boldsymbol{S}+\boldsymbol{j}\cdot\boldsymbol{E}=0, \tag{1.4.13}$$

the inhomogeneous continuity equation for energy, with the *energy density*

$$U=\frac{1}{8\pi}\left(\boldsymbol{E}^2+\boldsymbol{B}^2\right) \tag{1.4.14}$$

and the *energy current density*

$$\boldsymbol{S}=\frac{c}{4\pi}\boldsymbol{E}\times\boldsymbol{B}, \tag{1.4.15}$$

also known as the *Poynting** *vector*. The interpretation as energy density and energy current density is justified by the analog of the argument in Section 1.2. Upon integrating over volume $V$ with surface S, we have

$$\frac{\mathrm{d}}{\mathrm{d}t}\underbrace{\int_V(\mathrm{d}\boldsymbol{r})\,U}_{\substack{\text{electro-}\\\text{magnetic}\\\text{energy}\\\text{enclosed in}\\\text{volume } V}}+\underbrace{\int_S\mathrm{d}\boldsymbol{S}\cdot\boldsymbol{S}}_{\substack{\text{electro-}\\\text{magnetic}\\\text{energy flux}\\\text{through}\\\text{surface } S}}+\underbrace{\int_V(\mathrm{d}\boldsymbol{r})\,\boldsymbol{j}\cdot\boldsymbol{E}}_{\substack{\text{mechanical}\\\text{power trans-}\\\text{ferred to the}\\\text{charged par-}\\\text{ticles in } V}}=0, \tag{1.4.16}$$

which confirms that the continuity equation (1.4.13) states the local conservation of energy. Here it should perhaps be regretted that, by convention, we use $\mathrm{d}\boldsymbol{S}$ to denote the vectorial surface element and $\boldsymbol{S}$ for the energy current density. Yes, there is a finite number of letters in the alphabet.

## 1.5 Conservation of momentum

There should also be a statement about momentum conservation. We proceed from the expression for the Lorentz force density of (1.4.3),

---

*John Henry POYNTING (1852–1914)

$$\boldsymbol{f} = \rho\boldsymbol{E} + \frac{1}{c}\boldsymbol{j} \times \boldsymbol{B}, \tag{1.5.1}$$

which — when integrated over volume $V$ — gives the total force, that is: the rate of momentum change, on the particles enclosed by $V$. We use Gauss's law and Ampère's law to replace $\rho$ and $\boldsymbol{j}$ in terms of the fields $\boldsymbol{E}$ and $\boldsymbol{B}$,

$$\boldsymbol{f} = \frac{1}{4\pi}\boldsymbol{E}\,\boldsymbol{\nabla}\cdot\boldsymbol{E} - \frac{1}{4\pi}\boldsymbol{B} \times \left(\boldsymbol{\nabla}\times\boldsymbol{B} - \frac{1}{c}\frac{\partial}{\partial t}\boldsymbol{E}\right), \tag{1.5.2}$$

and, again, the objective is to recognize the right-hand side as a sum of a time derivative (of the momentum density) and a divergence (of the momentum flux density). First, the time derivative:

$$\begin{aligned}\boldsymbol{B} \times \left(-\frac{1}{c}\frac{\partial}{\partial t}\boldsymbol{E}\right) &= \frac{1}{c}\frac{\partial}{\partial t}(\boldsymbol{E}\times\boldsymbol{B}) - \boldsymbol{E} \times \frac{1}{c}\frac{\partial}{\partial t}\boldsymbol{B}\\ &= \frac{1}{c}\frac{\partial}{\partial t}(\boldsymbol{E}\times\boldsymbol{B}) + \boldsymbol{E}\times(\boldsymbol{\nabla}\times\boldsymbol{E}),\end{aligned} \tag{1.5.3}$$

which brings us to

$$\begin{aligned}\boldsymbol{f} = &-\frac{\partial}{\partial t}\left(\frac{1}{4\pi c}\boldsymbol{E}\times\boldsymbol{B}\right)\\ &+\frac{1}{4\pi}\left[\boldsymbol{E}\,\boldsymbol{\nabla}\cdot\boldsymbol{E} - \boldsymbol{E}\times(\boldsymbol{\nabla}\times\boldsymbol{E}) + \boldsymbol{B}\,\boldsymbol{\nabla}\cdot\boldsymbol{B} - \boldsymbol{B}\times(\boldsymbol{\nabla}\times\boldsymbol{B})\right]\end{aligned} \tag{1.5.4}$$

after adding, for restoring the symmetry between $\boldsymbol{E}$ and $\boldsymbol{B}$, the vanishing term $\boldsymbol{B}\,\boldsymbol{\nabla}\cdot\boldsymbol{B} = 0$. Second, the divergence:

$$\begin{aligned}\boldsymbol{E}\times(\boldsymbol{\nabla}\times\boldsymbol{E}) - \boldsymbol{E}\,\boldsymbol{\nabla}\cdot\boldsymbol{E} &= \boldsymbol{\nabla}\left(\frac{1}{2}\boldsymbol{E}^2\right) - \boldsymbol{E}\cdot\boldsymbol{\nabla}\boldsymbol{E} - \boldsymbol{E}\,\boldsymbol{\nabla}\cdot\boldsymbol{E}\\ &= \boldsymbol{\nabla}\cdot\left(\boldsymbol{1}\frac{1}{2}\boldsymbol{E}^2 - \boldsymbol{E}\,\boldsymbol{E}\right),\end{aligned} \tag{1.5.5}$$

where $\boldsymbol{1}$ is the unit dyadic and the factor of $\frac{1}{2}$ is necessary because on the left-hand side the gradient differential operator differentiates only one factor of $\boldsymbol{E}^2 = \boldsymbol{E}\cdot\boldsymbol{E}$, whereas it differentiates both factors on the right-hand side. Together with the analogous magnetic contribution we thus have

$$\boldsymbol{f} = -\frac{\partial}{\partial t}\left(\frac{1}{4\pi c}\boldsymbol{E}\times\boldsymbol{B}\right) - \boldsymbol{\nabla}\cdot\left(\boldsymbol{1}\frac{1}{8\pi}\left(\boldsymbol{E}^2 + \boldsymbol{B}^2\right) - \frac{1}{4\pi}\left(\boldsymbol{E}\,\boldsymbol{E} + \boldsymbol{B}\,\boldsymbol{B}\right)\right) \tag{1.5.6}$$

or

$$\frac{\partial}{\partial t}\boldsymbol{G} + \boldsymbol{\nabla}\cdot\mathsf{T} + \boldsymbol{f} = 0\,, \tag{1.5.7}$$

the inhomogeneous continuity equation for momentum, with the *momentum density* vector

$$\boldsymbol{G} = \frac{1}{4\pi c}\boldsymbol{E}\times\boldsymbol{B} \tag{1.5.8}$$

and the *momentum current density* dyadic

$$\mathsf{T} = \mathbf{1}U - \frac{1}{4\pi}\left(\boldsymbol{E}\,\boldsymbol{E} + \boldsymbol{B}\,\boldsymbol{B}\right), \tag{1.5.9}$$

also known as *Maxwell's stress tensor*, which involves the energy density $U$ of (1.4.14). It is also equal to the trace of the dyadic $\mathsf{T}$, see

$$\begin{aligned}\mathrm{tr}\{\mathsf{T}\} &= \mathrm{tr}\{\mathbf{1}\}\,U - \frac{1}{4\pi}\mathrm{tr}\{\boldsymbol{E}\,\boldsymbol{E} + \boldsymbol{B}\,\boldsymbol{B}\}\\ &= 3U - \underbrace{\frac{1}{4\pi}\left(\boldsymbol{E}^2 + \boldsymbol{B}^2\right)}_{=\,2U} = U\,.\end{aligned} \tag{1.5.10}$$

The justification of identifying $\boldsymbol{G}$ with the momentum density of the electromagnetic field and $\mathsf{T}$ with the momentum current density is fully analogous to the argument for $U$ and $\boldsymbol{S}$ in Section 1.4. It follows further that the continuity equation (1.5.7) states the local conservation of momentum.

We note the important intimate relation between the momentum density and the energy current density,

$$\boldsymbol{G} = \frac{1}{c^2}\boldsymbol{S}\,, \tag{1.5.11}$$

which we paraphrase as

$$\begin{aligned}&(\text{momentum density} = \text{mass density times velocity})\\ &= \frac{1}{c^2}(\text{energy current density} = \text{energy density times velocity})\,,\end{aligned} \tag{1.5.12}$$

very suggestive of mass = energy/$c^2$ and clearly reminiscent of Einstein's* famous energy-mass relation. More about this in Chapter 2.

*Albert EINSTEIN (1879–1955)

## 1.6 Conservation of angular momentum

We get the statement about conservation of angular momentum by considering the torque

$$\boldsymbol{\tau} = \int_V (\mathrm{d}\boldsymbol{r})\, \boldsymbol{r} \times \boldsymbol{f}\,, \tag{1.6.1}$$

the rate of change of the total angular momentum of the particles inside volume $V$. The torque density $\boldsymbol{r} \times \boldsymbol{f}$ is the last term in the equation that we get from the inhomogeneous continuity equation for momentum (1.5.7) upon taking the cross product with $\boldsymbol{r}$,

$$\boldsymbol{r} \times \frac{\partial}{\partial t}\boldsymbol{G} + \boldsymbol{r} \times (\boldsymbol{\nabla} \cdot \mathsf{T}) + \boldsymbol{r} \times \boldsymbol{f} = 0\,. \tag{1.6.2}$$

The first term is immediately recognized as the time derivative of the *angular momentum density* $\boldsymbol{r} \times \boldsymbol{G}$,

$$\boldsymbol{r} \times \frac{\partial}{\partial t}\boldsymbol{G} = \frac{\partial}{\partial t}(\boldsymbol{r} \times \boldsymbol{G})\,. \tag{1.6.3}$$

It is a bit more complicated to write the second term as a divergence. We begin by noting that in $\boldsymbol{r} \times \big(\boldsymbol{\nabla} \cdot \mathsf{T}\big)$ the dot product refers to the left vector in the dyadic $\mathsf{T}$, whereas the cross product refers to the right vector. For the $\boldsymbol{E}\,\boldsymbol{E}$ contribution to $\mathsf{T}$ this is more explicitly stated by

$$\begin{aligned} \boldsymbol{r} \times [\boldsymbol{\nabla} \cdot (\boldsymbol{E}\,\boldsymbol{E})] &= \boldsymbol{\nabla} \cdot (\boldsymbol{E}\,\boldsymbol{r} \times \boldsymbol{E}) - \boldsymbol{E} \cdot \underbrace{(\boldsymbol{\nabla}\,\boldsymbol{r})}_{=\mathbf{1}} \times \boldsymbol{E} \\ &= \boldsymbol{\nabla} \cdot (-\boldsymbol{E}\,\boldsymbol{E} \times \boldsymbol{r}) - \underbrace{\boldsymbol{E} \times \boldsymbol{E}}_{=0}\,, \end{aligned} \tag{1.6.4}$$

so that

$$\boldsymbol{r} \times [\boldsymbol{\nabla} \cdot (\boldsymbol{E}\,\boldsymbol{E} + \boldsymbol{B}\,\boldsymbol{B})] = \boldsymbol{\nabla} \cdot [-(\boldsymbol{E}\,\boldsymbol{E} + \boldsymbol{B}\,\boldsymbol{B}) \times \boldsymbol{r}]\,. \tag{1.6.5}$$

And for the $\mathbf{1}U$ term in $\mathsf{T}$, we have

$$\begin{aligned} \boldsymbol{r} \times \boldsymbol{\nabla} \cdot (\mathbf{1}U) = \boldsymbol{r} \times \boldsymbol{\nabla} U &= -\boldsymbol{\nabla} \times (\boldsymbol{r}\,U) + U\,\underbrace{\boldsymbol{\nabla} \times \boldsymbol{r}}_{=0} \\ &= -\boldsymbol{\nabla} \cdot (\mathbf{1} \times \boldsymbol{r}\,U) \end{aligned} \tag{1.6.6}$$

or

$$\boldsymbol{r} \times \boldsymbol{\nabla} \cdot (\mathbf{1}U) = \boldsymbol{\nabla} \cdot (-\mathbf{1}U \times \boldsymbol{r})\,. \tag{1.6.7}$$

In summary, then,

$$\boldsymbol{r} \times (\boldsymbol{\nabla} \cdot \mathsf{T}) = \boldsymbol{\nabla} \cdot (-\mathsf{T} \times \boldsymbol{r}), \tag{1.6.8}$$

and

$$\frac{\partial}{\partial t}(\boldsymbol{r} \times \boldsymbol{G}) + \boldsymbol{\nabla} \cdot (-\mathsf{T} \times \boldsymbol{r}) + \boldsymbol{r} \times \boldsymbol{f} = 0 \tag{1.6.9}$$

is implied. This is the inhomogeneous continuity equation for angular momentum, where, in full analogy to the corresponding equations for energy and momentum, we recognize that

| | | |
|---|---|---|
| $\boldsymbol{r} \times \boldsymbol{G}$ | is the angular momentum density; | |
| $-\mathsf{T} \times \boldsymbol{r}$ | is the angular momentum current density; | |
| and $\boldsymbol{r} \times \boldsymbol{f}$ | is the torque density that gives rise to the transfer of angular momentum between the particles and the electromagnetic field. | (1.6.10) |

The analogy also tells us that the continuity equation (1.6.9) states the local conservation of angular momentum.

## 1.7 Virial theorem

Not another conservation law, but rather a virial theorem is obtained if we take the scalar product of $\boldsymbol{r}$ and the momentum conservation equation (1.5.7),

$$\boldsymbol{r} \cdot \frac{\partial}{\partial t}\boldsymbol{G} + \boldsymbol{r} \cdot (\boldsymbol{\nabla} \cdot \mathsf{T}) + \boldsymbol{r} \cdot \boldsymbol{f} = 0. \tag{1.7.1}$$

Here, recalling that the basic ingredient of a dyadic is a dyadic product of two vectors, we consider

$$\begin{aligned} \boldsymbol{r} \cdot [\boldsymbol{\nabla} \cdot (\boldsymbol{Y}\,\boldsymbol{Z})] &= \boldsymbol{\nabla} \cdot (\boldsymbol{Y}\,\boldsymbol{r} \cdot \boldsymbol{Z}) - \boldsymbol{Y} \cdot \underbrace{(\boldsymbol{\nabla}\,\boldsymbol{r})}_{=\mathbf{1}} \cdot \boldsymbol{Z} \\ &= \boldsymbol{\nabla} \cdot (\boldsymbol{Y}\,\boldsymbol{Z} \cdot \boldsymbol{r}) - \boldsymbol{Y} \cdot \boldsymbol{Z} \\ &= \boldsymbol{\nabla} \cdot (\boldsymbol{Y}\,\boldsymbol{Z} \cdot \boldsymbol{r}) - \mathrm{tr}\{\boldsymbol{Y}\,\boldsymbol{Z}\}, \end{aligned} \tag{1.7.2}$$

and since this is true for any two vector fields $\boldsymbol{Y}$ and $\boldsymbol{Z}$, we also have

$$\begin{aligned} \boldsymbol{r} \cdot (\boldsymbol{\nabla} \cdot \mathsf{T}) &= \boldsymbol{\nabla} \cdot (\mathsf{T} \cdot \boldsymbol{r}) - \mathrm{tr}\{\mathsf{T}\} \\ &= \boldsymbol{\nabla} \cdot (\mathsf{T} \cdot \boldsymbol{r}) - U, \end{aligned} \tag{1.7.3}$$

in view of what we found for the trace of the momentum current density in (1.5.10).

We put the ingredients together and have

$$\frac{\partial}{\partial t}(\boldsymbol{r}\cdot\boldsymbol{G})+\boldsymbol{\nabla}\cdot(\mathbf{T}\cdot\boldsymbol{r})+\boldsymbol{r}\cdot\boldsymbol{f}=U\,, \tag{1.7.4}$$

a virial theorem for the electromagnetic field. We will have an application for it shortly; see (2.2.9) below.

According to Clausius,* we call $\sum_j \boldsymbol{r}_j\cdot\boldsymbol{p}_j$ the *virial* for a collection of particles with position vectors $\boldsymbol{r}_j$ and momentum vectors $\boldsymbol{p}_j$. In analogy, the quantity $\boldsymbol{r}\cdot\boldsymbol{G}$ is the *virial density* of the electromagnetic field and $\mathbf{T}\cdot\boldsymbol{r}$ is then the *virial current density* vector. Indeed, it is appropriate to call (1.7.4) a "virial theorem."

*Rudolf Julius Emanuel CLAUSIUS (1822–1888)

# Chapter 2

# Electromagnetic Pulses

As an application of the various conservation laws and in particular, the relation $\boldsymbol{S} = c^2\boldsymbol{G}$ between the energy current density and the momentum density, we now consider the situation of an *electromagnetic pulse*: purely electromagnetic energy enclosed in a volume $V$ that does not contain any charged particles. So, the geometry is such that

$$\begin{aligned} &\rho = 0,\ \boldsymbol{j} = 0 \text{ inside } V \text{ where } \boldsymbol{E},\ \boldsymbol{B} \text{ are nonzero,} \\ \text{and}\ \ &\boldsymbol{E} = 0,\ \boldsymbol{B} = 0 \text{ on the surface } S \text{ of } V. \end{aligned}$$

For instance, you could shoot a flash with a camera, then at some later time the flash will occupy some distant volume wherein there are no charged particles to which the electromagnetic field of the flash of light could transfer energy or momentum or angular momentum.

## 2.1 Conserved energy, momentum, angular momentum

Under these circumstances, the total energy

$$E = \int_V (\mathrm{d}\boldsymbol{r})\, U\,, \tag{2.1.1}$$

total momentum,

$$\boldsymbol{P} = \int_V (\mathrm{d}\boldsymbol{r})\, \boldsymbol{G}\,, \tag{2.1.2}$$

and total angular momentum of the pulse,

$$\boldsymbol{J} = \int_V (\mathrm{d}\boldsymbol{r})\, \boldsymbol{r} \times \boldsymbol{G}\,, \tag{2.1.3}$$

are conserved quantities. For example,

$$\frac{\mathrm{d}}{\mathrm{d}t}E = \int_V (\mathrm{d}\boldsymbol{r})\,\frac{\partial U}{\partial t} = -\int_S \mathrm{d}\boldsymbol{S}\cdot\boldsymbol{S} - \int_V (\mathrm{d}\boldsymbol{r})\,\boldsymbol{j}\cdot\boldsymbol{E}\,, \tag{2.1.4}$$

where the first term on the right-hand side, the surface integral of the energy current density, and the second term, the total power transferred to the charges, both vanish because $\boldsymbol{S} = \frac{c}{4\pi}\boldsymbol{E}\times\boldsymbol{B} = 0$ on the surface $S$, and $\boldsymbol{j} = 0$ inside the volume $V$. Accordingly,

$$\frac{\mathrm{d}}{\mathrm{d}t}E = 0\,, \tag{2.1.5}$$

and likewise

$$\frac{\mathrm{d}}{\mathrm{d}t}\boldsymbol{P} = 0 \quad\text{and}\quad \frac{\mathrm{d}}{\mathrm{d}t}\boldsymbol{J} = 0\,. \tag{2.1.6}$$

In other words, the mechanical properties of the pulse of light are just like those of a collection of particles on which no external forces are acting.

## 2.2 Energy velocity, momentum velocity

Now, let us exploit the relation (1.5.11), that is: $\boldsymbol{S} = c^2\boldsymbol{G}$, between the energy current density and the momentum density, by considering

$$\begin{aligned}\frac{\partial}{\partial t}\left(\boldsymbol{r}U - c^2t\boldsymbol{G}\right) &= \boldsymbol{r}\frac{\partial U}{\partial t} - c^2\boldsymbol{G} - c^2t\frac{\partial}{\partial t}\boldsymbol{G}\\ &= \boldsymbol{r}\left(-\boldsymbol{\nabla}\cdot\boldsymbol{S} - \boldsymbol{j}\cdot\boldsymbol{E}\right) - \boldsymbol{S} - c^2t\left(-\boldsymbol{\nabla}\cdot\mathsf{T} - \boldsymbol{f}\right)\\ &= -\boldsymbol{\nabla}\cdot\underbrace{\left(\boldsymbol{S}\,\boldsymbol{r} - c^2t\,\mathsf{T}\right)}_{=0\text{ on }S} \underbrace{- \boldsymbol{r}\,\boldsymbol{j}\cdot\boldsymbol{E} + c^2t\boldsymbol{f}}_{=0\text{ in }V}\,,\end{aligned} \tag{2.2.1}$$

so that, for the electromagnetic pulse,

$$\frac{\mathrm{d}}{\mathrm{d}t}\int_V (\mathrm{d}\boldsymbol{r})\left(\boldsymbol{r}U - c^2t\boldsymbol{G}\right) = 0\,. \tag{2.2.2}$$

Now, in view of $\int_V (\mathrm{d}\boldsymbol{r})\, U(\boldsymbol{r},t) = E = \text{constant}$, we can write

$$\int_V (\mathrm{d}\boldsymbol{r})\ \boldsymbol{r}\ U(\boldsymbol{r},t) = E\ \langle\boldsymbol{r}\rangle_E(t)\,, \tag{2.2.3}$$

thereby introducing the energy-weighted average position $\langle\boldsymbol{r}\rangle_E(t)$. This leads us to

$$\frac{\mathrm{d}}{\mathrm{d}t}\left(E\ \langle\boldsymbol{r}\rangle_E(t) - c^2 t\, \boldsymbol{P}\right) = 0 \tag{2.2.4}$$

or

$$\boldsymbol{v}_E = \frac{\mathrm{d}}{\mathrm{d}t}\langle\boldsymbol{r}\rangle_E = \frac{c^2}{E}\boldsymbol{P}\,, \tag{2.2.5}$$

which is the constant *energy velocity* of the pulse.

Since the second inequality in (2.3.4) below states that $c\left|\boldsymbol{P}\right| \le E$, it follows that $\left|\boldsymbol{v}_E\right| \le c$. The energy speed cannot exceed the speed of light.

When we denote by mass $m$ the proportionality factor between the momentum and velocity, $\boldsymbol{P} = m\,\boldsymbol{v}_E$, then we get Einstein's energy-mass relation,

$$E = mc^2\,, \tag{2.2.6}$$

here for electromagnetic pulses.

We can also introduce a momentum-weighted average position $\langle\boldsymbol{r}\rangle_P(t)$, which we define by the pair of equations

$$\langle\boldsymbol{r}\rangle_P(t)\cdot\boldsymbol{P} = \int_V (\mathrm{d}\boldsymbol{r})\ \boldsymbol{r}\cdot\boldsymbol{G}(\boldsymbol{r},t)\,,$$

$$\langle\boldsymbol{r}\rangle_P(t)\times\boldsymbol{P} = \int_V (\mathrm{d}\boldsymbol{r})\ \boldsymbol{r}\times\boldsymbol{G}(\boldsymbol{r},t) = \boldsymbol{J}\,. \tag{2.2.7}$$

We conclude that

$$\boldsymbol{v}_P\times\boldsymbol{P} = 0 \tag{2.2.8}$$

holds for $\boldsymbol{v}_P = \frac{\mathrm{d}}{\mathrm{d}t}\langle \boldsymbol{r}\rangle_P$, the *momentum velocity* of the pulse. In addition, there is the virial theorem of (1.7.4), which implies first

$$\begin{aligned}\frac{\mathrm{d}}{\mathrm{d}t}\left(\langle \boldsymbol{r}\rangle_P \cdot \boldsymbol{P}\right) &= \frac{\mathrm{d}}{\mathrm{d}t}\int\limits_V (\mathrm{d}\boldsymbol{r})\; \boldsymbol{r}\cdot\boldsymbol{G} \\ &= \int\limits_V (\mathrm{d}\boldsymbol{r})\; U = E \end{aligned} \tag{2.2.9}$$

and then

$$\boldsymbol{v}_P \cdot \boldsymbol{P} = E\,. \tag{2.2.10}$$

Since $\boldsymbol{P} = \frac{1}{c^2} E\, \boldsymbol{v}_E$, we have

$$\boldsymbol{v}_E \cdot \boldsymbol{v}_P = c^2 \tag{2.2.11}$$

for the scalar product of the energy velocity $\boldsymbol{v}_E$ and the momentum velocity $\boldsymbol{v}_P$.

The relation (2.2.8) tells us that $\boldsymbol{v}_P$ is parallel to $\boldsymbol{P}$, and is therefore parallel to $\boldsymbol{v}_E$. It follows that either both speeds $v_E = |\boldsymbol{v}_E|$ and $v_P = |\boldsymbol{v}_P|$ are equal to the speed of light, $v_E = v_P = c$, or one speed is superluminal (that is: exceeds the speed of light, $v > c$) and the other speed is subluminal ($v < c$). The latter must be the energy velocity. We can thus surmise that, if we had at hand a solution of the wave equation for the electromagnetic pulse as well as the group and phase velocity associated with it, the group velocity, which is typically subluminal, will agree with the energy velocity, and the phase velocity with the momentum velocity. Indeed, this correspondence is correct for pulses propagating in vacuum.

For an alternative way of looking at these matters, we start with

$$\begin{aligned}\frac{\partial}{\partial t}\left[\left(r^2-(ct)^2\right)U\right] &= \left(r^2-(ct)^2\right)\frac{\partial}{\partial t}U - 2c^2 t\,U \\ &= \left(r^2-(ct)^2\right)\left(-\boldsymbol{\nabla}\cdot\boldsymbol{S} - \boldsymbol{j}\cdot\boldsymbol{E}\right) - 2c^2 t\,U \\ &= -\boldsymbol{\nabla}\cdot\left[\left(r^2-(ct)^2\right)\boldsymbol{S}\right] + \underbrace{2\boldsymbol{r}\cdot\boldsymbol{S}}_{=2c^2\boldsymbol{r}\cdot\boldsymbol{G}} - 2c^2 t\,U \\ &\quad - \left(r^2-(ct)^2\right)\boldsymbol{j}\cdot\boldsymbol{E} \end{aligned} \tag{2.2.12}$$

or

$$\frac{\partial}{\partial t}\Big[\big(r^2-(ct)^2\big)\,U\Big]+\boldsymbol{\nabla}\cdot\big[\big(r^2-(ct)^2\big)\,\boldsymbol{S}\big]$$
$$=2c^2(\boldsymbol{r}\cdot\boldsymbol{G}-tU)-\big(r^2-(ct)^2\big)\,\boldsymbol{j}\cdot\boldsymbol{E}\,,\qquad(2.2.13)$$

which we integrate over the volume $V$ of the electromagnetic pulse (remembering that $\boldsymbol{j}=0$ in $V$ and $\boldsymbol{S}=0$ on its surface) to arrive at

$$\frac{\mathrm{d}}{\mathrm{d}t}\int\limits_V(\mathrm{d}\boldsymbol{r})\ \big(r^2-(ct)^2\big)\,U=2c^2\int\limits_V(\mathrm{d}\boldsymbol{r})\,(\boldsymbol{r}\cdot\boldsymbol{G}-t\,U)\,.\qquad(2.2.14)$$

Except for

$$\int\limits_V(\mathrm{d}\boldsymbol{r})\,r^2\,U=E\,\langle r^2\rangle_E\qquad(2.2.15)$$

all integrals have appeared above, and we have

$$\frac{\mathrm{d}}{\mathrm{d}t}\big(E\,\langle r^2\rangle_E-(ct)^2E\big)=2c^2\,\big(\langle\boldsymbol{r}\rangle_P\cdot\boldsymbol{P}-tE\big)\,.\qquad(2.2.16)$$

The right-hand side is, in fact, constant in time,

$$\frac{\mathrm{d}}{\mathrm{d}t}\big(\langle\boldsymbol{r}\rangle_P\cdot\boldsymbol{P}-tE\big)=\boldsymbol{v}_P\cdot\boldsymbol{P}-E=0\qquad(2.2.17)$$

as stated in (2.2.10). We, therefore, take its value at the initial time $t=0$

$$\langle\boldsymbol{r}\rangle_P(t)\cdot\boldsymbol{P}-tE=\langle\boldsymbol{r}\rangle_P(0)\cdot\boldsymbol{P}\,,\qquad(2.2.18)$$

and

$$\langle r^2\rangle_E(t)=\langle r^2\rangle_E(0)+2ct\langle\boldsymbol{r}\rangle_P(0)\cdot\frac{c\boldsymbol{P}}{E}+(ct)^2\qquad(2.2.19)$$

follows. Clearly, then, after a sufficiently long lapse of time, we have

$$\sqrt{\langle r^2\rangle_E(t)}\cong ct,\qquad(2.2.20)$$

the mean-square-energy distance grows at the rate of the speed of light.

This is *as if* we had little lumps of electromagnetic energy (Einstein's light quanta) making up the pulse, whereby each lump is moving with the speed of light $c$, but in varying directions,

$$\boldsymbol{r}_j(t)=\boldsymbol{r}_j(0)+\boldsymbol{v}_jt\quad\text{with}\quad|\boldsymbol{v}_j|=c\qquad(2.2.21)$$

for the $j$th lump. The ensemble average over all lumps,

$$\langle \boldsymbol{r}(t)^2 \rangle = \langle \boldsymbol{r}(0)^2 \rangle + 2\langle \boldsymbol{r}(0) \cdot \boldsymbol{v} \rangle t + c^2 t^2 , \tag{2.2.22}$$

has exactly the same structure as (2.2.19). The correspondence

$$\begin{aligned} \langle r^2 \rangle_E(t) &\longleftrightarrow \langle \boldsymbol{r}(t)^2 \rangle , \\ \text{and} \quad \langle \boldsymbol{r} \rangle_P(0) \cdot \frac{c\boldsymbol{P}}{E} &\longleftrightarrow \langle \boldsymbol{r}(0) \cdot \boldsymbol{v}/c \rangle , \end{aligned} \tag{2.2.23}$$

turns one equation into the other, with electromagnetic pulse averages on the left and ensemble averages on the right.

## 2.3 Unidirectional pulses

We close the discussion of the mechanical properties of electromagnetic pulses with a few remarks about *unidirectional* pulses, in which all motion is in the same direction. It is then expected that there is just one velocity $\boldsymbol{v}_E = \boldsymbol{v}_P \equiv \boldsymbol{v}$ with the speed equal to the speed of light, $v = |\boldsymbol{v}| = c$. The statement (2.2.5) then gives $c|\boldsymbol{P}| = E$, or

$$E = \int_V (\mathrm{d}\boldsymbol{r}) \, \frac{1}{8\pi}(\boldsymbol{E}^2 + \boldsymbol{B}^2) = c \left| \int_V (\mathrm{d}\boldsymbol{r}) \, \frac{1}{4\pi c} \boldsymbol{E} \times \boldsymbol{B} \right| = c|\boldsymbol{P}| , \tag{2.3.1}$$

implying

$$\int_V (\mathrm{d}\boldsymbol{r}) \, \frac{1}{2}(\boldsymbol{E}^2 + \boldsymbol{B}^2) = \left| \int_V (\mathrm{d}\boldsymbol{r}) \boldsymbol{E} \times \boldsymbol{B} \right| \leq \int_V (\mathrm{d}\boldsymbol{r}) |\boldsymbol{E} \times \boldsymbol{B}| \tag{2.3.2}$$

for such a unidirectional pulse.

On the other hand, however, there is the mathematical identity

$$\begin{aligned} |\boldsymbol{E} \times \boldsymbol{B}|^2 &= \boldsymbol{E}^2 \boldsymbol{B}^2 - (\boldsymbol{E} \cdot \boldsymbol{B})^2 \\ &= \left[ \frac{1}{2} (\boldsymbol{E}^2 + \boldsymbol{B}^2) \right]^2 - \left[ \frac{1}{2} (\boldsymbol{E}^2 - \boldsymbol{B}^2) \right]^2 - (\boldsymbol{E} \cdot \boldsymbol{B})^2 \\ &\leq \left[ \frac{1}{2} (\boldsymbol{E}^2 + \boldsymbol{B}^2) \right]^2 , \end{aligned} \tag{2.3.3}$$

where the equal sign holds only if $\boldsymbol{E}^2 = \boldsymbol{B}^2$ and $\boldsymbol{E} \cdot \boldsymbol{B} = 0$. Just this must be the case in view of the pair of inequalities that we obtain when combining (2.3.2) and (2.3.3),

$$\int_V (\mathrm{d}\boldsymbol{r})\, \frac{1}{2}\left(\boldsymbol{E}^2+\boldsymbol{B}^2\right) \le \int_V (\mathrm{d}\boldsymbol{r})\, |\boldsymbol{E}\times\boldsymbol{B}| \le \int_V (\mathrm{d}\boldsymbol{r})\, \frac{1}{2}\left(\boldsymbol{E}^2+\boldsymbol{B}^2\right). \quad (2.3.4)$$

Accordingly, we conclude that

$$|\boldsymbol{E}| = |\boldsymbol{B}| \quad \text{and} \quad \boldsymbol{E}\perp\boldsymbol{B} \quad (2.3.5)$$

in a unidirectional electromagnetic pulse. Further, for the inequality in (2.3.2) to be an equality, $\boldsymbol{E}\times\boldsymbol{B}$ must point in the direction of $\boldsymbol{v}$ everywhere in the pulse,

$$\boldsymbol{E}\times\boldsymbol{B} = |\boldsymbol{E}|\,|\boldsymbol{B}|\,\frac{\boldsymbol{v}}{c}. \quad (2.3.6)$$

In summary, $\boldsymbol{E}$, $\boldsymbol{B}$, and $\boldsymbol{v}$ form a right-handed orthogonal vector triplet at any location in the pulse:

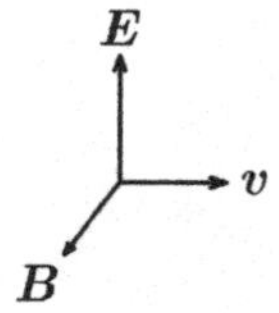


(2.3.7)

# Chapter 3

# Lorentz Transformation

We thus established that electromagnetic radiation, once emitted from a source of moving charged particles, propagates at the speed of light. That is: relativistic circumstances prevail. As a consequence, we now study how the various fields — $\rho$, $\boldsymbol{j}$, $\boldsymbol{E}$, $\boldsymbol{B}$, $\Phi$, $\boldsymbol{A}$ — change under Lorentz transformations.

## 3.1 Coordinate transformation, 4-vector, 4-dyadic

To set the stage, we consider two coordinate systems, with cartesian axes aligned in parallel, that are moving relative to each other with velocity $\boldsymbol{v} = v\boldsymbol{e}_z$:

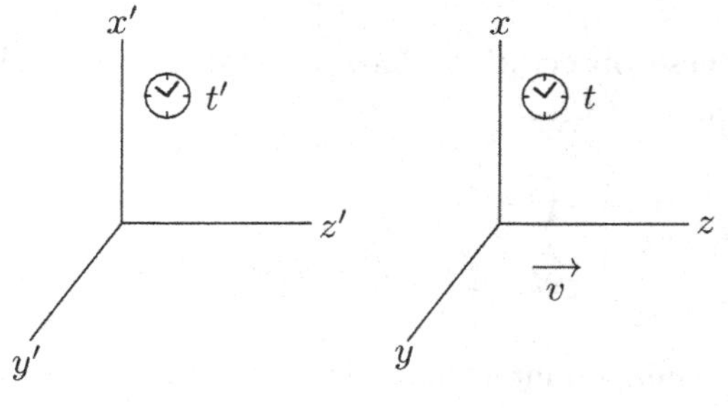


(3.1.1)

so that the origin $x = y = z = 0$ is described by $x' = y' = 0$, $z' = vt'$ in the primed coordinates. An event that occurs at time $t$ at position $\boldsymbol{r}$ with reference to the unprimed system, occurs at time $t'$ and position $\boldsymbol{r}'$, given by

$$t' = \gamma\left(t + \frac{vz}{c^2}\right), \quad \boldsymbol{r}' = \begin{pmatrix} x' \\ y' \\ z' \end{pmatrix} = \begin{pmatrix} x \\ y \\ \gamma(z + vt) \end{pmatrix}, \tag{3.1.2}$$

in the primed frame of reference, where

$$\gamma = \frac{1}{\sqrt{1-(v/c)^2}} = \frac{1}{\sqrt{1-\beta^2}} \quad \text{with} \quad \beta = \frac{v}{c}\,. \tag{3.1.3}$$

We split $\boldsymbol{r}'$ into the components that are parallel ($\boldsymbol{r}'_\parallel$) and perpendicular ($\boldsymbol{r}'_\perp$) to the $z$ direction of relative motion. Then

$$\boldsymbol{r}' = \boldsymbol{r}'_\perp + \boldsymbol{r}'_\parallel \tag{3.1.4}$$

with

$$\boldsymbol{r}'_\perp = \boldsymbol{r}_\perp \qquad (x \text{ and } y \text{ components}) \tag{3.1.5}$$

and

$$\boldsymbol{r}'_\parallel = \gamma(\boldsymbol{r}_\parallel + \boldsymbol{v}t) \qquad (z \text{ component}). \tag{3.1.6}$$

Here, the parallel component of $\boldsymbol{r}$ is

$$\boldsymbol{r}_\parallel = \frac{\boldsymbol{v}}{v}\frac{\boldsymbol{v}}{v}\cdot\boldsymbol{r} \tag{3.1.7}$$

and its perpendicular component is

$$\boldsymbol{r}_\perp = \boldsymbol{r} - \boldsymbol{r}_\parallel = \left(\mathbf{1} - \frac{\boldsymbol{v}\boldsymbol{v}}{v^2}\right)\cdot\boldsymbol{r} = -\frac{\boldsymbol{v}}{v}\times\left(\frac{\boldsymbol{v}}{v}\times\boldsymbol{r}\right). \tag{3.1.8}$$

We combine these observations into a compact vectorial form of the *Lorentz transformation* (3.1.2),

$$\begin{aligned} t' &= \gamma\left(t + \frac{\boldsymbol{v}\cdot\boldsymbol{r}}{c^2}\right),\\ \boldsymbol{r}' &= \gamma\boldsymbol{v}t + \left(\mathbf{1} + (\gamma-1)\frac{\boldsymbol{v}\,\boldsymbol{v}}{v^2}\right)\cdot\boldsymbol{r}\,. \end{aligned} \tag{3.1.9}$$

Clearly, space and time are interrelated here and, therefore, it is often convenient to switch from $t$ to $ct$, the distance covered by light during the lapse of $t$. We form a 4-component column out of $ct$ and $\boldsymbol{r}$, thereby regarding $\boldsymbol{r}$ as a 3-component column vector. Then the Lorentz transformation can be written as

$$\begin{pmatrix} ct' \\ \boldsymbol{r}' \end{pmatrix} = \begin{pmatrix} \gamma & \gamma\dfrac{\boldsymbol{v}^{\mathrm{T}}}{c} \\ \gamma\dfrac{\boldsymbol{v}}{c} & \mathbf{1} + (\gamma-1)\dfrac{\boldsymbol{v}\,\boldsymbol{v}^{\mathrm{T}}}{v^2} \end{pmatrix} \begin{pmatrix} ct \\ \boldsymbol{r} \end{pmatrix}, \tag{3.1.10}$$

where consistently row vectors are introduced as transposes of column vectors so that the scalar product between two column vectors appears as the multiplication of a row with a column, as illustrated by

$$\underbrace{\boldsymbol{v}\cdot\boldsymbol{r}}_{\text{two columns}} = \underbrace{\boldsymbol{v}^{\mathrm{T}}\boldsymbol{r}}_{\text{row times column}} . \tag{3.1.11}$$

The matrix above in (3.1.10) is a $4\times 4$ matrix, more specifically a *4-dyadic.* It multiplies the *4-vector* $\begin{pmatrix} ct \\ \boldsymbol{r} \end{pmatrix}$ and as a result of the multiplication we get $\begin{pmatrix} ct' \\ \boldsymbol{r}' \end{pmatrix}$, another *4-vector.*

More generally, any 4-component object that transforms under Lorentz transformations by the same 4-dyadic that applies to $\begin{pmatrix} ct \\ \boldsymbol{r} \end{pmatrix}$ is a 4-vector. Dyadic products of two 4-vectors give 4-dyadics, and so forth. One also speaks of tensors of 1st rank (4-vectors), tensors of 2nd rank (4-dyadics), tensors of 3rd rank, and so forth.

## 3.2 Infinitesimal transformations; rapidity

It is much easier to deal with infinitesimal Lorentz transformations than with finite ones. If the behavior under arbitrary infinitesimal transformations is understood, finite transformations can be constructed from very many successive infinitesimal transformations. So we put $\boldsymbol{v}\to\delta\boldsymbol{v}$ and discard all terms that are of second or higher order in $\delta\boldsymbol{v}$, and arrive at the *infinitesimal Lorentz transformation*

$$\begin{aligned} t' &= t+\delta t \quad \text{with} \quad \delta t = \frac{\delta\boldsymbol{v}}{c^2}\cdot\boldsymbol{r} , \\ \boldsymbol{r}' &= \boldsymbol{r}+\delta\boldsymbol{r} \quad \text{with} \quad \delta\boldsymbol{r} = \delta\boldsymbol{v}\, t . \end{aligned} \tag{3.2.1}$$

It is instructive to see how the finite transformations grow out of the infinitesimal ones. For this purpose, we return to the situation in which $\delta\boldsymbol{v}$ is in the $z$ direction, so that

$$\delta t = \frac{\delta v}{c^2}\, z , \qquad \delta x = 0 ,\ \delta y = 0 ,\ \delta z = \delta v\, t . \tag{3.2.2}$$

Now we introduce a parameter $\theta$, the so-called *rapidity*, by means of

$$\delta v = c\,\delta\theta . \tag{3.2.3}$$

Then

$$\frac{\partial t(\theta)}{\partial\theta} = \frac{1}{c} z(\theta)\,,$$
$$\frac{\partial x(\theta)}{\partial\theta} = 0\,,$$
$$\frac{\partial y(\theta)}{\partial\theta} = 0\,,$$
$$\frac{\partial z(\theta)}{\partial\theta} = c\,t(\theta)\,, \tag{3.2.4}$$

which are solved by

$$\begin{aligned} t(\theta) &= t(0)\cosh(\theta) + \frac{z(0)}{c}\sinh(\theta)\,, \\ x(\theta) &= x(0), \\ y(\theta) &= y(0), \\ z(\theta) &= z(0)\cosh(\theta) + ct(0)\sinh(\theta)\,. \end{aligned} \tag{3.2.5}$$

Upon identifying $t(0) = t$, $x(0) = x$, $y(0) = y$, $z(0) = z$ and $t(\theta) = t'$, $x(\theta) = x'$, $y(\theta) = y'$, $z(\theta) = z'$ as well as

$$\tanh(\theta) = \frac{v}{c} = \beta\,,\ \cosh(\theta) = \frac{1}{\sqrt{1-\tanh(\theta)^2}} = \frac{1}{\sqrt{1-(v/c)^2}} = \gamma\,,$$
$$\sinh(\theta) = \tanh(\theta)\cosh(\theta) = \beta\gamma\,, \tag{3.2.6}$$

we return to (3.1.2),

$$t' = \gamma\left(t + \frac{v}{c^2}z\right),$$
$$x' = x\,,\ y' = y\,,\ z' = \gamma(z + vt)\,, \tag{3.2.7}$$

as we should. For the record, we note further that the 4-dyadic of (3.1.10), expressed in terms of the rapidity $\theta$, appears as

$$\begin{pmatrix} \cosh\theta & 0 & 0 & \sinh\theta \\ 0 & 1 & 0 & 0 \\ 0 & 0 & 1 & 0 \\ \sinh\theta & 0 & 0 & \cosh\theta \end{pmatrix} \tag{3.2.8}$$

for $\boldsymbol{v}^{\mathrm{T}} = (0,\,0,\,v) = (0,\,0,\,\beta c)$.

### 3.3 Transformation laws for fields

A Lorentz *scalar field* is a function of $\boldsymbol{r}$ and $t$ that has the same form in all frames

$$\lambda(\boldsymbol{r}', t') = \lambda(\boldsymbol{r}, t)\,, \tag{3.3.1}$$

where, of course, $\boldsymbol{r}', t'$ refer to the same event as $\boldsymbol{r}, t$. If we express $\boldsymbol{r}', t'$ in terms of $\boldsymbol{r}, t$, we have, in the first place, another function of $\boldsymbol{r}, t$ on the left. For infinitesimal Lorentz transformation, we denote it by $\lambda + \delta\lambda$, so that

$$(\lambda + \delta\lambda)(\boldsymbol{r} + \delta\boldsymbol{r}, t + \delta t) = \lambda(\boldsymbol{r}, t)\,. \tag{3.3.2}$$

This implies

$$\delta\lambda(\boldsymbol{r}, t) = -\left(\delta t \frac{\partial}{\partial t} + \delta\boldsymbol{r} \cdot \boldsymbol{\nabla}\right) \lambda(\boldsymbol{r}, t) \tag{3.3.3}$$

or

$$\delta\lambda(\boldsymbol{r}, t) = -\delta_{\text{coor}} \lambda(\boldsymbol{r}, t) \tag{3.3.4}$$

with

$$\delta_{\text{coor}} = \frac{\delta\boldsymbol{v}}{c^2} \cdot \boldsymbol{r} \frac{\partial}{\partial t} + \delta\boldsymbol{v}\, t \cdot \boldsymbol{\nabla}\,. \tag{3.3.5}$$

The change recorded by $\delta_{\text{coor}}$ originates solely in the change of the coordinates; it is part and parcel of the transformation of all *fields*. In the case of a scalar field, it is the whole story.

The dependence of $\delta_{\text{coor}}$ on $\boldsymbol{r}$ and $t$ is important in the next step, when we establish the infinitesimal transformation law for the gradient and the time derivative of the scalar field $\lambda$,

$$\begin{aligned}
\delta(\boldsymbol{\nabla}\lambda) = \boldsymbol{\nabla}(\delta\lambda) &= -\delta_{\text{coor}}(\boldsymbol{\nabla}\lambda) - (\boldsymbol{\nabla}\delta_{\text{coor}})\lambda \\
&= -\delta_{\text{coor}}(\boldsymbol{\nabla}\lambda) - \frac{\delta\boldsymbol{v}}{c^2}\frac{\partial}{\partial t}\lambda\,, \\
\delta\Big(\frac{\partial}{\partial t}\lambda\Big) = \frac{\partial}{\partial t}(\delta\lambda) &= -\delta_{\text{coor}}\Big(\frac{\partial}{\partial t}\lambda\Big) - \Big(\frac{\partial}{\partial t}\delta_{\text{coor}}\Big)\lambda \\
&= -\delta_{\text{coor}}\Big(\frac{\partial}{\partial t}\lambda\Big) - \delta\boldsymbol{v} \cdot \boldsymbol{\nabla}\lambda\,,
\end{aligned} \tag{3.3.6}$$

or

$$\delta\Big(-\frac{1}{c}\frac{\partial}{\partial t}\lambda\Big) = -\delta_{\rm coor}\Big(-\frac{1}{c}\frac{\partial}{\partial t}\lambda\Big) + \frac{\delta\boldsymbol{v}}{c}\cdot\boldsymbol{\nabla}\lambda\,,$$
$$\delta(\boldsymbol{\nabla}\lambda) = -\delta_{\rm coor}(\boldsymbol{\nabla}\lambda) + \frac{\delta\boldsymbol{v}}{c}\Big(-\frac{1}{c}\frac{\partial}{\partial t}\lambda\Big). \qquad (3.3.7)$$

These equations state that, in addition to the field transformation of $\delta_{\rm coor}$, $\begin{pmatrix} -\frac{1}{c}\frac{\partial}{\partial t}\lambda \\ \boldsymbol{\nabla}\lambda \end{pmatrix}$ transforms by the same law as the event 4-vector $\begin{pmatrix} ct \\ \boldsymbol{r} \end{pmatrix}$. Therefore, we recognize that we have here a first example of a *4-vector field*,

$$\underbrace{\begin{pmatrix} -\frac{1}{c}\frac{\partial}{\partial t} \\ \boldsymbol{\nabla} \end{pmatrix}}_{\text{4-vector}} \underbrace{\lambda(\boldsymbol{r},t)}_{\text{scalar field}} = \underbrace{\begin{pmatrix} -\frac{1}{c}\frac{\partial}{\partial t}\lambda \\ \boldsymbol{\nabla}\lambda \end{pmatrix}}_{\text{4-vector field}}, \qquad (3.3.8)$$

where we note in passing that the differential operators form a 4-vector, the *4-gradient*. Do not miss the important detail of the minus sign carried by the time derivative.

We recall the gauge transformation of Section 1.3,

$$\Phi \to \Phi - \frac{1}{c}\frac{\partial}{\partial t}\lambda\,, \qquad \boldsymbol{A} \to \boldsymbol{A} + \boldsymbol{\nabla}\lambda\,, \qquad (3.3.9)$$

and conclude immediately that $\begin{pmatrix} \Phi \\ \boldsymbol{A} \end{pmatrix}$ is a 4-vector field, the *4-potential* so that

$$\delta\Phi = -\delta_{\rm coor}\Phi + \frac{\delta\boldsymbol{v}}{c}\cdot\boldsymbol{A}\,,$$
$$\delta\boldsymbol{A} = -\delta_{\rm coor}\boldsymbol{A} + \frac{\delta\boldsymbol{v}}{c}\Phi \qquad (3.3.10)$$

are the responses of $\Phi$ and $\boldsymbol{A}$ to infinitesimal Lorentz transformations. Further differentiation, in accordance with (1.3.1),

$$\boldsymbol{E} = -\frac{1}{c}\frac{\partial}{\partial t}\boldsymbol{A} - \boldsymbol{\nabla}\Phi\,, \qquad \boldsymbol{B} = \boldsymbol{\nabla}\times\boldsymbol{A}\,, \qquad (3.3.11)$$

will establish the transformation laws for $\boldsymbol{E}$ and $\boldsymbol{B}$.

We begin with the magnetic field,

$$\begin{aligned}\delta \boldsymbol{B} = \delta(\boldsymbol{\nabla} \times \boldsymbol{A}) = \boldsymbol{\nabla} \times (\delta \boldsymbol{A}) &= \boldsymbol{\nabla} \times \left(-\delta_{\text{coor}} \boldsymbol{A} + \frac{\delta \boldsymbol{v}}{c} \Phi\right) \\ &= -\delta_{\text{coor}}(\boldsymbol{\nabla} \times \boldsymbol{A}) - (\boldsymbol{\nabla} \delta_{\text{coor}}) \times \boldsymbol{A} - \frac{\delta \boldsymbol{v}}{c} \times \boldsymbol{\nabla} \Phi \\ &= -\delta_{\text{coor}} \boldsymbol{B} + \frac{\delta \boldsymbol{v}}{c} \times \left(-\frac{1}{c}\frac{\partial}{\partial t} \boldsymbol{A} - \boldsymbol{\nabla}\Phi\right)\end{aligned} \tag{3.3.12}$$

or

$$\delta \boldsymbol{B} = -\delta_{\text{coor}} \boldsymbol{B} + \frac{\delta \boldsymbol{v}}{c} \times \boldsymbol{E} \,. \tag{3.3.13}$$

Similarly, we get

$$\begin{aligned}\delta \boldsymbol{E} &= -\delta_{\text{coor}} \boldsymbol{E} + \frac{\delta \boldsymbol{v}}{c} \cdot \boldsymbol{\nabla} \boldsymbol{A} + \frac{\delta \boldsymbol{v}}{c^2} \frac{\partial}{\partial t} \Phi \\ &\qquad - \boldsymbol{\nabla}\left(\frac{\delta \boldsymbol{v}}{c} \cdot \boldsymbol{A}\right) - \frac{\delta \boldsymbol{v}}{c} \frac{1}{c} \frac{\partial}{\partial t} \Phi \\ &= -\delta_{\text{coor}} \boldsymbol{E} - \frac{\delta \boldsymbol{v}}{c} \times (\boldsymbol{\nabla} \times \boldsymbol{A})\end{aligned} \tag{3.3.14}$$

or

$$\delta \boldsymbol{E} = -\delta_{\text{coor}} \boldsymbol{E} - \frac{\delta \boldsymbol{v}}{c} \times \boldsymbol{B} \tag{3.3.15}$$

for the response of $\boldsymbol{E}$ to an infinitesimal Lorentz transformation. The pair of equations (3.3.13) and (3.3.15) is of a structure that we have not encountered before. We can conclude immediately that $\boldsymbol{E}$ and $\boldsymbol{B}$ are *not* the spatial parts of 4-vector fields because they do not transform the same way as $\Phi$ and $\boldsymbol{A}$ do.

The homogeneous Maxwell's equations are now found to be invariant under Lorentz transformation,

$$\begin{aligned}\delta(\boldsymbol{\nabla} \cdot \boldsymbol{B}) = \boldsymbol{\nabla} \cdot (\delta \boldsymbol{B}) &= -\delta_{\text{coor}} \underbrace{\boldsymbol{\nabla} \cdot \boldsymbol{B}}_{=0} - \frac{\delta \boldsymbol{v}}{c^2} \cdot \frac{\partial}{\partial t} \boldsymbol{B} + \boldsymbol{\nabla} \cdot \left(\frac{\delta \boldsymbol{v}}{c} \times \boldsymbol{E}\right) \\ &= -\frac{\delta \boldsymbol{v}}{c} \cdot \left(\frac{1}{c}\frac{\partial}{\partial t} \boldsymbol{B} + \boldsymbol{\nabla} \times \boldsymbol{E}\right) = 0 \,,\end{aligned} \tag{3.3.16}$$

and

$$\begin{aligned}\delta\left(\frac{1}{c}\frac{\partial}{\partial t}\boldsymbol{B}+\boldsymbol{\nabla}\times\boldsymbol{E}\right) &= -\,\delta_{\text{coor}}\left(\frac{1}{c}\frac{\partial}{\partial t}\boldsymbol{B}+\boldsymbol{\nabla}\times\boldsymbol{E}\right)\\ &\quad -\frac{1}{c}\delta\boldsymbol{v}\cdot\boldsymbol{\nabla}\boldsymbol{B}-\frac{\delta\boldsymbol{v}}{c^2}\times\frac{\partial}{\partial t}\boldsymbol{E}\\ &\quad +\frac{\delta\boldsymbol{v}}{c}\times\frac{1}{c}\frac{\partial}{\partial t}\boldsymbol{E}-\boldsymbol{\nabla}\times\left(\frac{\delta\boldsymbol{v}}{c}\times\boldsymbol{B}\right)\\ &= -\,\frac{\delta\boldsymbol{v}}{c}\,\boldsymbol{\nabla}\cdot\boldsymbol{B}=0\,,\end{aligned}\tag{3.3.17}$$

indeed.

We use the inhomogeneous Maxwell's equations to determine the transformation laws for the charge density $\rho$ and the current density $\boldsymbol{j}$,

$$\begin{aligned}\delta\rho &= \delta\Big(\frac{1}{4\pi}\boldsymbol{\nabla}\cdot\boldsymbol{E}\Big)=\frac{1}{4\pi}\boldsymbol{\nabla}\cdot(\delta\boldsymbol{E})\\ &= \frac{1}{4\pi}\boldsymbol{\nabla}\cdot\left(-\delta_{\text{coor}}\boldsymbol{E}-\frac{\delta\boldsymbol{v}}{c}\times\boldsymbol{B}\right)\\ &= -\delta_{\text{coor}}\Big(\frac{1}{4\pi}\boldsymbol{\nabla}\cdot\boldsymbol{E}\Big)-\frac{1}{4\pi}\frac{\delta\boldsymbol{v}}{c}\cdot\frac{1}{c}\frac{\partial}{\partial t}\boldsymbol{E}+\frac{1}{4\pi}\frac{\delta\boldsymbol{v}}{c}\cdot(\boldsymbol{\nabla}\times\boldsymbol{B})\\ &= -\delta_{\text{coor}}\rho+\frac{1}{4\pi}\frac{\delta\boldsymbol{v}}{c}\cdot\underbrace{\left(-\frac{1}{c}\frac{\partial}{\partial t}\boldsymbol{E}+\boldsymbol{\nabla}\times\boldsymbol{B}\right)}_{=\frac{4\pi}{c}\boldsymbol{j}}\\ &= -\delta_{\text{coor}}\rho+\frac{1}{c^2}\delta\boldsymbol{v}\cdot\boldsymbol{j}\,,\end{aligned}\tag{3.3.18}$$

and

$$\begin{aligned}\delta\boldsymbol{j} &= \delta\left(\frac{c}{4\pi}\boldsymbol{\nabla}\times\boldsymbol{B}-\frac{1}{4\pi}\frac{\partial}{\partial t}\boldsymbol{E}\right)\\ &= -\delta_{\text{coor}}\left(\frac{c}{4\pi}\boldsymbol{\nabla}\times\boldsymbol{B}-\frac{1}{4\pi}\frac{\partial}{\partial t}\boldsymbol{E}\right)\\ &\quad +\frac{c}{4\pi}\left(-\frac{\delta\boldsymbol{v}}{c^2}\times\frac{\partial}{\partial t}\boldsymbol{B}+\boldsymbol{\nabla}\times\left(\frac{\delta\boldsymbol{v}}{c}\times\boldsymbol{E}\right)+\frac{\delta\boldsymbol{v}}{c}\cdot\boldsymbol{\nabla}\boldsymbol{E}+\frac{\delta\boldsymbol{v}}{c}\times\frac{1}{c}\frac{\partial}{\partial t}\boldsymbol{B}\right)\\ &= -\delta_{\text{coor}}\boldsymbol{j}+\frac{c}{4\pi}\frac{\delta\boldsymbol{v}}{c}\,\boldsymbol{\nabla}\cdot\boldsymbol{E}\\ &= -\delta_{\text{coor}}\boldsymbol{j}+\delta\boldsymbol{v}\,\rho\,.\end{aligned}\tag{3.3.19}$$

We observe that $c\rho$ and $\boldsymbol{j}$ make up another 4-vector field, the *4-current density* $\begin{pmatrix} c\rho \\ \boldsymbol{j} \end{pmatrix}$,

$$\begin{aligned}
\delta(c\rho) &= -\delta_{\text{coor}}(c\rho) + \frac{\delta \boldsymbol{v}}{c} \cdot \boldsymbol{j}\,, \\
\delta \boldsymbol{j} &= -\delta_{\text{coor}}\boldsymbol{j} + \frac{\delta \boldsymbol{v}}{c}(c\rho)\,.
\end{aligned} \tag{3.3.20}$$

It is then clear that all four Maxwell's equations are invariant under Lorentz transformations. In fact, we just exploited the invariance to find the transformation laws for $\rho$ and $\boldsymbol{j}$, but we could have found those also by explicitly transforming the densities of a single point charge, for example.

Of the quantities that we have met in the various conservation laws for energy, momentum, and angular momentum, namely $U$, $\boldsymbol{S}$, $\boldsymbol{G}$, $\mathsf{T}$, and $\boldsymbol{j} \cdot \boldsymbol{E}$, $\boldsymbol{f}$, we first consider Joule's power density $\boldsymbol{j} \cdot \boldsymbol{E}$,

$$\begin{aligned}
\delta(\boldsymbol{j} \cdot \boldsymbol{E}) &= \delta \boldsymbol{j} \cdot \boldsymbol{E} + \boldsymbol{j} \cdot \delta \boldsymbol{E} \\
&= -\delta_{\text{coor}}(\boldsymbol{j} \cdot \boldsymbol{E}) + \delta \boldsymbol{v} \cdot \rho \boldsymbol{E} - \boldsymbol{j} \cdot \left(\frac{\delta \boldsymbol{v}}{c} \times \boldsymbol{B}\right) \\
&= -\delta_{\text{coor}}(\boldsymbol{j} \cdot \boldsymbol{E}) + \delta \boldsymbol{v} \cdot \left(\rho \boldsymbol{E} + \frac{1}{c}\boldsymbol{j} \times \boldsymbol{B}\right)
\end{aligned} \tag{3.3.21}$$

or

$$\delta\left(\frac{1}{c}\boldsymbol{j} \cdot \boldsymbol{E}\right) = -\delta_{\text{coor}}\left(\frac{1}{c}\boldsymbol{j} \cdot \boldsymbol{E}\right) + \frac{\delta \boldsymbol{v}}{c} \cdot \boldsymbol{f}, \tag{3.3.22}$$

and then the Lorentz force density,

$$\begin{aligned}
\delta \boldsymbol{f} &= \delta\rho\, \boldsymbol{E} + \rho\, \delta \boldsymbol{E} + \frac{1}{c}\delta \boldsymbol{j} \times \boldsymbol{B} + \frac{1}{c}\boldsymbol{j} \times \delta \boldsymbol{B} \\
&= -\delta_{\text{coor}}\boldsymbol{f} + \frac{\delta \boldsymbol{v}}{c} \cdot \boldsymbol{j}\, \boldsymbol{E} - \rho \frac{\delta \boldsymbol{v}}{c} \times \boldsymbol{B} + \frac{1}{c}\delta \boldsymbol{v} \rho \times \boldsymbol{B} + \frac{1}{c}\boldsymbol{j} \times \left(\frac{\delta \boldsymbol{v}}{c} \times \boldsymbol{E}\right) \\
&= -\delta_{\text{coor}}\boldsymbol{f} + \frac{\delta \boldsymbol{v}}{c}\frac{1}{c}\boldsymbol{j} \cdot \boldsymbol{E}\,.
\end{aligned} \tag{3.3.23}$$

Accordingly, the *4-force density* $\begin{pmatrix} \frac{1}{c}\boldsymbol{j} \cdot \boldsymbol{E} \\ \boldsymbol{f} \end{pmatrix}$ is yet another 4-vector field,

$$\begin{aligned}
\delta\left(\frac{1}{c}\boldsymbol{j} \cdot \boldsymbol{E}\right) &= -\delta_{\text{coor}}\left(\frac{1}{c}\boldsymbol{j} \cdot \boldsymbol{E}\right) + \frac{\delta \boldsymbol{v}}{c} \cdot \boldsymbol{f}, \\
\delta \boldsymbol{f} &= -\delta_{\text{coor}}\boldsymbol{f} + \frac{\delta \boldsymbol{v}}{c}\frac{1}{c}\boldsymbol{j} \cdot \boldsymbol{E}\,.
\end{aligned} \tag{3.3.24}$$

Upon making the components of the 4-force density explicit in the conservation laws for energy and momentum, (1.4.13) and (1.5.7), we have the pair of equations

$$\frac{1}{c}\frac{\partial}{\partial t}U + \boldsymbol{\nabla}\cdot\left(\frac{1}{c}\boldsymbol{S}\right) + \frac{1}{c}\boldsymbol{j}\cdot\boldsymbol{E} = 0\,,$$
$$\frac{1}{c}\frac{\partial}{\partial t}(c\boldsymbol{G}) + \boldsymbol{\nabla}\cdot\mathsf{T} + \boldsymbol{f} = 0\,, \tag{3.3.25}$$

and it follows that

$$\begin{pmatrix} \dfrac{1}{c}\dfrac{\partial}{\partial t}U + \boldsymbol{\nabla}\cdot\left(\dfrac{1}{c}\boldsymbol{S}\right) \\ \dfrac{1}{c}\dfrac{\partial}{\partial t}(c\boldsymbol{G}) + \boldsymbol{\nabla}\cdot\mathsf{T} \end{pmatrix} \tag{3.3.26}$$

is a 4-vector field as well. But what about $U$, $\frac{1}{c}\boldsymbol{S}$, $c\boldsymbol{G}$, and $\mathsf{T}$ themselves? As we will see, they constitute the $16 = 1 + 3 + 3 + 9$ components of a 4-dyadic.

## 3.4 4-columns, 4-rows

A clue is provided by the continuity equation for the electric charge,

$$\frac{\partial}{\partial t}\rho + \boldsymbol{\nabla}\cdot\boldsymbol{j} = 0\,, \tag{3.4.1}$$

which, in order to exhibit the 4-current explicitly, we rewrite as the product of a row with a column,

$$\left(\frac{1}{c}\frac{\partial}{\partial t},\ \boldsymbol{\nabla}^{\mathrm{T}}\right)\begin{pmatrix} c\rho \\ \boldsymbol{j} \end{pmatrix} = \frac{1}{c}\frac{\partial}{\partial t}(c\rho) + \boldsymbol{\nabla}^{\mathrm{T}}\boldsymbol{j} = 0\,, \tag{3.4.2}$$

where the row-times-column product $\boldsymbol{\nabla}^{\mathrm{T}}\boldsymbol{j}$ equals the dot product of the two columns: $\boldsymbol{\nabla}^{\mathrm{T}}\boldsymbol{j} = \boldsymbol{\nabla}\cdot\boldsymbol{j}$, in accordance with what we observed in (3.1.11). We recognize the ingredients of two 4-vectors, the 4-gradient $\begin{pmatrix} -\frac{1}{c}\frac{\partial}{\partial t} \\ \boldsymbol{\nabla} \end{pmatrix}$ and the 4-current $\begin{pmatrix} c\rho \\ \boldsymbol{j} \end{pmatrix}$, and it is tempting to regard the left-hand side in

(3.4.2) as the dot product of these two 4-vectors,

$$\underbrace{\begin{pmatrix} -\dfrac{1}{c}\dfrac{\partial}{\partial t} \\ \boldsymbol{\nabla} \end{pmatrix} \cdot \begin{pmatrix} c\rho \\ \boldsymbol{j} \end{pmatrix}}_{\text{dot product of two 4-vectors}} = \underbrace{\left( \frac{1}{c}\frac{\partial}{\partial t}, \; \boldsymbol{\nabla}^{\mathrm{T}} \right) \begin{pmatrix} c\rho \\ \boldsymbol{j} \end{pmatrix}}_{\text{4-row times 4-column}}, \tag{3.4.3}$$

which would require that a minus sign is included in the time-like component, when transposition turns a 4-column into a 4-row,

$$\begin{pmatrix} a_0 \\ \boldsymbol{a} \end{pmatrix} \xrightarrow[\text{transposition}]{} \left( -a_0, \; \boldsymbol{a}^{\mathrm{T}} \right). \tag{3.4.4}$$

This will only be a sensible concept if products of a 4-row and a 4-column result in a 4-scalar.

Let us see. Infinitesimal Lorentz transformations of a 4-vector of column type are of the familiar form

$$\delta \begin{pmatrix} a_0 \\ \boldsymbol{a} \end{pmatrix} = \begin{pmatrix} \dfrac{\delta \boldsymbol{v}}{c} \cdot \boldsymbol{a} \\ \dfrac{\delta \boldsymbol{v}}{c} a_0 \end{pmatrix} = \begin{pmatrix} 0 & \delta \boldsymbol{v}^{\mathrm{T}}/c \\ \delta \boldsymbol{v}/c & \mathbf{0} \end{pmatrix} \begin{pmatrix} a_0 \\ \boldsymbol{a} \end{pmatrix}, \tag{3.4.5}$$

where $\mathbf{0}$ is the null dyadic, and the response of 4-vectors of row type follows from first noting that

$$\left( -a_0, \; \boldsymbol{a}^{\mathrm{T}} \right) = \begin{pmatrix} -a_0 \\ \boldsymbol{a} \end{pmatrix}^{\mathrm{T}} = \left[ \begin{pmatrix} -1 & \boldsymbol{0}^{\mathrm{T}} \\ \boldsymbol{0} & \mathbf{1} \end{pmatrix} \begin{pmatrix} a_0 \\ \boldsymbol{a} \end{pmatrix} \right]^{\mathrm{T}}, \tag{3.4.6}$$

where $\boldsymbol{0}$ is the null vector and $^{\mathrm{T}}$ indicates the usual transposition without the minus sign in the time-like component. Then

$$\begin{aligned} \delta \left( -a_0, \; \boldsymbol{a}^{\mathrm{T}} \right) &= \left[ \begin{pmatrix} -1 & \boldsymbol{0}^{\mathrm{T}} \\ \boldsymbol{0} & \mathbf{1} \end{pmatrix} \begin{pmatrix} 0 & \delta \boldsymbol{v}^{\mathrm{T}}/c \\ \delta \boldsymbol{v}/c & \mathbf{0} \end{pmatrix} \begin{pmatrix} a_0 \\ \boldsymbol{a} \end{pmatrix} \right]^{\mathrm{T}} \\ &= \left( a_0, \; \boldsymbol{a}^{\mathrm{T}} \right) \begin{pmatrix} 0 & \delta \boldsymbol{v}^{\mathrm{T}}/c \\ \delta \boldsymbol{v}/c & \mathbf{0} \end{pmatrix} \begin{pmatrix} -1 & \boldsymbol{0}^{\mathrm{T}} \\ \boldsymbol{0} & \mathbf{1} \end{pmatrix} \\ &= \left( -a_0, \; \boldsymbol{a}^{\mathrm{T}} \right) \begin{pmatrix} -1 & \boldsymbol{0}^{\mathrm{T}} \\ \boldsymbol{0} & \mathbf{1} \end{pmatrix} \begin{pmatrix} 0 & \delta \boldsymbol{v}^{\mathrm{T}}/c \\ \delta \boldsymbol{v}/c & \mathbf{0} \end{pmatrix} \begin{pmatrix} -1 & \boldsymbol{0}^{\mathrm{T}} \\ \boldsymbol{0} & \mathbf{1} \end{pmatrix} \end{aligned} \tag{3.4.7}$$

or, finally,

$$\delta\left(-a_0,\ \boldsymbol{a}^{\mathrm{T}}\right)=\left(-a_0,\ \boldsymbol{a}^{\mathrm{T}}\right)\begin{pmatrix} 0 & -\delta\boldsymbol{v}^{\mathrm{T}}/c \\ -\delta\boldsymbol{v}/c & \boldsymbol{0}\end{pmatrix}. \tag{3.4.8}$$

In summary,

$$\text{4-column: } \delta\begin{pmatrix} a_0 \\ \boldsymbol{a}\end{pmatrix}=\delta L\begin{pmatrix} a_0 \\ \boldsymbol{a}\end{pmatrix},$$
$$\text{4-row: } \delta\left(-a_0,\ \boldsymbol{a}^{\mathrm{T}}\right)=-\left(-a_0,\ \boldsymbol{a}^{\mathrm{T}}\right)\delta L, \tag{3.4.9}$$

where

$$\delta L=\begin{pmatrix} 0 & \delta\boldsymbol{v}^{\mathrm{T}}/c \\ \delta\boldsymbol{v}/c & \boldsymbol{0}\end{pmatrix} \tag{3.4.10}$$

is the 4-dyadic of the infinitesimal Lorentz transformation in (3.4.5).

As a consequence, an infinitesimal Lorentz transformation has no effect on the 4-row times 4-column product,

$$\delta\left(\left(-a_0,\ \boldsymbol{a}^{\mathrm{T}}\right)\begin{pmatrix} b_0 \\ \boldsymbol{b}\end{pmatrix}\right)=\left(-a_0,\ \boldsymbol{a}^{\mathrm{T}}\right)(\underbrace{-\delta L}_{\text{transforms the row}}+\underbrace{\delta L}_{\text{transforms the column}})\begin{pmatrix} b_0 \\ \boldsymbol{b}\end{pmatrix}=0\,, \tag{3.4.11}$$

so that

$$-a_0b_0+\boldsymbol{a}\cdot\boldsymbol{b}=\begin{pmatrix} a_0 \\ \boldsymbol{a}\end{pmatrix}\cdot\begin{pmatrix} b_0 \\ \boldsymbol{b}\end{pmatrix}=\left(-a_0,\ \boldsymbol{a}^{\mathrm{T}}\right)\begin{pmatrix} b_0 \\ \boldsymbol{b}\end{pmatrix} \tag{3.4.12}$$

is a 4-scalar, indeed.

It follows that the vanishing right-hand side of the continuity equation (3.4.1) for electric charge is a 4-scalar: All observers agree that electric charge is absolutely conserved. Further, we recognize that the condition (1.3.7) of the Lorentz gauge is a scalar product,

$$\boldsymbol{\nabla}\cdot\boldsymbol{A}+\frac{1}{c}\frac{\partial}{\partial t}\Phi=\underbrace{\begin{pmatrix} -\dfrac{1}{c}\dfrac{\partial}{\partial t} \\ \boldsymbol{\nabla}\end{pmatrix}}_{\text{4-vector}}\cdot\underbrace{\begin{pmatrix} \Phi \\ \boldsymbol{A}\end{pmatrix}}_{\text{4-vector field}}=0\,, \tag{3.4.13}$$

where the final "0" is a vanishing 4-scalar field. Accordingly, the potentials are in the Lorentz gauge in *all* reference frames if they are in one. By contrast, the radiation gauge will not hold in other frames. One has to perform a gauge transformation if one wants to have the potentials in the radiation gauge in all reference frames.

Returning now to the pair of equations for the conservation of energy and momentum, (1.4.13) and (1.5.7), we combine them into the 4-vector form

$$\underbrace{\left(\frac{1}{c}\frac{\partial}{\partial t}, \boldsymbol{\nabla}^{\mathrm{T}}\right)}_{\text{4-gradient row vector}} \underbrace{\begin{pmatrix} -U & c\boldsymbol{G}^{\mathrm{T}} \\ -\frac{1}{c}\boldsymbol{S} & \mathsf{T} \end{pmatrix}}_{\text{4-dyadic field}} + \underbrace{\left(-\frac{1}{c}\boldsymbol{j}\cdot\boldsymbol{E}, \boldsymbol{f}^{\mathrm{T}}\right)}_{\text{4-force density row-vector field}} = 0\,, \tag{3.4.14}$$

where the final "0" is a vanishing 4-row vector field. The 4-dyadic field for energy and momentum in (3.4.14),

$$\begin{pmatrix} -U & c\boldsymbol{G}^{\mathrm{T}} \\ -\frac{1}{c}\boldsymbol{S} & \mathsf{T} \end{pmatrix} = \text{“energy-momentum tensor”}\,, \tag{3.4.15}$$

incorporates the minus sign in the first column that is required by the minus sign in the time-like power component of the 4-force in row form.

The prototype 4-dyadic is the product of a 4-column with a 4-row, which tells us that the response to an infinitesimal Lorentz transformation is given by

$$\begin{aligned} \delta\Bigl[\underbrace{\underbrace{\begin{pmatrix} a_0 \\ \boldsymbol{a} \end{pmatrix}}_{\text{4-column}} \underbrace{(-b_0,\, \boldsymbol{b}^{\mathrm{T}})}_{\text{4-row}}}_{\text{4-dyadic}}\Bigr] &= \left[\delta\begin{pmatrix} a_0 \\ \boldsymbol{a} \end{pmatrix}\right](-b_0,\, \boldsymbol{b}^{\mathrm{T}}) + \begin{pmatrix} a_0 \\ \boldsymbol{a} \end{pmatrix}\delta\,(-b_0,\, \boldsymbol{b}^{\mathrm{T}}) \\ &= \delta L\left[\begin{pmatrix} a_0 \\ \boldsymbol{a} \end{pmatrix}(-b_0,\, \boldsymbol{b}^{\mathrm{T}})\right] \\ &\quad - \left[\begin{pmatrix} a_0 \\ \boldsymbol{a} \end{pmatrix}(-b_0,\, \boldsymbol{b}^{\mathrm{T}})\right]\delta L\,, \end{aligned} \tag{3.4.16}$$

so that, for any 4-dyadic $D$,

$$\delta D = \delta L\, D - D\, \delta L \tag{3.4.17}$$

and, accordingly,

$$\delta D = -\delta_{\mathrm{coor}}\, D + \delta L\, D - D\, \delta L \tag{3.4.18}$$

applies to a 4-dyadic field.

Therefore, the infinitesimal Lorentz transformation of the 4-dyadic energy-momentum field is given by

$$
\begin{aligned}
\delta \begin{pmatrix} -U & c\boldsymbol{G}^{\mathrm{T}} \\ -\frac{1}{c}\boldsymbol{S} & \mathsf{T} \end{pmatrix} &= -\delta_{\mathrm{coor}} \begin{pmatrix} -U & c\boldsymbol{G}^{\mathrm{T}} \\ -\frac{1}{c}\boldsymbol{S} & \mathsf{T} \end{pmatrix} \\
&\quad + \begin{pmatrix} 0 & \delta\boldsymbol{v}^{\mathrm{T}}/c \\ \delta\boldsymbol{v}/c & \mathbf{0} \end{pmatrix} \begin{pmatrix} -U & c\boldsymbol{G}^{\mathrm{T}} \\ -\frac{1}{c}\boldsymbol{S} & \mathsf{T} \end{pmatrix} \\
&\quad - \begin{pmatrix} -U & c\boldsymbol{G}^{\mathrm{T}} \\ -\frac{1}{c}\boldsymbol{S} & \mathsf{T} \end{pmatrix} \begin{pmatrix} 0 & \delta\boldsymbol{v}^{\mathrm{T}}/c \\ \delta\boldsymbol{v}/c & \mathbf{0} \end{pmatrix} \\
&= -\delta_{\mathrm{coor}} \begin{pmatrix} -U & c\boldsymbol{G}^{\mathrm{T}} \\ -\frac{1}{c}\boldsymbol{S} & \mathsf{T} \end{pmatrix} \\
&\quad + \begin{pmatrix} -\frac{1}{c^2}\delta\boldsymbol{v}^{\mathrm{T}}\boldsymbol{S} - \boldsymbol{G}^{\mathrm{T}}\delta\boldsymbol{v} & \frac{\delta\boldsymbol{v}^{\mathrm{T}}}{c}\mathsf{T} + U\frac{\delta\boldsymbol{v}^{\mathrm{T}}}{c} \\ -\frac{\delta\boldsymbol{v}}{c}U - \mathsf{T}\frac{\delta\boldsymbol{v}}{c} & \delta\boldsymbol{v}\,\boldsymbol{G}^{\mathrm{T}} + \frac{1}{c^2}\boldsymbol{S}\delta\boldsymbol{v}^{\mathrm{T}} \end{pmatrix},
\end{aligned}
\tag{3.4.19}
$$

where we understand $\mathsf{T}$, $\mathbf{0}$, and also $\mathbf{1}$, as 3-dyadics that are of column type on the left and of row type on the right. Accordingly,

$$
\begin{aligned}
\delta U &= -\delta_{\mathrm{coor}} U + \frac{1}{c^2}\delta\boldsymbol{v}^{\mathrm{T}}\boldsymbol{S} + \boldsymbol{G}^{\mathrm{T}}\delta\boldsymbol{v}\,, \\
\delta\boldsymbol{S} &= -\delta_{\mathrm{coor}}\boldsymbol{S} + \delta\boldsymbol{v}\,U + \mathsf{T}\delta\boldsymbol{v}\,, \\
\delta\boldsymbol{G} &= -\delta_{\mathrm{coor}}\boldsymbol{G} + \frac{1}{c^2}\delta\boldsymbol{v}\,U + \frac{1}{c^2}\mathsf{T}\delta\boldsymbol{v}\,, \\
\delta\mathsf{T} &= -\delta_{\mathrm{coor}}\mathsf{T} + \delta\boldsymbol{v}\,\boldsymbol{G}^{\mathrm{T}} + \frac{1}{c^2}\boldsymbol{S}\delta\boldsymbol{v}^{\mathrm{T}},
\end{aligned}
\tag{3.4.20}
$$

or, without the pedantic distinction between rows and columns,

$$
\begin{aligned}
\delta U &= -\delta_{\mathrm{coor}} U + \frac{\delta\boldsymbol{v}}{c}\cdot\left(\frac{1}{c}\boldsymbol{S} + c\boldsymbol{G}\right), \\
\delta\boldsymbol{S} &= -\delta_{\mathrm{coor}}\boldsymbol{S} + \delta\boldsymbol{v}\,U + \mathsf{T}\cdot\delta\boldsymbol{v}\,, \\
\delta\boldsymbol{G} &= -\delta_{\mathrm{coor}}\boldsymbol{G} + \frac{1}{c^2}\delta\boldsymbol{v}\,U + \frac{1}{c^2}\mathsf{T}\cdot\delta\boldsymbol{v}\,, \\
\delta\mathsf{T} &= -\delta_{\mathrm{coor}}\mathsf{T} + \delta\boldsymbol{v}\,\boldsymbol{G} + \frac{1}{c^2}\boldsymbol{S}\delta\boldsymbol{v}\,.
\end{aligned}
\tag{3.4.21}
$$

As a simple check of consistency, we verify that the same transformation law for $U$, say, follows from transforming $\boldsymbol{E}$ and $\boldsymbol{B}$ in $U = \frac{1}{8\pi}(\boldsymbol{E}^2 + \boldsymbol{B}^2)$:

$$\begin{aligned}
\delta U &= \delta\left(\frac{1}{8\pi}(\boldsymbol{E}^2 + \boldsymbol{B}^2)\right) = \frac{1}{4\pi}(\boldsymbol{E}\cdot\delta\boldsymbol{E} + \boldsymbol{B}\cdot\delta\boldsymbol{B}) \\
&= -\delta_{\mathrm{coor}} U + \frac{1}{4\pi}\left[\boldsymbol{E}\cdot\left(-\frac{\delta\boldsymbol{v}}{c}\times\boldsymbol{B}\right) + \boldsymbol{B}\cdot\left(\frac{\delta\boldsymbol{v}}{c}\times\boldsymbol{E}\right)\right] \\
&= -\delta_{\mathrm{coor}} U + \frac{\delta\boldsymbol{v}}{c}\cdot\Bigg[\underbrace{\frac{1}{4\pi}\boldsymbol{E}\times\boldsymbol{B}}_{=\frac{1}{c}\boldsymbol{S}} + \underbrace{\frac{1}{4\pi}\boldsymbol{E}\times\boldsymbol{B}}_{=c\boldsymbol{G}}\Bigg]
\end{aligned} \tag{3.4.22}$$

is correct indeed, but there is quite some arbitrariness in assigning $\frac{1}{c}\boldsymbol{S}$ to one term $\frac{1}{4\pi}\boldsymbol{E}\times\boldsymbol{B}$ and $c\boldsymbol{G}$ to the other. This ambiguity can be removed by extending the analysis from electrodynamics in vacuum to electrodynamics in media. Then the displacement vector field $\boldsymbol{D}(\boldsymbol{r},t)$ and the magnetic induction field $\boldsymbol{H}(\boldsymbol{r},t)$ appear in $\frac{1}{c}\boldsymbol{S} = \boldsymbol{E}\times\boldsymbol{H}$ and $c\boldsymbol{G} = \boldsymbol{D}\times\boldsymbol{B}$. (Confession: I am not sure about the prevailing terminology. Perhaps it is more common to call $\boldsymbol{H}$ the magnetic field and $\boldsymbol{B}$ the magnetic induction. I have also come across the weird term "magnetizing field" for $\boldsymbol{H}$, and you can meet "magnetic flux density" as well, either for $\boldsymbol{B}$ or for $\boldsymbol{H}$. Never mind the terminology though: The fundamental field surely is $\boldsymbol{B}$.)

So, $U$, $\frac{1}{c}\boldsymbol{S}$, $c\boldsymbol{G}$, and $\mathsf{T}$ are the ingredients of a 4-dyadic, that for energy and momentum of the fields $\boldsymbol{E}$ and $\boldsymbol{B}$, but what about the electric and magnetic fields themselves? We recall once more their expressions in terms of the potentials,

$$\boldsymbol{E} = -\frac{1}{c}\frac{\partial}{\partial t}\boldsymbol{A} - \boldsymbol{\nabla}\Phi\,, \qquad \boldsymbol{B} = \boldsymbol{\nabla}\times\boldsymbol{A}\,, \tag{3.4.23}$$

and note that these involve the components of the 4-potential $\begin{pmatrix}\Phi\\ \boldsymbol{A}\end{pmatrix}$ and the 4-gradient $\begin{pmatrix}-\frac{1}{c}\frac{\partial}{\partial t}\\ \boldsymbol{\nabla}\end{pmatrix}$.

The scalar product of these two 4-vectors appears in the Lorentz gauge condition (3.4.13), so we now consider the dyadic product,

$$\begin{pmatrix}-\frac{1}{c}\frac{\partial}{\partial t}\\ \boldsymbol{\nabla}\end{pmatrix}(-\Phi,\ \boldsymbol{A}^{\mathrm{T}}) = \begin{pmatrix}\frac{1}{c}\frac{\partial}{\partial t}\Phi & -\frac{1}{c}\frac{\partial}{\partial t}\boldsymbol{A}^{\mathrm{T}}\\ -\boldsymbol{\nabla}\Phi & \boldsymbol{\nabla}\boldsymbol{A}^{\mathrm{T}}\end{pmatrix} \tag{3.4.24}$$

and its Lorentz transpose

$$\begin{pmatrix} -1 & \boldsymbol{0}^{\mathrm{T}} \\ \boldsymbol{0} & \mathbf{1} \end{pmatrix} \begin{pmatrix} \frac{1}{c}\frac{\partial}{\partial t}\Phi & -\frac{1}{c}\frac{\partial}{\partial t}\boldsymbol{A}^{\mathrm{T}} \\ -\boldsymbol{\nabla}\Phi & \boldsymbol{\nabla}\boldsymbol{A}^{\mathrm{T}} \end{pmatrix}^{\mathrm{T}} \begin{pmatrix} -1 & \boldsymbol{0}^{\mathrm{T}} \\ \boldsymbol{0} & \mathbf{1} \end{pmatrix} = \begin{pmatrix} \frac{1}{c}\frac{\partial}{\partial t}\Phi & \boldsymbol{\nabla}^{\mathrm{T}}\Phi \\ \frac{1}{c}\frac{\partial}{\partial t}\boldsymbol{A} & (\boldsymbol{\nabla}\boldsymbol{A}^{\mathrm{T}})^{\mathrm{T}} \end{pmatrix}, \tag{3.4.25}$$

where, in addition to the usual matrix transposition, we must include the sign changing dyadics on the left (for the column part of the dyadic) and on the right (for the row part), just as we did in (3.4.7).

The difference of these two dyadic products of the 4-gradient and the 4-potential is

$$\begin{pmatrix} 0 & -\frac{1}{c}\frac{\partial}{\partial t}\boldsymbol{A}^{\mathrm{T}} - \boldsymbol{\nabla}^{\mathrm{T}}\Phi \\ -\boldsymbol{\nabla}\Phi - \frac{1}{c}\frac{\partial}{\partial t}\boldsymbol{A} & \boldsymbol{\nabla}\boldsymbol{A}^{\mathrm{T}} - (\boldsymbol{\nabla}\boldsymbol{A}^{\mathrm{T}})^{\mathrm{T}} \end{pmatrix} = \begin{pmatrix} 0 & \boldsymbol{E}^{\mathrm{T}} \\ \boldsymbol{E} & -\boldsymbol{B}\times\mathbf{1} \end{pmatrix}, \tag{3.4.26}$$

as we realize upon writing out the cartesian components of the 3-dyadic, beginning with

$$\boldsymbol{\nabla}\boldsymbol{A}^{\mathrm{T}} = \begin{pmatrix} \frac{\partial}{\partial x}A_x & \frac{\partial}{\partial x}A_y & \frac{\partial}{\partial x}A_z \\ \frac{\partial}{\partial y}A_x & \frac{\partial}{\partial y}A_y & \frac{\partial}{\partial y}A_z \\ \frac{\partial}{\partial z}A_x & \frac{\partial}{\partial z}A_y & \frac{\partial}{\partial z}A_z \end{pmatrix} \tag{3.4.27}$$

and then continuing to arrive at

$$\begin{aligned} \boldsymbol{\nabla}\boldsymbol{A}^{\mathrm{T}} - (\boldsymbol{\nabla}\boldsymbol{A}^{\mathrm{T}})^{\mathrm{T}} &= \begin{pmatrix} 0 & \frac{\partial}{\partial x}A_y - \frac{\partial}{\partial y}A_x & \frac{\partial}{\partial x}A_z - \frac{\partial}{\partial z}A_x \\ \frac{\partial}{\partial y}A_x - \frac{\partial}{\partial x}A_y & 0 & \frac{\partial}{\partial y}A_z - \frac{\partial}{\partial z}A_y \\ \frac{\partial}{\partial z}A_x - \frac{\partial}{\partial x}A_z & \frac{\partial}{\partial z}A_y - \frac{\partial}{\partial y}A_z & 0 \end{pmatrix} \\ &= \begin{pmatrix} 0 & B_z & -B_y \\ -B_z & 0 & B_x \\ B_y & -B_x & 0 \end{pmatrix} = -\begin{pmatrix} B_x \\ B_y \\ B_z \end{pmatrix} \times \begin{pmatrix} 1\,0\,0 \\ 0\,1\,0 \\ 0\,0\,1 \end{pmatrix} \\ &= -\boldsymbol{B}\times\mathbf{1}\,. \end{aligned} \tag{3.4.28}$$

Alternatively, we could recognize that

$$\boldsymbol{\nabla}\boldsymbol{A}^{\mathrm{T}} - \left(\boldsymbol{\nabla}\boldsymbol{A}^{\mathrm{T}}\right)^{\mathrm{T}} = -\mathbf{1} \times \boldsymbol{B}^{\mathrm{T}}, \tag{3.4.29}$$

and since these dyadics are antisymmetric — the difference on the left-hand side clearly changes sign under transposition — we also learn that

$$-\boldsymbol{B} \times \mathbf{1} = \left(\boldsymbol{B} \times \mathbf{1}\right)^{\mathrm{T}} = -\mathbf{1} \times \boldsymbol{B}^{\mathrm{T}}. \tag{3.4.30}$$

The resulting *field 4-dyadic* is antisymmetric under Lorentz transposition

$$\begin{aligned} \begin{pmatrix} 0 & \boldsymbol{E}^{\mathrm{T}} \\ \boldsymbol{E} & -\boldsymbol{B} \times \mathbf{1} \end{pmatrix} &\to \begin{pmatrix} -1 & \boldsymbol{0}^{\mathrm{T}} \\ \boldsymbol{0} & \mathbf{1} \end{pmatrix} \begin{pmatrix} 0 & \boldsymbol{E}^{\mathrm{T}} \\ \boldsymbol{E} & -\boldsymbol{B} \times \mathbf{1} \end{pmatrix}^{\mathrm{T}} \begin{pmatrix} -1 & \boldsymbol{0}^{\mathrm{T}} \\ \boldsymbol{0} & \mathbf{1} \end{pmatrix} \\ &= \begin{pmatrix} 0 & -\boldsymbol{E}^{\mathrm{T}} \\ -\boldsymbol{E} & \boldsymbol{B} \times \mathbf{1} \end{pmatrix}, \end{aligned} \tag{3.4.31}$$

and so we recognize that $\boldsymbol{E}$ and $\boldsymbol{B}$ make up the components of an antisymmetric 4-dyadic, often called the "field tensor."

The dot product with the 4-gradient,

$$\begin{aligned} \left(\frac{1}{c}\frac{\partial}{\partial t}, \boldsymbol{\nabla}^{\mathrm{T}}\right) \begin{pmatrix} 0 & -\boldsymbol{E}^{\mathrm{T}} \\ -\boldsymbol{E} & \boldsymbol{B} \times \mathbf{1} \end{pmatrix} &= \left(-\boldsymbol{\nabla}\cdot\boldsymbol{E},\; -\frac{1}{c}\frac{\partial}{\partial t}\boldsymbol{E}^{\mathrm{T}} + \boldsymbol{\nabla}^{\mathrm{T}}\boldsymbol{B} \times \mathbf{1}\right) \\ &= \left(-\boldsymbol{\nabla}\cdot\boldsymbol{E},\; -\frac{1}{c}\frac{\partial}{\partial t}\boldsymbol{E}^{\mathrm{T}} + \left(\boldsymbol{\nabla} \times \boldsymbol{B}\right)^{\mathrm{T}}\right) \\ &= \frac{4\pi}{c}\left(-c\rho,\; \boldsymbol{j}^{\mathrm{T}}\right), \end{aligned} \tag{3.4.32}$$

gives the row version of the 4-current density and thus states the inhomogeneous Maxwell's equations in compact 4-vector notation.

To deal with the homogeneous Maxwell's equations, we note that the infinitesimal transformation equations for $\boldsymbol{E}$ and $\boldsymbol{B}$, see (3.3.13) and (3.3.15),

$$\begin{aligned} \delta\boldsymbol{E} &= -\delta_{\mathrm{coor}}\boldsymbol{E} - \frac{\delta\boldsymbol{v}}{c} \times \boldsymbol{B}, \\ \delta\boldsymbol{B} &= -\delta_{\mathrm{coor}}\boldsymbol{B} + \frac{\delta\boldsymbol{v}}{c} \times \boldsymbol{E}, \end{aligned} \tag{3.4.33}$$

are themselves invariant under the replacements

$$\boldsymbol{E} \to \boldsymbol{B}, \qquad \boldsymbol{B} \to -\boldsymbol{E}. \tag{3.4.34}$$

It follows that the field dyadic that results from this replacement, that is

$$\begin{pmatrix} 0 & \boldsymbol{B}^{\mathrm{T}} \\ \boldsymbol{B} & \boldsymbol{E}\times\mathbf{1} \end{pmatrix} \tag{3.4.35}$$

is also a proper 4-dyadic. The dot product with the 4-gradient,

$$\begin{aligned} \left(\frac{1}{c}\frac{\partial}{\partial t},\ \boldsymbol{\nabla}^{\mathrm{T}}\right)\begin{pmatrix} 0 & \boldsymbol{B}^{\mathrm{T}} \\ \boldsymbol{B} & \boldsymbol{E}\times\mathbf{1} \end{pmatrix} &= \left(\boldsymbol{\nabla}\cdot\boldsymbol{B},\ \left(\frac{1}{c}\frac{\partial}{\partial t}\boldsymbol{B}+\boldsymbol{\nabla}\times\boldsymbol{E}\right)^{\mathrm{T}}\right) \\ &= (0,\ \boldsymbol{o}^{\mathrm{T}}), \end{aligned} \tag{3.4.36}$$

gives us the time-like component and the space-like component of a 4-row, precisely the differential expressions in the homogeneous Maxwell's equations. The right-hand sides of them, vanishing everywhere and at all times, compose of course a null 4-vector.

We note that Maxwell's equations as a whole are *not* invariant under $\boldsymbol{E}\to\boldsymbol{B}$, $\boldsymbol{B}\to-\boldsymbol{E}$, because there is no corresponding transformation for the charge and current densities, unless we introduce also magnetic charges and currents. There is, however, no experimental evidence that magnetic charges exist in Nature. It is worth noting, though, that — as was first observed by Dirac* — the existence of magnetic charges would explain, in a very natural way, why electric charge occurs only in multiples of a charge unit, the elementary charge $e_0$, which equals $4.80320\times10^{-10}$ Fr (or $1.60218\times 10^{-19}$ C in SI units). This is so because the electric Coulomb field of an electric point charge and the magnetic Coulomb field of a magnetic point charge together form an electromagnetic field with a nonzero total angular momentum; see Section 3.6 below. Now, quantum mechanics demands that angular momentum is quantized in units of $\frac{1}{2}\hbar$ and, as a consequence, the product of the two charges can take on only certain discrete values. Therefore, a single magnetic charge somewhere in the universe would force all electric charges to be multiples of a unit charge.

In the absence of charges and currents, Maxwell's equations (1.1.1) read

$$\begin{aligned} \boldsymbol{\nabla}\cdot\boldsymbol{E} &= 0\,, & \boldsymbol{\nabla}\cdot\boldsymbol{B} &= 0\,, \\ \boldsymbol{\nabla}\times\boldsymbol{B}-\frac{1}{c}\frac{\partial}{\partial t}\boldsymbol{E} &= 0\,, & \boldsymbol{\nabla}\times\boldsymbol{E}+\frac{1}{c}\frac{\partial}{\partial t}\boldsymbol{B} &= 0\,, \end{aligned} \tag{3.4.37}$$

and this set *is* invariant under the replacements (3.4.34) or, more generally, under the so-called electromagnetic rotation $(\boldsymbol{E}+\mathrm{i}\boldsymbol{B})\to\mathrm{e}^{\mathrm{i}\varphi}(\boldsymbol{E}+\mathrm{i}\boldsymbol{B})$ with arbitrary angle $\varphi$. It follows that we need charges to tell apart the electric

*Paul Adrien Maurice DIRAC (1902–1984)

and magnetic fields. Indeed, the fields play very different roles in the force density

$$\boldsymbol{f} = \rho \boldsymbol{E} + \frac{1}{c} \boldsymbol{j} \times \boldsymbol{B}, \tag{3.4.38}$$

where you cannot simply replace $\boldsymbol{E}$ by $\boldsymbol{B}$ and $\boldsymbol{B}$ by $-\boldsymbol{E}$. A general expression for the force density, one that includes the corresponding terms for magnetic charge and current densities, would be invariant provided that the charge and current densities are transformed likewise, of course.

## 3.5 Finite Lorentz transformations

Having established the response of the potentials $\Phi$, $\boldsymbol{A}$, of the charge and current distributions $\rho$, $\boldsymbol{j}$, and of the fields $\boldsymbol{E}$, $\boldsymbol{B}$ to infinitesimal Lorentz transformations, let us now find how they behave under finite transformation. For the 4-vector fields $\begin{pmatrix} \Phi \\ \boldsymbol{A} \end{pmatrix}$ and $\begin{pmatrix} c\rho \\ \boldsymbol{j} \end{pmatrix}$, we can just rely on our knowledge about the 4-vector of time and position, the *event* 4-vector $\begin{pmatrix} ct \\ \boldsymbol{r} \end{pmatrix}$, for which we have (3.1.10),

$$\begin{aligned} \begin{pmatrix} ct' \\ \boldsymbol{r}' \end{pmatrix} &= \begin{pmatrix} \gamma & \gamma \boldsymbol{v}^{\mathrm{T}}/c \\ \gamma \boldsymbol{v}/c & \mathbf{1} + (\gamma - 1)\dfrac{\boldsymbol{v}\boldsymbol{v}^{\mathrm{T}}}{v^2} \end{pmatrix} \begin{pmatrix} ct \\ \boldsymbol{r} \end{pmatrix} \\ &= L(\boldsymbol{v}) \begin{pmatrix} ct \\ \boldsymbol{r} \end{pmatrix}. \end{aligned} \tag{3.5.1}$$

Therefore,

$$\begin{pmatrix} \Phi'(\boldsymbol{r}', t') \\ \boldsymbol{A}'(\boldsymbol{r}', t') \end{pmatrix} = L(\boldsymbol{v}) \begin{pmatrix} \Phi(\boldsymbol{r}, t) \\ \boldsymbol{A}(\boldsymbol{r}, t) \end{pmatrix} \Bigg|_{\begin{pmatrix} ct \\ \boldsymbol{r} \end{pmatrix} = L(\boldsymbol{v})^{-1} \begin{pmatrix} ct' \\ \boldsymbol{r}' \end{pmatrix}}, \tag{3.5.2}$$

where we must remember that the application of the Lorentz-transformation 4-dyadic $L(\boldsymbol{v})$ is not the whole story, we must also transform the coordinates, thereby expressing $ct$ and $\boldsymbol{r}$ in terms of $ct'$ and $\boldsymbol{r}'$, the event coordinates after the transformation. This coordinate transformation is, of course, what is accounted for by $\delta_{\text{coor}}$ in the infinitesimal transformations of Section 3.3.

As an example, we consider the Coulomb potential of a point charge $e$ at rest in the $t$, $\boldsymbol{r}$ frame,

$$\Phi(\boldsymbol{r},t) = \frac{e}{|\boldsymbol{r}|} = \frac{e}{r}\,,$$
$$\boldsymbol{A}(\boldsymbol{r},t) = 0\,. \tag{3.5.3}$$

Then, in the $t'$, $\boldsymbol{r}'$ frame, we have

$$\Phi'(\boldsymbol{r}',t') = \gamma\frac{e}{r}\,,$$
$$\boldsymbol{A}'(\boldsymbol{r}',t') = \gamma\frac{\boldsymbol{v}}{c}\frac{e}{r}\,, \tag{3.5.4}$$

where $r = |\boldsymbol{r}|$ is to be expressed in terms of $t'$ and $\boldsymbol{r}'$. For $\boldsymbol{v} = v\boldsymbol{e}_z$, we have

$$x = x'\,,$$
$$y = y'\,,$$
$$z = \gamma(z' - vt')\,,$$
$$t = \gamma\left(t' - \frac{v}{c^2}z'\right)\,, \tag{3.5.5}$$

so that

$$\begin{aligned} r &= \sqrt{x^2 + y^2 + z^2} \\ &= \sqrt{x'^2 + y'^2 + \gamma^2(z' - vt')^2} \\ &= \gamma\sqrt{\frac{x'^2 + y'^2}{\gamma^2} + (z' - vt')^2}\,, \end{aligned} \tag{3.5.6}$$

and

$$\Phi'(\boldsymbol{r}',t') = \frac{e}{\sqrt{\dfrac{x'^2 + y'^2}{\gamma^2} + (z' - vt')^2}}\,,$$
$$\boldsymbol{A}'(\boldsymbol{r}',t') = \frac{\boldsymbol{v}}{c}\Phi'(\boldsymbol{r}',t') \tag{3.5.7}$$

are the potentials in the primed frame in which the charge is at $(x', y', z') = (0, 0, vt')$ at time $t'$.

For the 4-current density $\begin{pmatrix} c\rho \\ \boldsymbol{j} \end{pmatrix}$ we have likewise

$$\begin{pmatrix} c\rho'(\boldsymbol{r}',t') \\ \boldsymbol{j}'(\boldsymbol{r}',t') \end{pmatrix} = L(\boldsymbol{v}) \begin{pmatrix} c\rho(\boldsymbol{r},t) \\ \boldsymbol{j}(\boldsymbol{r},t) \end{pmatrix} \Bigg|_{\begin{pmatrix} ct \\ \boldsymbol{r} \end{pmatrix} = L(\boldsymbol{v})^{-1} \begin{pmatrix} ct' \\ \boldsymbol{r}' \end{pmatrix}} . \tag{3.5.8}$$

The corresponding example of

$$\begin{aligned} \rho(\boldsymbol{r},t) &= e\delta(\boldsymbol{r}) , \\ \boldsymbol{j}(\boldsymbol{r},t) &= 0 , \end{aligned} \tag{3.5.9}$$

that is: point charge at rest at $\boldsymbol{r} = 0$ again, gives

$$\begin{aligned} \rho'(\boldsymbol{r}',t') &= \gamma e\delta(\boldsymbol{r}) , \\ \boldsymbol{j}'(\boldsymbol{r}',t') &= \boldsymbol{v}\rho'(\boldsymbol{r}',t') , \end{aligned} \tag{3.5.10}$$

where

$$\delta(\boldsymbol{r}) = \delta(x)\delta(y)\delta(z) \tag{3.5.11}$$

has a three-dimensional delta function on the left and the product of three one-dimensional delta functions on the right, so that

$$\begin{aligned} \delta(\boldsymbol{r}) &= \delta(x')\delta(y')\delta\big(\gamma(z' - vt')\big) \\ &= \frac{1}{\gamma}\delta(x')\delta(y')\delta(z' - vt') \\ &= \frac{1}{\gamma}\delta(\boldsymbol{r}' - \boldsymbol{v}t') \end{aligned} \tag{3.5.12}$$

for $\boldsymbol{v} = v\boldsymbol{e}_z$. In summary, then,

$$\begin{aligned} \rho'(\boldsymbol{r}',t') &= e\delta(\boldsymbol{r}' - \boldsymbol{v}t') , \\ \boldsymbol{j}'(\boldsymbol{r}',t') &= \boldsymbol{v}\rho'(\boldsymbol{r}',t') , \end{aligned} \tag{3.5.13}$$

as we could have expected for charge $e$ at position $\boldsymbol{v}t'$ at time $t'$.

In order to obtain the finite Lorentz transformation for $\boldsymbol{E}$ and $\boldsymbol{B}$ we could exploit the 4-dyadic that they form,

$$\begin{pmatrix} 0 & \boldsymbol{E}'^{\mathrm{T}} \\ \boldsymbol{E}' & -\boldsymbol{B}'\times\mathbf{1} \end{pmatrix}(\boldsymbol{r}',t') = L(\boldsymbol{v})\begin{pmatrix} 0 & \boldsymbol{E}^{\mathrm{T}} \\ \boldsymbol{E} & -\boldsymbol{B}\times\mathbf{1} \end{pmatrix}(\boldsymbol{r},t)L(\boldsymbol{v})^{-1}\Bigg|_{\binom{ct}{\boldsymbol{r}} = L(\boldsymbol{v})^{-1}\binom{ct'}{\boldsymbol{r}'}}, \tag{3.5.14}$$

and you should regard this as an exercise for yourself. For a change, however, let us return to the infinitesimal transformation

$$\delta\boldsymbol{E} = -\delta_{\mathrm{coor}}\boldsymbol{E} - \frac{\delta\boldsymbol{v}}{c}\times\boldsymbol{B}\,, \qquad \delta\boldsymbol{B} = -\delta_{\mathrm{coor}}\boldsymbol{B} + \frac{\delta\boldsymbol{v}}{c}\times\boldsymbol{E}\,, \tag{3.5.15}$$

which we consider for $\delta\boldsymbol{v} = c\delta\theta\,\boldsymbol{e}_z$ with the rapidity parameter $\theta$ that we first met in Section 3.2. We remember that $\delta_{\mathrm{coor}}$ accounts for the coordinate transformation that we need for all fields ($\equiv$ functions of $\boldsymbol{r}$ and $t$), and keep that implicit. Then

$$\begin{aligned} \delta E_x &= \delta\theta\, B_y\,, & \delta B_x &= -\delta\theta\, E_y\,, \\ \delta E_y &= -\delta\theta\, B_x\,, & \delta B_y &= \delta\theta\, E_x\,, \\ \delta E_z &= 0\,, & \delta B_z &= 0\,, \end{aligned} \tag{3.5.16}$$

which are the pairs

$$\begin{aligned} \frac{\partial}{\partial\theta}(E_x, B_y) &= (B_y, E_x)\,, \\ \frac{\partial}{\partial\theta}(E_y, B_x) &= (-B_x, -E_y)\,, \\ \frac{\partial}{\partial\theta}(E_z, B_z) &= 0\,. \end{aligned} \tag{3.5.17}$$

These differential equations are solved by

$$\begin{aligned} E_x(\theta) &= E_x(0)\cosh\theta + B_y(0)\sinh\theta\,, \\ E_y(\theta) &= E_y(0)\cosh\theta - B_x(0)\sinh\theta\,, \\ E_z(\theta) &= E_z(0)\,; \\ B_x(\theta) &= B_x(0)\cosh\theta - E_y(0)\sinh\theta\,, \\ B_y(\theta) &= B_y(0)\cosh\theta + E_x(0)\sinh\theta\,, \\ B_z(\theta) &= B_z(0)\,, \end{aligned} \tag{3.5.18}$$

or, after identifying $\boldsymbol{E}(\theta=0)$ with $\boldsymbol{E}(\boldsymbol{r},t)$ and $\boldsymbol{E}(\theta>0)$ with $\boldsymbol{E}'(\boldsymbol{r}',t')$ and likewise for the magnetic fields, and recalling the relations (3.2.6),

$$\begin{aligned}
\boldsymbol{E}'_{\perp}(\boldsymbol{r}',t') &= \gamma\boldsymbol{E}_{\perp}(\boldsymbol{r},t)-\gamma\frac{\boldsymbol{v}}{c}\times\boldsymbol{B}_{\perp}(\boldsymbol{r},t)\,,\\
\boldsymbol{E}'_{\parallel}(\boldsymbol{r}',t') &= \boldsymbol{E}_{\parallel}(\boldsymbol{r},t)\,,\\
\boldsymbol{B}'_{\perp}(\boldsymbol{r}',t') &= \gamma\boldsymbol{B}_{\perp}(\boldsymbol{r},t)+\gamma\frac{\boldsymbol{v}}{c}\times\boldsymbol{E}_{\perp}(\boldsymbol{r},t)\,,\\
\boldsymbol{B}'_{\parallel}(\boldsymbol{r}',t') &= \boldsymbol{B}_{\parallel}(\boldsymbol{r},t)\,,
\end{aligned}\tag{3.5.19}$$

where, for instance, $\boldsymbol{E}_{\parallel}=\dfrac{\boldsymbol{v}\,\boldsymbol{v}}{v^2}\cdot\boldsymbol{E}$ and $\boldsymbol{E}_{\perp}=\boldsymbol{E}-\boldsymbol{E}_{\parallel}=\dfrac{\boldsymbol{v}}{v}\times\left(\boldsymbol{E}\times\dfrac{\boldsymbol{v}}{v}\right)$ are the parallel and perpendicular components of the electric field, respectively.

Since $\boldsymbol{v}\times\boldsymbol{B}_{\perp}=\boldsymbol{v}\times\boldsymbol{B}$ and $\boldsymbol{v}\times\boldsymbol{E}_{\perp}=\boldsymbol{v}\times\boldsymbol{E}$, we get

$$\begin{aligned}
\boldsymbol{E}' &= \underbrace{\gamma\Bigl(\mathbf{1}-\frac{\boldsymbol{v}\,\boldsymbol{v}}{v^2}\Bigr)\cdot\boldsymbol{E}-\gamma\frac{\boldsymbol{v}}{c}\times\boldsymbol{B}}_{=\boldsymbol{E}'_{\perp}}+\underbrace{\frac{\boldsymbol{v}\,\boldsymbol{v}}{v^2}\cdot\boldsymbol{E}}_{=\boldsymbol{E}'_{\parallel}}\\
&= \Bigl(\gamma\mathbf{1}-(\gamma-1)\frac{\boldsymbol{v}\,\boldsymbol{v}}{v^2}\Bigr)\cdot\boldsymbol{E}-\gamma\frac{\boldsymbol{v}}{c}\times\boldsymbol{B}
\end{aligned}\tag{3.5.20}$$

and

$$\boldsymbol{B}'=\Bigl(\gamma\mathbf{1}-(\gamma-1)\frac{\boldsymbol{v}\,\boldsymbol{v}}{v^2}\Bigr)\cdot\boldsymbol{B}+\gamma\frac{\boldsymbol{v}}{c}\times\boldsymbol{E}\,,\tag{3.5.21}$$

where we remember that $\boldsymbol{E}'=\boldsymbol{E}'(\boldsymbol{r}',t')$ and $\boldsymbol{B}'=\boldsymbol{B}'(\boldsymbol{r}',t')$ on the left and $\boldsymbol{E}=\boldsymbol{E}(\boldsymbol{r},t)$ and $\boldsymbol{B}=\boldsymbol{B}(\boldsymbol{r},t)$ on the right, with

$$\begin{pmatrix} ct\\ \boldsymbol{r}\end{pmatrix}=L(\boldsymbol{v})^{-1}\begin{pmatrix} ct'\\ \boldsymbol{r}'\end{pmatrix}\tag{3.5.22}$$

being understood. As they should, the relations (3.5.20) and (3.5.21) are clearly turned into each other when we interchange $\boldsymbol{E}$ and $\boldsymbol{B}$ in accordance with (3.4.34), that is: $(\boldsymbol{E},\boldsymbol{B})\to(\boldsymbol{B},-\boldsymbol{E})$.

## 3.6 A digression: Dirac's magnetic monopole

As mentioned in passing between (3.4.36) and (3.4.37), Dirac entertained the speculation that, in addition to the magnetic dipoles of paired north and south poles and their divergenceless magnetic fields, there could also be magnetic monopoles — single magnetic charges that act as isolated north poles or isolated south poles if the charge is positive or negative, respectively. There is no experimental evidence at all for such magnetic charges, no violation of Gilbert's law $\boldsymbol{\nabla}\cdot\boldsymbol{B}=0$ has ever been observed.

Nevertheless, let us follow Dirac's guidance and consider the electromagnetic field associated with an electric charge $e$ at rest at position $\boldsymbol{a}$ and a magnetic charge $g$ at rest at position $\boldsymbol{b}$. In this static situation, the electric field is the Coulomb field

$$\boldsymbol{E}(\boldsymbol{r}) = e\frac{\boldsymbol{r}-\boldsymbol{a}}{|\boldsymbol{r}-\boldsymbol{a}|^3} = -\boldsymbol{\nabla}\frac{e}{|\boldsymbol{r}-\boldsymbol{a}|} \tag{3.6.1}$$

and the magnetic field is

$$\boldsymbol{B}(\boldsymbol{r}) = g\frac{\boldsymbol{r}-\boldsymbol{b}}{|\boldsymbol{r}-\boldsymbol{b}|^3} = -\boldsymbol{\nabla}\frac{g}{|\boldsymbol{r}-\boldsymbol{b}|}\,. \tag{3.6.2}$$

The momentum density

$$\boldsymbol{G}(\boldsymbol{r}) = \frac{1}{4\pi c}\boldsymbol{E}(\boldsymbol{r})\times\boldsymbol{B}(\boldsymbol{r}) \tag{3.6.3}$$

is the cross product of two gradient fields and, therefore, it is a curl field,

$$\boldsymbol{G}(\boldsymbol{r}) = \frac{eg}{4\pi c}\boldsymbol{\nabla}\times\left(\frac{1}{|\boldsymbol{r}-\boldsymbol{a}|}\boldsymbol{\nabla}\frac{1}{|\boldsymbol{r}-\boldsymbol{b}|}\right). \tag{3.6.4}$$

As a consequence, the total momentum vanishes,

$$\int(\mathrm{d}\boldsymbol{r})\,\boldsymbol{G}(\boldsymbol{r}) = 0\,, \tag{3.6.5}$$

and the point of reference for the angular momentum is irrelevant,

$$\begin{aligned}\boldsymbol{J} &= \int(\mathrm{d}\boldsymbol{r})\,\boldsymbol{r}\times\boldsymbol{G}(\boldsymbol{r}) = \int(\mathrm{d}\boldsymbol{r})\,(\boldsymbol{r}-\boldsymbol{a})\times\boldsymbol{G}(\boldsymbol{r})\\ &= \int(\mathrm{d}\boldsymbol{r})\,\boldsymbol{r}\times\boldsymbol{G}(\boldsymbol{r}+\boldsymbol{a})\\ &= \frac{eg}{4\pi c}\int(\mathrm{d}\boldsymbol{r})\,\boldsymbol{r}\times\left[\boldsymbol{\nabla}\times\left(\frac{1}{r}\boldsymbol{\nabla}\frac{1}{|\boldsymbol{r}-(\boldsymbol{b}-\boldsymbol{a})|}\right)\right].\end{aligned} \tag{3.6.6}$$

This integrand is of the form

$$\boldsymbol{r}\times\Bigl(\boldsymbol{\nabla}\times\boldsymbol{F}\Bigr) = \boldsymbol{\nabla}\Bigl(\boldsymbol{r}\cdot\boldsymbol{F}\Bigr) - \boldsymbol{\nabla}\cdot\Bigl(\boldsymbol{r}\,\boldsymbol{F}\Bigr) + 2\boldsymbol{F}\,, \tag{3.6.7}$$

for which

$$\int(\mathrm{d}\boldsymbol{r})\,\boldsymbol{r}\times\Bigl(\boldsymbol{\nabla}\times\boldsymbol{F}\Bigr) = 2\int(\mathrm{d}\boldsymbol{r})\,\boldsymbol{F}\,, \tag{3.6.8}$$

so that

$$J = \frac{eg}{2\pi c}\int(\mathrm{d}\boldsymbol{r})\,\frac{1}{r}\boldsymbol{\nabla}\frac{1}{|\boldsymbol{r}-\boldsymbol{s}|}$$

$$= \frac{eg}{2\pi c}\int(\mathrm{d}\boldsymbol{r})\,\frac{1}{r}\frac{\boldsymbol{s}-\boldsymbol{r}}{|\boldsymbol{s}-\boldsymbol{r}|^3} \quad \text{with} \quad \boldsymbol{s}=\boldsymbol{b}-\boldsymbol{a}\,. \tag{3.6.9}$$

Here we recall a lesson from electrostatics, namely that the electric field of a purely radial charge density $\rho(r)$ is given by

$$\boldsymbol{E}(\boldsymbol{r}) = \int(\mathrm{d}\boldsymbol{r}')\,\rho(r')\frac{\boldsymbol{r}-\boldsymbol{r}'}{|\boldsymbol{r}-\boldsymbol{r}'|^3} = Q(r)\frac{\boldsymbol{r}}{r^3} \tag{3.6.10}$$

where

$$Q(r) = 4\pi\int_0^r \mathrm{d}r'\,r'^2\rho(r') \tag{3.6.11}$$

is the charge inside the sphere of radius $r$. (The analogous statement about the gravitational field of a spherically symmetric mass distribution was discovered by Newton.*) In the context of (3.6.9) this tells us that

$$\boldsymbol{J} = \frac{eg}{2\pi c}\frac{\boldsymbol{s}}{s^3}\underbrace{4\pi\int_0^s \mathrm{d}r\,r^2\frac{1}{r}}_{=2\pi s^2} = \frac{eg}{c}\boldsymbol{n}\,, \tag{3.6.12}$$

where $\boldsymbol{n} = \dfrac{\boldsymbol{s}}{s}$ is the unit vector pointing from the electric charge $e$ to the magnetic charge $g$. Quite remarkably, the angular momentum $\boldsymbol{J}$ does not depend on the distance between the two charges, they could be close to each other or separated by an astronomical distance.

In quantum physics, a component of the angular momentum, such as $\boldsymbol{n}\cdot\boldsymbol{J} = eg/c$, can only take on a value that is an integer multiple of $\frac{1}{2}\hbar$, whereby $2\pi\hbar = 6.62607\times10^{-27}\,\mathrm{erg\,s}$ (cgs units) $= 0.658212$ eV fs (atomic units) is Planck's† constant, of course. This implies that

$$eg = \frac{\ell}{2}\hbar c \quad \text{with} \quad \ell = 0, \pm1, \pm2, \ldots\,, \tag{3.6.13}$$

so that the existence of a single magnetic charge $g$ anywhere in the universe would require all electric charges to be integer multiples of $\frac{1}{2}\hbar c/g$. This is Dirac's insight of 1931. It appears that Pierre Curie‡ hypothesized the

*Sir Isaac NEWTON (1643–1727) †Max Karl Ernst Ludwig PLANCK (1858–1947)
‡Pierre CURIE (1859–1906)

existence of magnetic charges already in 1894, but at that time the quantum properties of angular momentum were not known and the implication about the discreteness of electric charge could not be recognized.

It is worth noting that the value of the observed unit of electric charge $e_0$ is such that the ratio $e_0^2/(\hbar c)$ is Sommerfeld's* fine structure constant $\alpha$,

$$\frac{e_0^2}{\hbar c} = \alpha = \frac{1}{137.036}, \tag{3.6.14}$$

which is a rather small number. In marked contrast, the corresponding value for the unit of magnetic charge, $g_0 = e_0/(2\alpha)$,

$$\frac{g_0^2}{\hbar c} = \frac{1}{4\alpha} = 34.26, \tag{3.6.15}$$

is a rather large number. This says that the force between two unit magnetic charges is larger by a factor of

$$\frac{1}{4\alpha^2} \simeq 4700 \tag{3.6.16}$$

than the force between two unit electric charges at the same distance. Magnetic charges, should they exist, would interact very strongly.

*Arnold SOMMERFELD (1868–1951)

## Chapter 4

# 3+1–Dimensional Notation

### 4.1 Contravariant and covariant components

Whereas we managed to study, and exploit, Lorentz transformations with notational conventions that derive directly from the familiar vector formalism of three-dimensional space, the systematic appearance of 4-component vectors, both of column-type and of row-type, along with $4\times 4$ dyadic quantities, is a reminder that there is a formalism at hand that emphasizes the four-dimensional space-time structure more directly. It is clear that one must carefully distinguish between the 4-vectors of column-type and those of row-type, which one does by using upper and lower indices, as illustrated by the vectors of both kinds for an event,

$$\text{column}: \begin{pmatrix} ct \\ \boldsymbol{r} \end{pmatrix} = \begin{pmatrix} ct \\ x \\ y \\ z \end{pmatrix} = \begin{pmatrix} x^0 \\ x^1 \\ x^2 \\ x^3 \end{pmatrix};$$

$$\text{row}: (-ct, \boldsymbol{r}^{\mathrm{T}}) = (-ct, x, y, z) = (x_0, x_1, x_2, x_3). \qquad (4.1.1)$$

Common terminology calls the $x^\mu$ ($\mu = 0, 1, 2, 3$) components the *contravariant* components, and the $x_\mu$ components are called *covariant.* So, what was earlier referred to as a vector would now be called a "contravariant 4-vector," and the vector is a "covariant 4-vector."

Terminology aside, what we need to remember is the sign change $x_0 = -ct = -x^0$ in the time-like component. A compact way of switching from one kind of coordinates to the other is by means of the so-called *metrical tensor* $g_{\mu\nu}$,

$$x_\mu = \sum_{\nu=0}^{3} g_{\mu\nu} x^\nu \tag{4.1.2}$$

where

$$g_{\mu\nu} = \begin{cases} -1 & \text{if} \quad \mu = \nu = 0 \\ +1 & \text{if} \quad \mu = \nu = 1,2,3 \\ 0 & \text{if} \quad \mu \neq \nu \end{cases} \tag{4.1.3}$$

and it is common convenient practice not to write the summation sign,

$$x_\mu = g_{\mu\nu} x^\nu \,, \tag{4.1.4}$$

where it is *Einstein's summation convention* to understand that all repeated indices are summed over (here: $_\nu$ and $^\nu$), whereby one index has to be covariant (lower) and the other contravariant (upper). For example,

$$x^\mu = g^\mu{}_\nu x^\nu, \tag{4.1.5}$$

where, of course,

$$g^\mu{}_\nu = \begin{cases} 1 & \text{if} \quad \mu = \nu \\ 0 & \text{if} \quad \mu \neq \nu \end{cases} \tag{4.1.6}$$

or

$$x^\mu = g^{\mu\nu} x_\nu \tag{4.1.7}$$

with

$$g^{\mu\nu} = \begin{cases} -1 & \text{if} \quad \mu = \nu = 0 \\ +1 & \text{if} \quad \mu = \nu = 1,2,3 \\ 0 & \text{if} \quad \mu \neq \nu \end{cases} \tag{4.1.8}$$

but we must *never* write $g^{\mu\nu} = g_{\mu\nu}$ because a 4-dyadic that has two contravariant indices cannot equal one that has two covariant indices, although in this particular case the numerical values of the 16 components are the same.

The product of a 4-row vector with a 4-column vector is a 4-scalar, with

$$(-ct, \boldsymbol{r}^{\mathrm{T}}) \begin{pmatrix} ct \\ \boldsymbol{r} \end{pmatrix} = -(ct)^2 + \boldsymbol{r}^2 \tag{4.1.9}$$

illustrating the basic case, where the product is invariant indeed,

$$\begin{aligned}\delta\big[-(ct)^2+\boldsymbol{r}^2\big] &= -2c^2t\,\delta t+2\boldsymbol{r}\cdot\delta\boldsymbol{r}\\ &= -2c^2t\frac{\delta\boldsymbol{v}\cdot\boldsymbol{r}}{c^2}+2\boldsymbol{r}\cdot\delta\boldsymbol{v}\,t=0\,.\end{aligned}\tag{4.1.10}$$

In the new notation we have

$$(-ct,\boldsymbol{r}^{\mathrm{T}})\begin{pmatrix}ct\\ \boldsymbol{r}\end{pmatrix}=x_\mu x^\mu\,,\tag{4.1.11}$$

and with no index left over, the notation suggests strongly that we have a scalar quantity, which is true as we know.

This example is indeed suggestive and indicates correctly the power of the mixed 4-vector notation with contravariant and covariant quantities: In an equation the free indices have to match.

The contravariant 4-gradient is, recall (3.3.8),

$$\partial\equiv\begin{pmatrix}-\dfrac{1}{c}\dfrac{\partial}{\partial t}\\ \boldsymbol{\nabla}\end{pmatrix}=\begin{pmatrix}\partial^0\\ \partial^1\\ \partial^2\\ \partial^3\end{pmatrix},\tag{4.1.12}$$

so that

$$\partial^\mu=\frac{\partial}{\partial x_\mu}\tag{4.1.13}$$

and

$$\partial_\mu=\frac{\partial}{\partial x^\mu}\,,\tag{4.1.14}$$

and we get

$$\partial_\mu x^\mu=\sum_{\mu=0}^{3}\frac{\partial}{\partial x^\mu}x^\mu=4\tag{4.1.15}$$

as the analog of $\boldsymbol{\nabla}\cdot\boldsymbol{r}=3$.

According to (3.3.20), the 4-current density is

$$j=\begin{pmatrix}c\rho\\ \boldsymbol{j}\end{pmatrix},\qquad j^0=c\rho\,,\ j^1=j_x\,,\ j^2=j_y\,,\ j^3=j_z\,,\tag{4.1.16}$$

and

$$\partial_\mu j^\mu=\frac{1}{c}\frac{\partial}{\partial t}c\rho+\boldsymbol{\nabla}\cdot\boldsymbol{j}=\frac{\partial}{\partial t}\rho+\boldsymbol{\nabla}\cdot\boldsymbol{j}=0,\tag{4.1.17}$$

states the continuity equation for electric charge in the 4-notation,

$$\partial_\mu j^\mu = 0 \quad \text{or} \quad \partial^\mu j_\mu = 0\,. \tag{4.1.18}$$

Another 4-vector field is $\begin{pmatrix} \Phi \\ \boldsymbol{A} \end{pmatrix} = \begin{pmatrix} A^0 \\ \boldsymbol{A} \end{pmatrix}$, the 4-potential, for which

$$\partial_\mu A^\mu = 0 \tag{4.1.19}$$

is the Lorentz gauge condition (1.3.7), obviously with a vanishing 4-scalar field on the right-hand side.

## 4.2 Field dyadic; energy-momentum dyadic

For the $\boldsymbol{E}$, $\boldsymbol{B}$ fields, we found the field 4-dyadic

$$\begin{pmatrix} 0 & \boldsymbol{E}^{\mathrm{T}} \\ \boldsymbol{E} & -\boldsymbol{B}\times\mathbf{1} \end{pmatrix} \tag{4.2.1}$$

in (3.4.26), which connects to 4-rows (on the left) and 4-columns (on the right) by ordinary matrix multiplication, so that it must be a mixed contravariant-covariant 4-dyadic with components $F^\mu{}_\nu$. Let us see what we get by translating the construction of (3.4.24)–(3.4.26),

$$F^\mu{}_\nu = \partial^\mu A_\nu - \partial_\nu A^\mu\,, \tag{4.2.2}$$

where $\partial^\mu A_\nu$ corresponds to $\begin{pmatrix} -\dfrac{1}{c}\dfrac{\partial}{\partial t} \\ \boldsymbol{\nabla} \end{pmatrix} \left(-\Phi, \boldsymbol{A}^{\mathrm{T}}\right)$ and $\partial_\nu A^\mu$ to its Lorentz transpose with correct consistent positioning of the contravariant and covariant indices. The gauge transformations (1.3.9) now appears as

$$A^\mu \to A^\mu + \partial^\mu \lambda\,, \tag{4.2.3}$$

and it is immediately obvious that they have no net effect in (4.2.2).

Quite explicitly we have

$$F^0{}_0 = \partial^0 A_0 - \partial_0 A^0 = \left(-\frac{1}{c}\frac{\partial}{\partial t}\right)(-\Phi) - \left(\frac{1}{c}\frac{\partial}{\partial t}\right)\Phi = 0\,, \tag{4.2.4}$$

as well as

$$F^0{}_1 = \partial^0 A_1 - \partial_1 A^0 = \left(-\frac{1}{c}\frac{\partial}{\partial t}\right)A_x - \frac{\partial}{\partial x}\Phi = E_x\,, \tag{4.2.5}$$

and likewise

$$F^0{}_2 = E_y\,, \quad F^0{}_3 = E_z\,, \tag{4.2.6}$$

which are supplemented by

$$F^1{}_0 = \partial^1 A_0 - \partial_0 A^1 = \frac{\partial}{\partial x}(-\Phi) - \frac{1}{c}\frac{\partial}{\partial t}A_x = E_x \tag{4.2.7}$$

and

$$F^2{}_0 = E_y\,, \quad F^3{}_0 = E_z\,. \tag{4.2.8}$$

Further, when both indices are space-like ($\mu, \nu = 1, 2, 3$),

$$F^1{}_1 = \partial^1 A_1 - \partial_1 A^1 = \frac{\partial}{\partial x}A_x - \frac{\partial}{\partial x}A_x = 0 \quad \text{and} \quad F^2{}_2 = F^3{}_3 = 0\,, \tag{4.2.9}$$

as well as

$$F^1{}_2 = \partial^1 A_2 - \partial_2 A^1 = \frac{\partial}{\partial x}A_y - \frac{\partial}{\partial y}A_x = B_z \quad \text{and} \quad F^2{}_1 = -B_z\,, \tag{4.2.10}$$

and likewise

$$F^2{}_3 = B_x\,, \quad F^3{}_1 = B_y\,, \qquad F^3{}_2 = -B_x\,, \quad F^1{}_3 = -B_y\,. \tag{4.2.11}$$

In summary, then,

$$(F^\mu{}_\nu)_{\mu,\nu=0,1,2,3} = \begin{pmatrix} 0 & \boldsymbol{E}^{\mathrm{T}} \\ \boldsymbol{E} & -\boldsymbol{B} \times \mathbf{1} \end{pmatrix}, \tag{4.2.12}$$

as anticipated.

It is clear that $F^{\mu\nu} = \partial^\mu A^\nu - \partial^\nu A^\mu = -F^{\nu\mu}$ is antisymmetric, and this carries over to $F^\mu{}_\nu$,

$$F^\mu{}_\nu = g_{\nu\lambda} F^{\mu\lambda} = -g_{\nu\lambda} F^{\lambda\mu} = -F_\nu{}^\mu\,. \tag{4.2.13}$$

But, whereas it is all right and systematic to regard $F^\mu{}_\nu$ as a $4 \times 4$ matrix — a column of rows or row of columns — one should not think of $F^{\mu\nu}$ as a $4 \times 4$ matrix, rather it stands for a "column of columns," that is: columns on top of each other. Likewise, $F_{\mu\nu}$ is a "row of rows," with rows aligned behind each other.

The combination of the gradient with the field dyadic should give us the current vector,

$$\partial_\nu F^{\mu\nu} = \frac{4\pi}{c} j^\mu\,. \tag{4.2.14}$$

Indeed it does, as we can see by either writing out the components, or by evaluating

$$\partial_\nu F^{\mu\nu} = \partial_\nu \partial^\mu A^\nu - \partial_\nu \partial^\nu A^\mu \,, \tag{4.2.15}$$

where we can invoke the Lorentz gauge and recognize $\partial_\nu A^\nu = 0$ as well as

$$-\partial_\nu \partial^\nu A^\mu = \left[ \left( \frac{1}{c} \frac{\partial}{\partial t} \right)^2 - \boldsymbol{\nabla}^2 \right] A^\mu = \frac{4\pi}{c} j^\mu \,, \tag{4.2.16}$$

which is (1.3.12), so that (4.2.14) is correct, indeed. It follows that (4.2.14) are the inhomogeneous Maxwell's equations.

What about the homogeneous Maxwell's equations? They are contained in

$$\partial_\lambda F_{\mu\nu} + \partial_\mu F_{\nu\lambda} + \partial_\nu F_{\lambda\mu} = 0 \,. \tag{4.2.17}$$

To justify this statement, we just note that the left-hand side vanishes, if any two of the indices are the same, say $\lambda = \mu$,

$$\partial_\mu F_{\mu\nu} + \partial_\mu \underbrace{F_{\nu\mu}}_{=-F_{\mu\nu}} + \partial_\nu \underbrace{F_{\mu\mu}}_{=0} = 0 \,. \qquad \text{(No summation!)} \tag{4.2.18}$$

Second, a cyclic permutation of the indices $\lambda \to \mu \to \nu \to \lambda$ does not change the left-hand side. Third, an interchange of two indices changes the sign of the left-hand side, see for $\mu \leftrightarrow \nu$:

$$\begin{aligned} \partial_\lambda F_{\mu\nu} + \partial_\mu F_{\nu\lambda} + \partial_\nu F_{\lambda\mu} &\to \partial_\lambda F_{\nu\mu} + \partial_\nu F_{\mu\lambda} + \partial_\mu F_{\lambda\nu} \\ &= -\partial_\lambda F_{\mu\nu} - \partial_\nu F_{\lambda\mu} - \partial_\mu F_{\nu\lambda} \\ &= -(\partial_\lambda F_{\mu\nu} + \partial_\mu F_{\nu\lambda} + \partial_\nu F_{\lambda\mu}) \,. \end{aligned} \tag{4.2.19}$$

Therefore, of the $4^3 = 64$ values on the left-hand side of (4.2.18), only four sets of six each are possibly nonzero, namely those for which $\lambda\mu\nu$ is a permutation of 012, or of 123, or of 230, or of 301.

We consider $\lambda\mu\nu = 012$,

$$\begin{aligned} \partial_0 F_{12} + \partial_1 F_{20} + \partial_2 F_{01} &= \frac{1}{c} \frac{\partial}{\partial t} B_z + \frac{\partial}{\partial x} E_y + \frac{\partial}{\partial y} (-E_x) \\ &= \left( \frac{1}{c} \frac{\partial}{\partial t} \boldsymbol{B} + \boldsymbol{\nabla} \times \boldsymbol{E} \right)_z = 0 \,, \end{aligned} \tag{4.2.20}$$

one component of the homogeneous equation of motion; likewise $\lambda\mu\nu = 230$ and $\lambda\mu\nu = 301$ give the other two components. And for $\lambda\mu\nu = 123$, we get the homogeneous equation of constraint,

$$\begin{aligned}\partial_1 F_{23} + \partial_2 F_{31} + \partial_3 F_{12} &= \frac{\partial}{\partial x} B_x + \frac{\partial}{\partial y} B_y + \frac{\partial}{\partial z} B_z \\ &= \boldsymbol{\nabla} \cdot \boldsymbol{B} = 0 \,. \end{aligned} \tag{4.2.21}$$

In summary, (4.2.17) indeed states the homogeneous Maxwell's equations.

The energy-momentum tensor of (3.4.15) is a 4-dyadic that is composed of the field tensor with components $F_{\mu\nu}$. The building blocks that we have available are $F^{\mu}{}_{\lambda} F^{\lambda}{}_{\nu}$ and $g^{\mu}{}_{\nu} F^{\kappa}{}_{\lambda} F^{\lambda}{}_{\kappa}$. Let us see:

$$\begin{aligned}\underbrace{F^{\mu}{}_{\lambda} F^{\lambda}{}_{\nu}}_{\text{ordinary } 4\times 4 \text{ matrix multiplication}} &\mathrel{\widehat{=}} \begin{pmatrix} 0 & \boldsymbol{E}^{\mathrm{T}} \\ \boldsymbol{E} & -\boldsymbol{B}\times\boldsymbol{1} \end{pmatrix} \begin{pmatrix} 0 & \boldsymbol{E}^{\mathrm{T}} \\ \boldsymbol{E} & -\boldsymbol{B}\times\boldsymbol{1} \end{pmatrix} \\ &= \begin{pmatrix} \boldsymbol{E}^2 & -(\boldsymbol{E}\times\boldsymbol{B})^{\mathrm{T}} \\ \boldsymbol{E}\times\boldsymbol{B} & \boldsymbol{E}\,\boldsymbol{E}^{\mathrm{T}} + \boldsymbol{B}\,\boldsymbol{B}^{\mathrm{T}} - \boldsymbol{1}\boldsymbol{B}^2 \end{pmatrix} , \end{aligned} \tag{4.2.22}$$

where

$$(\boldsymbol{B}\times\boldsymbol{1})^2 = \boldsymbol{B}\,\boldsymbol{B}^{\mathrm{T}} - \boldsymbol{1}\boldsymbol{B}^2 \tag{4.2.23}$$

is used. We can establish this identity either by squaring the $3\times 3$ matrix of (3.4.28) or by recalling that $\boldsymbol{B}\times\boldsymbol{1} = \boldsymbol{1}\times\boldsymbol{B}^{\mathrm{T}}$ and employing two arbitrary 3-vectors,

$$\begin{aligned}\boldsymbol{a}^{\mathrm{T}}(\boldsymbol{B}\times\boldsymbol{1})^2\boldsymbol{b} &= \boldsymbol{a}^{\mathrm{T}}\,\boldsymbol{1}\times\boldsymbol{B}^{\mathrm{T}}\,\boldsymbol{B}\times\boldsymbol{1}\,\boldsymbol{b} \\ &= (\boldsymbol{a}\times\boldsymbol{B})^{\mathrm{T}}(\boldsymbol{B}\times\boldsymbol{b}) = \boldsymbol{a}^{\mathrm{T}}\boldsymbol{B}\,\boldsymbol{B}^{\mathrm{T}}\boldsymbol{b} - \boldsymbol{a}^{\mathrm{T}}\boldsymbol{b}\,\boldsymbol{B}^{\mathrm{T}}\boldsymbol{B} \\ &= \boldsymbol{a}^{\mathrm{T}}\big(\boldsymbol{B}\,\boldsymbol{B}^{\mathrm{T}} - \boldsymbol{1}\boldsymbol{B}^2\big)\,\boldsymbol{b} \,. \end{aligned} \tag{4.2.24}$$

Since $F^{\kappa}{}_{\lambda} F^{\lambda}{}_{\kappa}$ is the 4-trace of $F^{\mu}{}_{\lambda} F^{\lambda}{}_{\nu}$, we have

$$F^{\kappa}{}_{\lambda} F^{\lambda}{}_{\kappa} = 2(\boldsymbol{E}^2 - \boldsymbol{B}^2) \,, \tag{4.2.25}$$

and a glance at the spatial $3\times 3$ sector tells us that

$$\begin{aligned}\mathsf{T} &= \boldsymbol{1}\frac{1}{8\pi}(\boldsymbol{E}^2 + \boldsymbol{B}^2) - \frac{1}{4\pi}(\boldsymbol{E}\,\boldsymbol{E}^{\mathrm{T}} + \boldsymbol{B}\,\boldsymbol{B}^{\mathrm{T}}) \\ &= \frac{1}{16\pi} F^{\kappa}{}_{\lambda} F^{\lambda}{}_{\kappa}\boldsymbol{1} - \frac{1}{4\pi}(F^{\mu}{}_{\lambda} F^{\lambda}{}_{\nu})_{3\times 3} \,, \end{aligned} \tag{4.2.26}$$

which suggests that the contravariant-covariant components of the energy-momentum tensor are given by

$$T^{\mu}{}_{\nu} = \frac{1}{4\pi} F^{\mu\lambda} F_{\nu\lambda} - \frac{1}{16\pi} g^{\mu}{}_{\nu} F^{\kappa\lambda} F_{\kappa\lambda} \,, \tag{4.2.27}$$

where the asymmetry of $F^{\mu\nu}$ is used twice. The purely spatial components are correct as constructed, but we need to verify the $T^0{}_0$ and $T^0{}_j$ ($j = 1,2,3$) components, as well as $T^j{}_0$. See

$$T^0{}_0 = \frac{1}{8\pi}(\boldsymbol{E}^2 - \boldsymbol{B}^2) - \frac{1}{4\pi}\boldsymbol{E}^2 = -\frac{1}{8\pi}(\boldsymbol{E}^2 + \boldsymbol{B}^2) = -U \tag{4.2.28}$$

for $\mu = \nu = 0$, and

$$T^0{}_1 = \frac{1}{4\pi}(\boldsymbol{E}\times\boldsymbol{B})_x\,, \quad T^0{}_2 = \frac{1}{4\pi}(\boldsymbol{E}\times\boldsymbol{B})_y\,, \quad T^0{}_3 = \frac{1}{4\pi}(\boldsymbol{E}\times\boldsymbol{B})_z \tag{4.2.29}$$

for $\mu = 0$ and $\nu = 1,2,3$ as well as

$$T^1{}_0 = -\frac{1}{4\pi}(\boldsymbol{E}\times\boldsymbol{B})_x\,, \quad T^2{}_0 = -\frac{1}{4\pi}(\boldsymbol{E}\times\boldsymbol{B})_y\,, \quad T^3{}_0 = -\frac{1}{4\pi}(\boldsymbol{E}\times\boldsymbol{B})_z \tag{4.2.30}$$

for $\mu = 1,2,3$ and $\nu = 0$. Indeed, we recognize that

$$(T^\mu{}_\nu)_{\mu,\nu=0,1,2,3} = \begin{pmatrix} -U & c\,\boldsymbol{G}^{\mathrm{T}} \\ -\frac{1}{c}\boldsymbol{S} & \mathsf{T} \end{pmatrix}, \tag{4.2.31}$$

after recalling that $\frac{1}{4\pi}\boldsymbol{E}\times\boldsymbol{B} = \frac{1}{c}\boldsymbol{S} = c\,\boldsymbol{G}$. We note that

$$\begin{aligned} T^{\mu\nu} &= \frac{1}{4\pi}F^{\mu\lambda}F^\nu{}_\lambda - \frac{1}{16\pi}g^{\mu\nu}F^{\kappa\lambda}F_{\kappa\lambda} \\ &= \frac{1}{4\pi}F^{\nu\lambda}F^\mu{}_\lambda - \frac{1}{16\pi}g^{\nu\mu}F^{\kappa\lambda}F_{\kappa\lambda} = T^{\nu\mu}\,, \end{aligned} \tag{4.2.32}$$

stating that $T^{\mu\nu}$ is a symmetric 4-dyadic. The verification of

$$\partial_\nu T^{\mu\nu} = -F^{\mu\nu}\frac{1}{c}j_\nu\,, \tag{4.2.33}$$

which is (3.4.14) in 4-notation is the subject of Exercise 37.

## 4.3 Wave 4-vector; 4-velocity; Doppler effect

Here is an application that shows the usefulness of 4-vector notation beyond, as we have done so far, the simple rewriting of already familiar statements. We ask the following question: Given that a conductor at rest scatters light without changing the frequency, what is the frequency change when the scatterer moves with velocity $\boldsymbol{u}$?

For the light wave, we consider a scalar, monochromatic, plane wave of the generic form

$$\lambda(\boldsymbol{r},t) = \mathrm{e}^{\mathrm{i}(\boldsymbol{k}\cdot\boldsymbol{r} - \omega t)}\,, \tag{4.3.1}$$

for which all observers agree on the value of the phase difference

$$\Delta\phi(1,2) = \boldsymbol{k}\cdot(\boldsymbol{r}_1 - \boldsymbol{r}_2) - \omega(t_1 - t_2) \tag{4.3.2}$$

between events '1' and '2' with space-time coordinates $\begin{pmatrix} ct_1 \\ \boldsymbol{r}_1 \end{pmatrix}$ and $\begin{pmatrix} ct_2 \\ \boldsymbol{r}_2 \end{pmatrix}$ in the unprimed frame of reference. It follows that the wave vector $\boldsymbol{k}$ and the (circular) frequency $\omega$ make up another 4-vector,

$$k = \begin{pmatrix} \omega/c \\ \boldsymbol{k} \end{pmatrix}, \tag{4.3.3}$$

the 4-vector version of the wave vector. For a wave propagating at the speed of light, that is: $c|\boldsymbol{k}| = \omega$, it is a null 4-vector,

$$k_\mu k^\mu = -(\omega/c)^2 + \boldsymbol{k}^2 = 0\,. \tag{4.3.4}$$

The conductor is a material body moving with velocity $\boldsymbol{u} = \dfrac{\mathrm{d}\boldsymbol{r}}{\mathrm{d}t}$, whose response to infinitesimal Lorentz transformations is given by

$$\begin{aligned}\delta\boldsymbol{u} &= \frac{\mathrm{d}\delta\boldsymbol{r}}{\mathrm{d}t} - \frac{\mathrm{d}\boldsymbol{r}\,\mathrm{d}\delta t}{(\mathrm{d}t)^2} = \frac{\mathrm{d}(\delta\boldsymbol{v}\,t)}{\mathrm{d}t} - \frac{\mathrm{d}\boldsymbol{r}\,\mathrm{d}(\delta\boldsymbol{v}\cdot\boldsymbol{r}/c^2)}{(\mathrm{d}t)^2} \\ &= \delta\boldsymbol{v} - \frac{\boldsymbol{u}}{c}\,\delta\boldsymbol{v}\cdot\frac{\boldsymbol{u}}{c}\,,\end{aligned} \tag{4.3.5}$$

which is the infinitesimal version of the relativistic velocity addition. The resulting infinitesimal change of the $\gamma$-factor for $\boldsymbol{u}$ is then

$$\delta\gamma_u = \delta\frac{1}{\sqrt{1-(u/c)^2}} = \gamma_u^3\frac{\boldsymbol{u}\cdot\delta\boldsymbol{u}}{c^2} = \frac{\delta\boldsymbol{v}}{c}\cdot\gamma_u\frac{\boldsymbol{u}}{c} \tag{4.3.6}$$

since $\gamma_u^2\boldsymbol{u}\cdot\delta\boldsymbol{u} = \boldsymbol{u}\cdot\delta\boldsymbol{v}$. This looks like the response of the time-like component $\gamma_u c$ of a 4-vector with the 3-vector $\gamma_u\boldsymbol{u}$. Indeed, we have the correct corresponding expression for the infinitesimal transformation of $\gamma_u\boldsymbol{u}$,

$$\delta(\gamma_u\boldsymbol{u}) = \left(\frac{\delta\boldsymbol{v}}{c}\cdot\gamma_u\frac{\boldsymbol{u}}{c}\right)\boldsymbol{u} + \gamma_u\left(\delta\boldsymbol{v} - \frac{\boldsymbol{u}}{c}\,\delta\boldsymbol{v}\cdot\frac{\boldsymbol{u}}{c}\right) = \delta\boldsymbol{v}\,\gamma_u\,. \tag{4.3.7}$$

We have thus established that the velocity 4-vector, or simply the 4-velocity, of a material object moving at velocity $\boldsymbol{u}$ is given by

$$\left(u^\mu\right)_{\mu=0,1,2,3} = \begin{pmatrix} \gamma_u c \\ \gamma_u \boldsymbol{u} \end{pmatrix} \quad \text{with} \quad \gamma_u = \frac{1}{\sqrt{1-u^2/c^2}}\,. \tag{4.3.8}$$

Its Lorentz square

$$u_\mu u^\mu = -{\gamma_u}^2(c^2-u^2) = -c^2 \tag{4.3.9}$$

is clearly a 4-scalar.

Now returning to the question posed above, we note that, for the scatterer at rest, its 4-velocity is $\begin{pmatrix} c \\ \boldsymbol{0} \end{pmatrix}$, and we have $k_{\text{in}} = \begin{pmatrix} \omega/c \\ \boldsymbol{k} \end{pmatrix}$ and $k_{\text{out}} = \begin{pmatrix} \omega'/c \\ \boldsymbol{k}' \end{pmatrix}$ for the wave 4-vectors of the incoming and outgoing light, respectively,

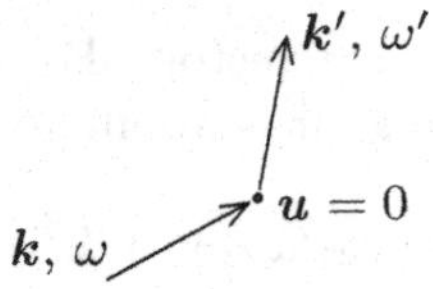


(4.3.10)

The statement $\omega = \omega'$ is given a Lorentz-invariant form by writing

$$u_\mu k^\mu_{\text{in}} = u_\mu k^\mu_{\text{out}}\,, \tag{4.3.11}$$

and this must now be equally true in a frame of reference in which the scatterer has velocity $\boldsymbol{u} \neq \boldsymbol{0}$. We conclude that

$$-\gamma_u\omega + \gamma_u \boldsymbol{u}\cdot\boldsymbol{k} = -\gamma_u\omega' + \gamma_u\boldsymbol{u}\cdot\boldsymbol{k}', \tag{4.3.12}$$

and with

$$\boldsymbol{u}\cdot\boldsymbol{k} = u\frac{\omega}{c}\cos\vartheta\,,$$

$$\boldsymbol{u}\cdot\boldsymbol{k}' = u\frac{\omega'}{c}\cos\vartheta'\,,$$

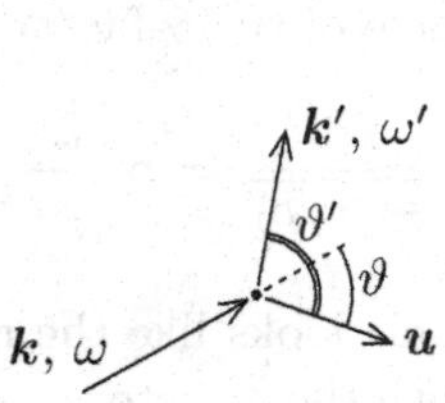


(4.3.13)

we get

$$\omega\left(1-\frac{u}{c}\cos\vartheta\right) = \omega'\left(1-\frac{u}{c}\cos\vartheta'\right). \tag{4.3.14}$$

The frequency of the scattered light,

$$\omega' = \frac{1 - \dfrac{u}{c}\cos\vartheta}{1 - \dfrac{u}{c}\cos\vartheta'}\,\omega\,, \tag{4.3.15}$$

depends on the direction of the incoming and the scattered light, and for $u = 0$ we get $\omega = \omega'$, as we should.

Physically speaking, this is a two-fold Doppler* effect. Once for the incoming frequency and once for the outgoing frequency, both are Doppler shifted in the rest frame of the scatterer, with each shift depending on the angle between the velocity $\boldsymbol{u}$ and the respective propagation direction as specified by $\boldsymbol{k}$ and $\boldsymbol{k}'$.

With this example, we leave the 4-vector notation until we have another use for it — as in Exercises 38–42 and 63, for instance.

*Christian Andreas DOPPLER (1803–1853)

## Chapter 5

# Action, Reaction — Interaction

## 5.1 Action principles of classical mechanics

### 5.1.1 *Lagrange's formulation*

The familiar *Lagrange* action principle* of classical mechanics — here: for a nonrelativistic particle of mass $m$ that moves under the influence of a force $\boldsymbol{F} = -\boldsymbol{\nabla}V$ with a potential energy $V(\boldsymbol{r},t)$ that could have a parametric time dependence — is based on the action

$$W_{12} = \int_2^1 \mathrm{d}t\, L \tag{5.1.1}$$

with the *Lagrange function*

$$L = \frac{m}{2}\left(\frac{\mathrm{d}\boldsymbol{r}}{\mathrm{d}t}\right)^2 - V(\boldsymbol{r},t) \tag{5.1.2}$$

for the particle trajectory $t \mapsto \boldsymbol{r}(t)$. The labels '1' and '2' stand for the final and the initial configuration, respectively, specified by $(t,\boldsymbol{r})_1 = (t_1,\boldsymbol{r}_1)$ and $(t,\boldsymbol{r})_2 = (t_2,\boldsymbol{r}_2)$.

We consider infinitesimal variations about the actual trajectory,

$$\boldsymbol{r} \to \boldsymbol{r} + \delta\boldsymbol{r}\,, \qquad t \to t + \delta t\,, \tag{5.1.3}$$

for which

$$\delta\,\mathrm{d}\boldsymbol{r} = \mathrm{d}\,\delta\boldsymbol{r}\,, \quad \delta\,\mathrm{d}t = \mathrm{d}\,\delta t\,, \quad \delta\frac{1}{\mathrm{d}t} = -\frac{\mathrm{d}\,\delta t}{(\mathrm{d}t)^2} \tag{5.1.4}$$

*Joseph Louis de LAGRANGE (1736–1813)

are needed for the evaluation of $\delta W$, the infinitesimal change of the action,

$$\delta W_{12} = \delta \int_2^1 \left[\frac{m}{2}\frac{(\mathrm{d}\boldsymbol{r})^2}{\mathrm{d}t} - \mathrm{d}t\,V(\boldsymbol{r},t)\right]. \tag{5.1.5}$$

This gives

$$\delta W_{12} = \int_2^1 \Bigg[m\frac{\mathrm{d}\boldsymbol{r}\cdot\mathrm{d}\delta\boldsymbol{r}}{\mathrm{d}t} - \mathrm{d}t\,\delta\boldsymbol{r}\cdot\boldsymbol{\nabla}V$$
$$- \frac{m}{2}\left(\frac{\mathrm{d}\boldsymbol{r}}{\mathrm{d}t}\right)^2\mathrm{d}\delta t - \mathrm{d}\delta t\,V - \mathrm{d}t\,\frac{\partial V}{\partial t}\delta t\Bigg] \tag{5.1.6}$$

or, after integrating by parts,

$$\delta W_{12} = \int_2^1 \mathrm{d}\Bigg[\underbrace{m\frac{\mathrm{d}\boldsymbol{r}}{\mathrm{d}t}}_{=\boldsymbol{p}}\cdot\delta\boldsymbol{r} - \underbrace{\left(\frac{m}{2}\left(\frac{\mathrm{d}\boldsymbol{r}}{\mathrm{d}t}\right)^2 + V\right)}_{=E}\delta t\Bigg]$$
$$+ \int_2^1 \mathrm{d}t\left[-\delta\boldsymbol{r}\cdot\left(m\frac{\mathrm{d}^2\boldsymbol{r}}{\mathrm{d}t^2} + \boldsymbol{\nabla}V\right) + \delta t\left(\frac{\mathrm{d}E}{\mathrm{d}t} - \frac{\partial E}{\partial t}\right)\right]$$
$$= (\boldsymbol{p}\cdot\delta\boldsymbol{r} - E\,\delta t)\Big|_2^1 = G_1 - G_2 \tag{5.1.7}$$

where

$$G = \boldsymbol{p}\cdot\delta\boldsymbol{r} - E\,\delta t\,, \tag{5.1.8}$$

the *generator* for infinitesimal variations of the initial and final configuration, exhibits the *momentum* $\boldsymbol{p} = m\dfrac{\mathrm{d}\boldsymbol{r}}{\mathrm{d}t}$, which accounts for changes in position, and the *energy* $E = \dfrac{m}{2}\left(\dfrac{\mathrm{d}\boldsymbol{r}}{\mathrm{d}t}\right)^2 + V(\boldsymbol{r},t)$, which accounts for changes in time.

The Principle of Stationary Action (PSA) thus states that the response of the action to infinitesimal variations about the actual trajectory originates solely in the change of the terminal configurations,

$$\delta W_{12} = G_1 - G_2\,, \tag{5.1.9}$$

whereas there is no contribution to $\delta W_{12}$ from the intermediate times. We can either argue that this follows from Newton's equation of motion

$$m\left(\frac{\mathrm{d}\boldsymbol{r}}{\mathrm{d}t}\right)^2 = -\boldsymbol{\nabla}V \tag{5.1.10}$$

and its implication

$$\frac{\mathrm{d}}{\mathrm{d}t}E = \frac{\partial}{\partial t}E\,, \tag{5.1.11}$$

or we can adopt the more useful point of view that these equations are implied by the PSA.

### 5.1.2 *Hamilton's formulation*

Hamilton's* version of the action principle also has the form of the PSA in (5.1.9), but now the action is given by

$$\begin{aligned} W_{12} &= \int_2^1 \mathrm{d}t \left[\boldsymbol{p}\cdot\frac{\mathrm{d}\boldsymbol{r}}{\mathrm{d}t} - H(\boldsymbol{r},\boldsymbol{p},t)\right] \\ &= \int_2^1 \Bigl(\boldsymbol{p}\cdot\mathrm{d}\boldsymbol{r} - \mathrm{d}t\,H\Bigr) \end{aligned} \tag{5.1.12}$$

where

$$H(\boldsymbol{r},\boldsymbol{p},t) = \frac{\boldsymbol{p}^2}{2m} + V(\boldsymbol{r},t) \tag{5.1.13}$$

is the *Hamilton function* (vulgo "the Hamiltonian"). Here we get

$$\delta W_{12} = \int_2^1 \Bigl(\delta\boldsymbol{p}\cdot\mathrm{d}\boldsymbol{r} + \boldsymbol{p}\cdot\mathrm{d}\delta\boldsymbol{r} - \mathrm{d}\delta t\,H - \mathrm{d}t\,\delta H\Bigr) \tag{5.1.14}$$

or, upon taking

$$\delta H = \delta\boldsymbol{r}\cdot\boldsymbol{\nabla}V + \delta\boldsymbol{p}\cdot\frac{\boldsymbol{p}}{m} + \delta t\frac{\partial V}{\partial t} \tag{5.1.15}$$

into account and integrating by parts,

$$\begin{aligned} \delta W_{12} = \int_2^1 \biggl[&\delta\boldsymbol{p}\cdot\Bigl(\mathrm{d}\boldsymbol{r} - \mathrm{d}t\frac{\boldsymbol{p}}{m}\Bigr) + \delta\boldsymbol{r}\cdot\Bigl(-\mathrm{d}\boldsymbol{p} - \mathrm{d}t\,\boldsymbol{\nabla}V\Bigr) \\ &+ \delta t\Bigl(\mathrm{d}H - \mathrm{d}t\frac{\partial V}{\partial t}\Bigr) + \mathrm{d}\Bigl(\boldsymbol{p}\cdot\delta\boldsymbol{r} - H\,\delta t\Bigr)\biggr] \\ = G_1 - G_2 \qquad& \end{aligned} \tag{5.1.16}$$

with the generator

$$G = \boldsymbol{p}\cdot\delta\boldsymbol{r} - H\,\delta t\,. \tag{5.1.17}$$

---

*Sir William Rowan HAMILTON (1805–1865)

The PSA requires that there is no contribution from intermediate times, and this implies the familiar *Hamilton's equations of motion*,

$$\begin{aligned}
\delta\boldsymbol{p}: &\quad \frac{\mathrm{d}}{\mathrm{d}t}\boldsymbol{r} = \frac{\boldsymbol{p}}{m} = \frac{\partial H}{\partial \boldsymbol{p}}\,,\\
\delta\boldsymbol{r}: &\quad \frac{\mathrm{d}}{\mathrm{d}t}\boldsymbol{p} = -\boldsymbol{\nabla}V = -\frac{\partial H}{\partial \boldsymbol{r}}\,,\\
\delta t: &\quad \frac{\mathrm{d}}{\mathrm{d}t}H = \frac{\partial H}{\partial t}\,,
\end{aligned} \tag{5.1.18}$$

the third of which follows also from the other two. This pair of first-order differential equations is, of course, equivalent to Newton's second-order differential equation of motion, inasmuch as

$$m\frac{\mathrm{d}^2}{\mathrm{d}t^2}\boldsymbol{r} = \frac{\mathrm{d}}{\mathrm{d}t}\boldsymbol{p} = -\boldsymbol{\nabla}V\,, \tag{5.1.19}$$

and the numerical value of the Hamilton function, evaluated along the actual trajectory, equals the energy $E$ of (5.1.7) and (5.1.8),

$$H = \frac{\boldsymbol{p}^2}{2m} + V \rightarrow \frac{m}{2}\left(\frac{\mathrm{d}\boldsymbol{r}}{\mathrm{d}t}\right)^2 + V = E\,. \tag{5.1.20}$$

Hamilton's point of view, therefore, is equivalent to that entertained by Lagrange, but the great advantage of Hamilton's formulation results from the appearance of the basic quantities $\boldsymbol{p}$ and $H$ in the generator $G$ of (5.1.17). In quantum mechanics this generator tells us the fundamental commutation relations.

### 5.1.3 *Schwinger's formulation*

Yet another point of view is Schwinger's.* Here the Lagrange function has the Hamilton-type form

$$L = \boldsymbol{p}\cdot\frac{\mathrm{d}\boldsymbol{r}}{\mathrm{d}t} - H(\boldsymbol{r},\boldsymbol{v},\boldsymbol{p},t)\,, \tag{5.1.21}$$

but now we have the velocity $\boldsymbol{v}$ as an independent variable in the Hamilton function

$$H(\boldsymbol{r},\boldsymbol{v},\boldsymbol{p},t) = \boldsymbol{p}\cdot\boldsymbol{v} - \frac{1}{2}m\boldsymbol{v}^2 + V(\boldsymbol{r},t)\,. \tag{5.1.22}$$

---

*Julian SCHWINGER (1918–1994)

Upon considering infinitesimal variations of $\boldsymbol{r}$, $\boldsymbol{v}$, $\boldsymbol{p}$, and $t$, the PSA establishes the generator

$$G = \boldsymbol{p} \cdot \delta\boldsymbol{r} - H\,\delta t \tag{5.1.23}$$

of Hamilton form, and the various implications for intermediate times are

$$\begin{aligned}
\delta\boldsymbol{r}: &\quad \frac{\mathrm{d}}{\mathrm{d}t}\boldsymbol{p} = -\frac{\partial H}{\partial \boldsymbol{r}} = -\boldsymbol{\nabla} V\,,\\
\delta\boldsymbol{v}: &\quad 0 = \frac{\partial H}{\partial \boldsymbol{v}} = \boldsymbol{p} - m\boldsymbol{v}\,,\\
\delta\boldsymbol{p}: &\quad \frac{\mathrm{d}}{\mathrm{d}t}\boldsymbol{r} = \frac{\partial H}{\partial \boldsymbol{p}} = \boldsymbol{v}\,,\\
\delta t: &\quad \frac{\mathrm{d}}{\mathrm{d}t}H = \frac{\partial}{\partial t}H\,.
\end{aligned} \tag{5.1.24}$$

We recognize that there are equations of motion for $\boldsymbol{r}$ and $\boldsymbol{p}$, the variables paired in $\boldsymbol{p} \cdot \mathrm{d}\boldsymbol{r} = \mathrm{d}t\, \boldsymbol{p} \cdot \frac{\mathrm{d}\boldsymbol{r}}{\mathrm{d}t}$, whereas the variation $\delta\boldsymbol{v}$ of the velocity yields $\boldsymbol{p} = m\boldsymbol{v}$, which is not an equation of motion but a *constraint*.

If we use this constraint to accept $\boldsymbol{v} = \frac{1}{m}\boldsymbol{p}$ as the definition of $\boldsymbol{v}$, then the Hamilton function becomes

$$H(\boldsymbol{r}, \boldsymbol{v}, \boldsymbol{p}, t)\Big|_{\boldsymbol{v} = \boldsymbol{p}/m} = \frac{\boldsymbol{p}^2}{2m} + V(\boldsymbol{r}, t)\,, \tag{5.1.25}$$

which is Hamilton's formulation. Alternatively, we can accept $\boldsymbol{v} = \frac{\mathrm{d}\boldsymbol{r}}{\mathrm{d}t}$ as the definition of $\boldsymbol{v}$ and $\boldsymbol{p} = m\boldsymbol{v}$ as the definition of $\boldsymbol{p}$, and then the Lagrange function acquires Lagrange's form,

$$\begin{aligned}
L &= \left[\boldsymbol{p} \cdot \left(\frac{\mathrm{d}\boldsymbol{r}}{\mathrm{d}t} - \boldsymbol{v}\right) + \frac{1}{2}m\boldsymbol{v}^2 - V(\boldsymbol{r}, t)\right]\Bigg|_{\boldsymbol{v} = \mathrm{d}\boldsymbol{r}/\mathrm{d}t,\ \boldsymbol{p} = m\boldsymbol{v}}\\
&= \frac{m}{2}\left(\frac{\mathrm{d}\boldsymbol{r}}{\mathrm{d}t}\right)^2 - V(\boldsymbol{r}, t)\,.
\end{aligned} \tag{5.1.26}$$

It should be clear, then, that Schwinger's formulation of the action principle is intermediate between the viewpoints of Lagrange and Hamilton and contains both as special cases.

### 5.1.4 *Velocity-dependent forces*

The option of having the velocity as an independent variable is particularly useful when one has to deal with velocity-dependent forces, as is the case for the Lorentz force. We proceed from Newton's equation of motion in the presence of the Lorentz force,

$$\frac{\mathrm{d}}{\mathrm{d}t}(m\boldsymbol{v}) = e\boldsymbol{E}(\boldsymbol{r},t) + e\frac{\boldsymbol{v}}{c} \times \boldsymbol{B}(\boldsymbol{r},t) \tag{5.1.27}$$

with $\boldsymbol{r} = \boldsymbol{r}(t)$ and $\boldsymbol{v}(t) = \dfrac{\mathrm{d}\boldsymbol{r}(t)}{\mathrm{d}t}$, and introduce the electromagnetic potentials,

$$\begin{aligned}\frac{\mathrm{d}}{\mathrm{d}t}(m\boldsymbol{v}) &= e\left(-\frac{1}{c}\frac{\partial}{\partial t}\boldsymbol{A} - \boldsymbol{\nabla}\Phi + \frac{\boldsymbol{v}}{c} \times (\boldsymbol{\nabla} \times \boldsymbol{A})\right) \\ &= e\left(-\frac{1}{c}\frac{\partial}{\partial t}\boldsymbol{A} - \frac{\boldsymbol{v}}{c}\cdot\boldsymbol{\nabla}\boldsymbol{A} - \boldsymbol{\nabla}\Phi + \boldsymbol{\nabla}\frac{\boldsymbol{v}}{c}\cdot\boldsymbol{A}\right).\end{aligned} \tag{5.1.28}$$

Here we recognize a total time derivative,

$$\frac{\partial}{\partial t}\boldsymbol{A} + \boldsymbol{v}\cdot\boldsymbol{\nabla}\boldsymbol{A} = \frac{\mathrm{d}}{\mathrm{d}t}\boldsymbol{A}, \tag{5.1.29}$$

and a gradient, and so arrive at

$$\frac{\mathrm{d}}{\mathrm{d}t}\left(m\boldsymbol{v} + \frac{e}{c}\boldsymbol{A}\right) = -\boldsymbol{\nabla}\left(e\left(\Phi - \frac{\boldsymbol{v}}{c}\cdot\boldsymbol{A}\right)\right) \tag{5.1.30}$$

or

$$\frac{\mathrm{d}}{\mathrm{d}t}\boldsymbol{p} = -\boldsymbol{\nabla}\left(e\Phi - e\frac{\boldsymbol{v}}{c}\cdot\boldsymbol{A}\right) \tag{5.1.31}$$

after identifying the momentum $\boldsymbol{p}$ in accordance with

$$\boldsymbol{p} = m\boldsymbol{v} + \frac{e}{c}\boldsymbol{A}, \tag{5.1.32}$$

the sum of kinetic momentum and potential momentum, quite analogous to the sum of kinetic energy and potential energy in (5.1.20). The resulting equation of motion has Newton's form $\dfrac{\mathrm{d}}{\mathrm{d}t}\boldsymbol{p} = -\boldsymbol{\nabla}V$, and this invites the use of the Schwinger-type Lagrange function

$$L = \boldsymbol{p}\cdot\left(\frac{\mathrm{d}\boldsymbol{r}}{\mathrm{d}t} - \boldsymbol{v}\right) + \frac{1}{2}mv^2 - e\Phi + e\frac{\boldsymbol{v}}{c}\cdot\boldsymbol{A}. \tag{5.1.33}$$

Let us verify that the respective variations give us the correct set of equations. The responses to variations of $\boldsymbol{p}(t)$, $\boldsymbol{r}(t)$, and $\boldsymbol{v}(t)$ are

$$\begin{aligned}
\delta\boldsymbol{p}: &\quad \frac{\mathrm{d}\boldsymbol{r}}{\mathrm{d}t} = \boldsymbol{v}\,,\\
\delta\boldsymbol{r}: &\quad \frac{\mathrm{d}\boldsymbol{p}}{\mathrm{d}t} = -\boldsymbol{\nabla}\Bigl(e\Phi - e\frac{\boldsymbol{v}}{c}\cdot\boldsymbol{A}\Bigr)\,,\\
\delta\boldsymbol{v}: &\quad \boldsymbol{p} = m\boldsymbol{v} + \frac{e}{c}\boldsymbol{A}\,,
\end{aligned}\tag{5.1.34}$$

which are indeed correct.

The last equation in (5.1.34), which is (5.1.32), is a constraint that we can use to eliminate the velocity $\boldsymbol{v}$ to arrive at the Hamilton-type Lagrange function

$$L = \boldsymbol{p}\cdot\frac{\mathrm{d}\boldsymbol{r}}{\mathrm{d}t} - H(\boldsymbol{r},\boldsymbol{p},t) \tag{5.1.35}$$

with the Hamilton function

$$\begin{aligned}
H &= \Bigl(\boldsymbol{p}\cdot\boldsymbol{v} - \frac{1}{2}m\boldsymbol{v}^2 + e\Phi(\boldsymbol{r},t) - e\frac{\boldsymbol{v}}{c}\cdot\boldsymbol{A}(\boldsymbol{r},t)\Bigr)\Bigg|_{\boldsymbol{v}=\frac{1}{m}(\boldsymbol{p}-(e/c)\boldsymbol{A})}\\
&= \Bigl(\frac{1}{2}m\boldsymbol{v}^2 + e\Phi(\boldsymbol{r},t)\Bigr)\Bigg|_{\boldsymbol{v}=\frac{1}{m}(\boldsymbol{p}-(e/c)\boldsymbol{A})}\,,
\end{aligned}\tag{5.1.36}$$

that is

$$H(\boldsymbol{r},\boldsymbol{p},t) = \frac{1}{2m}\Bigl(\boldsymbol{p} - \frac{e}{c}\boldsymbol{A}(\boldsymbol{r},t)\Bigr)^2 + e\Phi(\boldsymbol{r},t)\,. \tag{5.1.37}$$

As a Hamilton operator rather than a Hamilton function, this $H(\boldsymbol{r},\boldsymbol{p},t)$ is very often taken as the starting point for studying the quantum-mechanical aspects of a charged particle in nonrelativistic motion under the influence of an electromagnetic field. Please note that the first term on the right is simply the kinetic energy $\frac{1}{2}m\boldsymbol{v}^2$ and that there is a square of the vector potential $\boldsymbol{A}(\boldsymbol{r},t)$ in $H(\boldsymbol{r},\boldsymbol{p},t)$ although the Lagrange function of (5.1.33) is linear in $\boldsymbol{A}$.

The generalization of (5.1.33) to a collection of charged massive particles is immediate,

$$\begin{aligned}
L &= \sum_a \Bigl[\boldsymbol{p}_a\cdot\Bigl(\frac{\mathrm{d}\boldsymbol{r}_a}{\mathrm{d}t} - \boldsymbol{v}_a\Bigr) + \frac{1}{2}m_a\boldsymbol{v}_a^2 - e_a\Phi(\boldsymbol{r}_a,t) + \frac{e_a}{c}\boldsymbol{v}_a\cdot\boldsymbol{A}(\boldsymbol{r}_a,t)\Bigr]\\
&= L_{\text{kin}} + L_{\text{int}}
\end{aligned}\tag{5.1.38}$$

with

$$L_{\text{kin}} = \sum_a \left[ \boldsymbol{p} \cdot \left( \frac{\mathrm{d}\boldsymbol{r}}{\mathrm{d}t} - \boldsymbol{v} \right) + \frac{1}{2} m v^2 \right]_a \tag{5.1.39}$$

and

$$L_{\text{int}} = \int (\mathrm{d}\boldsymbol{r}) \left[ -\rho(\boldsymbol{r},t)\Phi(\boldsymbol{r},t) + \frac{1}{c}\boldsymbol{j}(\boldsymbol{r},t)\cdot \boldsymbol{A}(\boldsymbol{r},t) \right], \tag{5.1.40}$$

where

$$\begin{aligned} \rho(\boldsymbol{r},t) &= \sum_a e_a \delta\bigl(\boldsymbol{r} - \boldsymbol{r}_a(t)\bigr), \\ \boldsymbol{j}(\boldsymbol{r},t) &= \sum_a e_a \boldsymbol{v}_a(t)\, \delta\bigl(\boldsymbol{r} - \boldsymbol{r}_a(t)\bigr) \end{aligned} \tag{5.1.41}$$

are the charge and current densities associated with the moving charges; see (1.4.4) and (1.4.5). The split

$$L = L_{\text{kin}} + L_{\text{int}} \tag{5.1.42}$$

of the Lagrange function identifies the two contributions: $L_{\text{kin}}$ for the kinetic energy of the particles and $L_{\text{int}}$ for the interaction of the particles with the electromagnetic field that exerts the Lorentz force on the charged particles.

## 5.2 Lagrange function of the electromagnetic field

Now turning to the electromagnetic field by itself, that is: for $\rho = 0$ and $\boldsymbol{j} = 0$, we have the two equations of motion

$$\boldsymbol{E} = -\frac{1}{c}\frac{\partial}{\partial t}\boldsymbol{A} - \boldsymbol{\nabla}\Phi, \qquad \boldsymbol{\nabla}\times\boldsymbol{B} - \frac{1}{c}\frac{\partial}{\partial t}\boldsymbol{E} = 0 \tag{5.2.1}$$

and the two constraints

$$\boldsymbol{B} = \boldsymbol{\nabla}\times\boldsymbol{A}, \qquad \boldsymbol{\nabla}\cdot\boldsymbol{E} = 0. \tag{5.2.2}$$

The fields $\boldsymbol{E}(\boldsymbol{r},t)$ and $\boldsymbol{A}(\boldsymbol{r},t)$ have equations of motion and, therefore, the analog of $\boldsymbol{p}\cdot\frac{\mathrm{d}\boldsymbol{r}}{\mathrm{d}t}$ is a term proportional to $\int(\mathrm{d}\boldsymbol{r})\, \boldsymbol{E}\cdot\frac{1}{c}\frac{\partial}{\partial t}\boldsymbol{A}$.

With the correct factors, the Lagrange function for the electromagnetic field is

$$L_{\text{emf}} = \int (\mathrm{d}\boldsymbol{r}) \frac{1}{4\pi} \left[ \boldsymbol{E} \cdot \left( -\frac{1}{c}\frac{\partial}{\partial t}\boldsymbol{A} - \boldsymbol{\nabla}\Phi \right) - \boldsymbol{B} \cdot (\boldsymbol{\nabla} \times \boldsymbol{A}) - \frac{1}{2}(\boldsymbol{E}^2 - \boldsymbol{B}^2) \right], \tag{5.2.3}$$

as we verify by considering infinitesimal variations of $\boldsymbol{E}$, $\boldsymbol{B}$, $\Phi$, and $\boldsymbol{A}$,

$$\begin{aligned} \delta L_{\text{emf}} = \int (\mathrm{d}\boldsymbol{r}) \frac{1}{4\pi} \Bigg[ & \delta\boldsymbol{E} \cdot \left( -\frac{1}{c}\frac{\partial}{\partial t}\boldsymbol{A} - \boldsymbol{\nabla}\Phi - \boldsymbol{E} \right) \\ & + \delta\boldsymbol{B} \cdot (-\boldsymbol{\nabla} \times \boldsymbol{A} + \boldsymbol{B}) \\ & + \delta\boldsymbol{A} \cdot \left( \frac{1}{c}\frac{\partial}{\partial t}\boldsymbol{E} - \boldsymbol{\nabla} \times \boldsymbol{B} \right) \\ & + \delta\Phi\, \boldsymbol{\nabla} \cdot \boldsymbol{E} \Bigg] \\ + \int (\mathrm{d}\boldsymbol{r}) & \frac{1}{4\pi} \boldsymbol{\nabla} \cdot (-\delta\Phi\boldsymbol{E} + \boldsymbol{B} \times \delta\boldsymbol{A}) \\ - \frac{1}{c}\frac{\mathrm{d}}{\mathrm{d}t} \int & (\mathrm{d}\boldsymbol{r}) \frac{1}{4\pi} \boldsymbol{E} \cdot \delta\boldsymbol{A}\,. \end{aligned} \tag{5.2.4}$$

The integrated divergence of $\boldsymbol{B}\times\delta\boldsymbol{A}-\delta\Phi\boldsymbol{E}$ vanishes (convert it to a surface integral over an infinitely remote surface), and the total time derivative contributes

$$-\frac{1}{c}\int (\mathrm{d}\boldsymbol{r}) \frac{1}{4\pi} \boldsymbol{E} \cdot \delta\boldsymbol{A} \tag{5.2.5}$$

to the generator of infinitesimal endpoint variations, the field analog of $\boldsymbol{p} \cdot \delta\boldsymbol{r}$, and the terms that refer to intermediate times must vanish irrespective of the variations $\delta\boldsymbol{E}$, $\delta\boldsymbol{B}$, $\delta\boldsymbol{A}$, and $\delta\Phi$, so that

$$\begin{aligned} \delta\boldsymbol{E}: &\quad \boldsymbol{E} = -\frac{1}{c}\frac{\partial}{\partial t}\boldsymbol{A} - \boldsymbol{\nabla}\Phi\,, \\ \delta\boldsymbol{B}: &\quad \boldsymbol{B} = \boldsymbol{\nabla} \times \boldsymbol{A}\,, \\ \delta\boldsymbol{A}: &\quad \boldsymbol{\nabla} \times \boldsymbol{B} - \frac{1}{c}\frac{\partial}{\partial t}\boldsymbol{E} = 0\,, \\ \delta\Phi: &\quad \boldsymbol{\nabla} \cdot \boldsymbol{E} = 0\,, \end{aligned} \tag{5.2.6}$$

the four equations of (5.2.1) and (5.2.2), indeed.

If we accept $\boldsymbol{B} = \boldsymbol{\nabla} \times \boldsymbol{A}$ as the definition of $\boldsymbol{B}$ and $\boldsymbol{\nabla} \cdot \boldsymbol{E} = 0$ as a restriction on $\boldsymbol{E}$, then

$$\begin{aligned} \boldsymbol{E} \cdot (-\boldsymbol{\nabla}\Phi) &= -\boldsymbol{\nabla} \cdot (\Phi\boldsymbol{E}) + \Phi\, \boldsymbol{\nabla} \cdot \boldsymbol{E} \\ &= -\boldsymbol{\nabla} \cdot (\Phi\boldsymbol{E}) \end{aligned} \tag{5.2.7}$$

does not contribute to $L_{\mathrm{emf}}$ and

$$-\boldsymbol{B} \cdot (\boldsymbol{\nabla} \times \boldsymbol{A}) - \frac{1}{2}(\boldsymbol{E}^2 - \boldsymbol{B}^2) = -\frac{1}{2}\left[\boldsymbol{E}^2 + (\boldsymbol{\nabla} \times \boldsymbol{A})^2\right] \tag{5.2.8}$$

gives

$$L_{\mathrm{emf}} = \int (\mathrm{d}\boldsymbol{r}) \Big(-\frac{1}{4\pi c}\Big)\, \boldsymbol{E} \cdot \frac{\partial}{\partial t}\boldsymbol{A} - H_{\mathrm{emf}} \tag{5.2.9}$$

with the Hamilton function of the electromagnetic field given by

$$H_{\mathrm{emf}} = \int (\mathrm{d}\boldsymbol{r})\, \frac{1}{8\pi}\left[\boldsymbol{E}^2 + (\boldsymbol{\nabla} \times \boldsymbol{A})^2\right] = \int (\mathrm{d}\boldsymbol{r})\, U, \tag{5.2.10}$$

where

$$U = \frac{1}{8\pi}\left[\boldsymbol{E}^2 + (\boldsymbol{\nabla} \times \boldsymbol{A})^2\right] \tag{5.2.11}$$

is the familiar energy density of (1.4.14).

If we now add $L_{\mathrm{int}}$ of (5.1.40) to $L_{\mathrm{emf}}$, that is:

$$L = L_{\mathrm{emf}} \to L = L_{\mathrm{emf}} + L_{\mathrm{int}}\,, \tag{5.2.12}$$

the response of $L_{\mathrm{int}}$ to variations of $\Phi$ and $\boldsymbol{A}$ adds the terms

$$\delta L_{\mathrm{int}} = \int (\mathrm{d}\boldsymbol{r}) \left[-\rho\, \delta\Phi + \frac{1}{c}\boldsymbol{j} \cdot \delta\boldsymbol{A}\right] \tag{5.2.13}$$

to $\delta L_{\mathrm{emf}}$ of (5.2.4), so that now

$$\begin{aligned} \delta\boldsymbol{A}: &\quad \boldsymbol{\nabla} \times \boldsymbol{B} - \frac{1}{c}\frac{\partial}{\partial t}\boldsymbol{E} = \frac{4\pi}{c}\boldsymbol{j}\,, \\ \delta\Phi: &\quad \boldsymbol{\nabla} \cdot \boldsymbol{E} = 4\pi\rho\,. \end{aligned} \tag{5.2.14}$$

These are the inhomogeneous Maxwell's equations, and this outcome finally explains why we included the factor $\frac{1}{4\pi}$ in the definition of $L_{\mathrm{emf}}$ in (5.2.3).

## 5.3 Particles and fields in interaction

The Lagrange function for the combined physical system of charged particles and the electromagnetic field is

$$L = L_{\text{kin}} + L_{\text{int}} + L_{\text{emf}} \tag{5.3.1}$$

where $L_{\text{kin}}$ of (5.1.38) describes the particles all by themselves, $L_{\text{emf}}$ is the part for the electromagnetic field by itself, and $L_{\text{int}}$ accounts for the interaction of the charges with the field. This interaction is both the action of the charges on the field — $\rho$ and $\boldsymbol{j}$ are the sources for $\boldsymbol{E}$ and $\boldsymbol{B}$ in Maxwell's equations — and the reaction of the field on the charges by means of the Lorentz force.

That there is a single interaction term in the Lagrange function that accounts for the action and also for the reaction is a manifestation of Newton's Third Law, the law of reciprocal actions: *Actioni contrariam semper et aequalem esse reactionem.* (To every action there is always opposed an equal reaction.) The particular form of the Lorentz force of (1.4.1) or (1.4.3) follows from Maxwell's equations and the law of reciprocal actions. A different force law would be inconsistent with Maxwell's equations.

## 5.4 Disposing of the gauge-dependent terms

It may seem odd that the vector potential $\boldsymbol{A}$ with its gauge dependence is paired with the electric field $\boldsymbol{E}$ in the term

$$\int (\mathrm{d}\boldsymbol{r}) \left(-\frac{1}{4\pi c}\right) \boldsymbol{E} \cdot \frac{\partial}{\partial t} \boldsymbol{A} \tag{5.4.1}$$

that gives the equations of motion for the electromagnetic field. Upon splitting the vector potential into its transverse part $\boldsymbol{A}_\perp$ and its longitudinal part $\boldsymbol{A}_\parallel$ (more about this splitting in Section 7.4),

$$\boldsymbol{A} = \boldsymbol{A}_\perp + \boldsymbol{A}_\parallel \quad \text{with} \quad \boldsymbol{\nabla} \cdot \boldsymbol{A}_\perp = 0 \quad \text{and} \quad \boldsymbol{\nabla} \times \boldsymbol{A}_\parallel = 0\,, \tag{5.4.2}$$

it becomes clear that a gauge transformation $\boldsymbol{A} \to \boldsymbol{A} + \boldsymbol{\nabla}\lambda$ solely affects the longitudinal part $\boldsymbol{A}_\parallel$ while the transverse part $\boldsymbol{A}_\perp$ is gauge independent. It is this gauge-independent component that is of physical significance.

We shall, therefore, adopt the radiation gauge, $\boldsymbol{\nabla} \cdot \boldsymbol{A} = 0$, which ensures that the vector potential is transverse and has no longitudinal component, $\boldsymbol{A} = \boldsymbol{A}_\perp$ and $\boldsymbol{A}_\parallel = 0$. Then, upon realizing that the integral of the product of a transverse field and a longitudinal field vanishes,

$$\begin{aligned}\int(\mathrm{d}\boldsymbol{r})\,\boldsymbol{F}_\perp^{(1)}\cdot\boldsymbol{F}_\parallel^{(2)} &= \int(\mathrm{d}\boldsymbol{r})\,\boldsymbol{F}_\perp^{(1)}\cdot\boldsymbol{\nabla}\lambda^{(2)}\\ &= \int(\mathrm{d}\boldsymbol{r})\left[\boldsymbol{\nabla}\cdot\left(\lambda^{(2)}\boldsymbol{F}_\perp^{(1)}\right)-\lambda^{(2)}\boldsymbol{\nabla}\cdot\boldsymbol{F}_\perp^{(1)}\right]\\ &= 0\,,\end{aligned}\tag{5.4.3}$$

we observe that

$$\begin{aligned}\int(\mathrm{d}\boldsymbol{r})\left(-\frac{1}{4\pi c}\right)\boldsymbol{E}\cdot\frac{\partial}{\partial t}\boldsymbol{A} &= \int(\mathrm{d}\boldsymbol{r})\left(-\frac{1}{4\pi c}\right)\left(\boldsymbol{E}_\parallel+\boldsymbol{E}_\perp\right)\cdot\frac{\partial}{\partial t}\boldsymbol{A}\\ &= \int(\mathrm{d}\boldsymbol{r})\left(-\frac{1}{4\pi c}\right)\boldsymbol{E}_\perp\cdot\frac{\partial}{\partial t}\boldsymbol{A}\,,\end{aligned}\tag{5.4.4}$$

which tells us that only the transverse part of the electric field is a fundamental dynamical variable whereas the longitudinal part is not. In view of the radiation gauge, we have

$$\boldsymbol{E}=\boldsymbol{E}_\perp+\boldsymbol{E}_\parallel \quad\text{with}\quad \boldsymbol{E}_\perp=-\frac{1}{c}\frac{\partial}{\partial t}\boldsymbol{A} \quad\text{and}\quad \boldsymbol{E}_\parallel=-\boldsymbol{\nabla}\Phi\,,\tag{5.4.5}$$

which is an invitation to eliminate $\Phi$ from the formalism. This is quite natural because there is no equation of motion for $\Phi$. There is also no equation of motion for $\boldsymbol{B}$ and, therefore, we accept $\boldsymbol{B}=\boldsymbol{\nabla}\times\boldsymbol{A}$ as the definition of $\boldsymbol{B}$ and so eliminate $\boldsymbol{B}$ as well.

For the $\boldsymbol{B}$ terms in $L_{\mathrm{emf}}$ we thus write

$$-\boldsymbol{B}\cdot(\boldsymbol{\nabla}\times\boldsymbol{A})+\frac{1}{2}\boldsymbol{B}^2=-\frac{1}{2}(\boldsymbol{\nabla}\times\boldsymbol{A})^2\,.\tag{5.4.6}$$

And for the $\boldsymbol{E}_\parallel$ terms we have

$$\begin{aligned}&\int(\mathrm{d}\boldsymbol{r})\,\frac{1}{4\pi}\left[\boldsymbol{E}_\parallel\cdot(-\boldsymbol{\nabla}\Phi)-\frac{1}{2}\boldsymbol{E}_\parallel^2\right]\\ &=\int(\mathrm{d}\boldsymbol{r})\,\frac{1}{8\pi}(\boldsymbol{\nabla}\Phi)^2=\int(\mathrm{d}\boldsymbol{r})\,\frac{1}{8\pi}\left[\boldsymbol{\nabla}\cdot(\Phi\boldsymbol{\nabla}\Phi)-\Phi\boldsymbol{\nabla}^2\Phi\right]\\ &=\frac{1}{2}\int(\mathrm{d}\boldsymbol{r})\,\rho\,\Phi=\frac{1}{2}\int(\mathrm{d}\boldsymbol{r})\,\rho(\boldsymbol{r},t)\int(\mathrm{d}\boldsymbol{r}')\,\frac{\rho(\boldsymbol{r}',t)}{|\boldsymbol{r}-\boldsymbol{r}'|}\,,\end{aligned}\tag{5.4.7}$$

where

$$-\boldsymbol{\nabla}^2\Phi=\boldsymbol{\nabla}\cdot\boldsymbol{E}_\parallel=4\pi\rho\tag{5.4.8}$$

is used together with the electrostatic relation

$$\Phi(\boldsymbol{r},t) = \int (\mathrm{d}\boldsymbol{r}') \frac{\rho(\boldsymbol{r}',t)}{|\boldsymbol{r}-\boldsymbol{r}'|} . \tag{5.4.9}$$

As mentioned in Section 1.3, this is the reason why some people call the radiation gauge the "Coulomb gauge." Perhaps one should indeed call it the "transverse gauge" and so emphasize its defining property.

In its final form, the sum of the various $\boldsymbol{E}_{\|}$ contributions is expressed in terms of the charge density of (5.1.41),

$$\rho(\boldsymbol{r},t) = \sum_a e_a \delta\big(\boldsymbol{r} - \boldsymbol{r}_a(t)\big) , \tag{5.4.10}$$

which is a function of the particle variables. The $-\rho\Phi$ term in $L_{\mathrm{int}}$ is of the same form — it is a particle term as well.

In summary, then, we have

$$L = L_{\mathrm{p}} + \int (\mathrm{d}\boldsymbol{r}) \, \frac{1}{c} \boldsymbol{j} \cdot \boldsymbol{A} + L_{\mathrm{f}} \tag{5.4.11}$$

with the particle Lagrange function

$$\begin{aligned} L_{\mathrm{p}} &= \sum_a \left[ \boldsymbol{p}_a \cdot \left( \frac{\mathrm{d}\boldsymbol{r}_a}{\mathrm{d}t} - \boldsymbol{v}_a \right) + \frac{1}{2} m_a \boldsymbol{v}_a^2 \right] - \frac{1}{2} \int (\mathrm{d}\boldsymbol{r})(\mathrm{d}\boldsymbol{r}') \frac{\rho(\boldsymbol{r},t)\rho(\boldsymbol{r}',t)}{|\boldsymbol{r}-\boldsymbol{r}'|} \\ &= \sum_a \left[ \boldsymbol{p}_a \cdot \left( \frac{\mathrm{d}\boldsymbol{r}_a}{\mathrm{d}t} - \boldsymbol{v}_a \right) + \frac{1}{2} m_a \boldsymbol{v}_a^2 \right] - \frac{1}{2} \sum_{a \neq b} \frac{e_a e_b}{|\boldsymbol{r}_a - \boldsymbol{r}_b|} , \end{aligned} \tag{5.4.12}$$

where we remove the self-interaction terms with $a = b$ from the double summation, and the field Lagrange function

$$L_{\mathrm{f}} = \int (\mathrm{d}\boldsymbol{r}) \, \frac{1}{4\pi} \left[ -\boldsymbol{E}_{\perp} \cdot \frac{1}{c} \frac{\partial}{\partial t} \boldsymbol{A} - \frac{1}{2} \Big( \boldsymbol{E}_{\perp}^2 + (\boldsymbol{\nabla} \times \boldsymbol{A})^2 \Big) \right] . \tag{5.4.13}$$

The $\boldsymbol{j} \cdot \boldsymbol{A}$ term in (5.4.11) accounts for the interaction between the charges and the field, where in fact only the transverse part of $\boldsymbol{j}$ matters,

$$\int (\mathrm{d}\boldsymbol{r}) \, \frac{1}{c} \boldsymbol{j} \cdot \boldsymbol{A} = \int (\mathrm{d}\boldsymbol{r}) \, \frac{1}{c} \boldsymbol{j}_{\perp} \cdot \boldsymbol{A} \tag{5.4.14}$$

because $\boldsymbol{A} = \boldsymbol{A}_{\perp}$ in the radiation gauge. We have met $\boldsymbol{j}_{\perp}$ before, on the right-hand side of (1.3.14), in the explicit form

$$\boldsymbol{j}_{\perp} = \boldsymbol{j} - \frac{1}{4\pi} \frac{\partial}{\partial t} \boldsymbol{\nabla}\Phi \tag{5.4.15}$$

with $\Phi(\boldsymbol{r},t)$ from (5.4.9), and have verified, in Exercise 47, that $\boldsymbol{\nabla}\cdot\boldsymbol{j}_{\perp}=0$ as a consequence of the continuity equation (1.2.1).

The Lagrange function $L_{\mathrm{f}}$, and the action associated with it, with its consistent focus on the gauge-invariant transverse fields, is the starting point for the first formulation of quantum electrodynamics. This matter is, however, beyond the scope of these lectures on classical electrodynamics.

# Chapter 6

# Retarded Potentials

The Lorentz transformation of Chapter 3, also in the contravariant-covariant formulation of Chapter 4, tells us how to find the electric and magnetic fields, and the potentials, in any other Lorentz frame if we know them in one frame. For $\boldsymbol{E}(\boldsymbol{r},t)$ and $\boldsymbol{B}(\boldsymbol{r},t)$ we can, in particular, rely on the transformation formulas in (3.5.20)–(3.5.22). But how do we get the potentials and the fields in the first place from the given charge and current densities?

## 6.1 Green's function

To this end, we need to solve the equations that determine $\Phi$ and $\boldsymbol{A}$ in terms of $\rho$ and $\boldsymbol{j}$. It is simplest to employ the Lorentz gauge, and then we are challenged to solve the inhomogeneous wave equations in (1.3.12), that is

$$\left(\frac{1}{c^2}\frac{\partial^2}{\partial t^2} - \boldsymbol{\nabla}^2\right)\begin{pmatrix}\Phi(\boldsymbol{r},t)\\ \boldsymbol{A}(\boldsymbol{r},t)\end{pmatrix} = \frac{4\pi}{c}\begin{pmatrix}c\rho(\boldsymbol{r},t)\\ \boldsymbol{j}(\boldsymbol{r},t)\end{pmatrix}, \tag{6.1.1}$$

where the 4-potential on the left is differentiated by the scalar differential operator

$$-\left(\frac{1}{c}\frac{\partial}{\partial t}, \boldsymbol{\nabla}^{\mathrm{T}}\right)\begin{pmatrix}-\dfrac{1}{c}\dfrac{\partial}{\partial t}\\ \boldsymbol{\nabla}\end{pmatrix} = -\left(-\left(\frac{1}{c}\frac{\partial}{\partial t}\right)^2 + \boldsymbol{\nabla}^2\right), \tag{6.1.2}$$

the product of the row with the column of the 4-gradient, to give the 4-current density on the right; recall (4.2.15) and (4.2.16) in this context.

Since the four components of $\begin{pmatrix}\Phi\\ \boldsymbol{A}\end{pmatrix}$ are not coupled in this equation, we can first just consider the time-like component,

$$\left(\frac{1}{c^2}\frac{\partial^2}{\partial t^2} - \boldsymbol{\nabla}^2\right)\Phi(\boldsymbol{r},t) = 4\pi\rho(\boldsymbol{r},t)\,, \tag{6.1.3}$$

and infer the solution for $\boldsymbol{A}$ once we have the solution for $\Phi$ at hand. Upon expressing $\rho$ and $\Phi$ in terms of their Fourier* transforms, in time and in position,

$$\begin{aligned}\rho(\boldsymbol{r},t) &= \int\frac{\mathrm{d}\omega}{2\pi}\,\mathrm{e}^{-\mathrm{i}\omega t}\int\frac{(\mathrm{d}\boldsymbol{k})}{(2\pi)^3}\,\mathrm{e}^{\mathrm{i}\boldsymbol{k}\cdot\boldsymbol{r}}\rho(\boldsymbol{k},\omega)\,,\\ \Phi(\boldsymbol{r},t) &= \int\frac{\mathrm{d}\omega}{2\pi}\,\mathrm{e}^{-\mathrm{i}\omega t}\int\frac{(\mathrm{d}\boldsymbol{k})}{(2\pi)^3}\,\mathrm{e}^{\mathrm{i}\boldsymbol{k}\cdot\boldsymbol{r}}\Phi(\boldsymbol{k},\omega)\,,\end{aligned} \tag{6.1.4}$$

the application of the differential operator to the exponential factors,

$$\left(\frac{1}{c^2}\frac{\partial^2}{\partial t^2} - \boldsymbol{\nabla}^2\right)\mathrm{e}^{-\mathrm{i}\omega t+\mathrm{i}\boldsymbol{k}\cdot\boldsymbol{r}} = \left(-\left(\frac{\omega}{c}\right)^2 + \boldsymbol{k}^2\right)\mathrm{e}^{-\mathrm{i}\omega t+\mathrm{i}\boldsymbol{k}\cdot\boldsymbol{r}}, \tag{6.1.5}$$

turns the differential equation into an algebraic equation

$$\left(k^2 - \left(\frac{\omega}{c}\right)^2\right)\Phi(\boldsymbol{k},\omega) = 4\pi\rho(\boldsymbol{k},\omega)\,. \tag{6.1.6}$$

Since $k^2-(\omega/c)^2$ vanishes when $|\omega/c| = k = |\boldsymbol{k}|$, and such values for $\omega$ and $\boldsymbol{k}$ are not excluded, we cannot just divide by $k^2-(\omega/c)^2$ to solve this equation for $\Phi(\boldsymbol{k},\omega)$. Rather, we must take two precaution measures: first, we have to allow for solutions of the homogeneous equation, $\Phi(\boldsymbol{k},\omega) \propto \delta(k^2-(\omega/c)^2)$; second, we must opt for *one* particular solution of the inhomogeneous equation, which is achieved by giving a small imaginary part to $\omega$, whose sign will be decided later.

Thus,

$$\begin{aligned}\Phi(\boldsymbol{k},\omega) = {} & \left.\frac{4\pi}{k^2-\left(\frac{\omega}{c}\pm\mathrm{i}\epsilon\right)^2}\,\rho(\boldsymbol{k},\omega)\right|_{0<\epsilon\to 0}\\ & + a(\boldsymbol{k},\omega)\delta\big(k^2-(\omega/c)^2\big)\,,\end{aligned} \tag{6.1.7}$$

where the limit $\epsilon\to 0$ is taken at a convenient later stage, and $a(\boldsymbol{k},\omega)$ is an arbitrary function that is reasonably smooth in the vicinity of $\omega = c|\boldsymbol{k}|$

*Jean Baptiste Joseph FOURIER (1768–1830)

and $\omega = -c|\boldsymbol{k}|$. The latter part contributes

$$\int \frac{d\omega}{2\pi} e^{-i\omega t} \int \frac{(d\boldsymbol{k})}{(2\pi)^3} e^{i\boldsymbol{k}\cdot\boldsymbol{r}} a(\boldsymbol{k},\omega) \underbrace{\delta(k^2-(\omega/c)^2)}_{=\frac{c}{2k}[\delta(ck-\omega)+\delta(ck+\omega)]}$$

$$= \int \frac{(d\boldsymbol{k})}{(2\pi)^3} \frac{c}{4\pi k} \left[ e^{i(\boldsymbol{k}\cdot\boldsymbol{r}-ckt)} a(\boldsymbol{k},ck) + e^{i(\boldsymbol{k}\cdot\boldsymbol{r}+ckt)} a(\boldsymbol{k},-ck) \right]$$

$$= \int \frac{(d\boldsymbol{k})}{(2\pi)^3} \frac{c}{4\pi k} \left[ e^{i(\boldsymbol{k}\cdot\boldsymbol{r}-ckt)} a(\boldsymbol{k},ck) + e^{-i(\boldsymbol{k}\cdot\boldsymbol{r}-ckt)} a(-\boldsymbol{k},-ck) \right] \tag{6.1.8}$$

to $\Phi(\boldsymbol{r},t)$, which is clearly a superposition of plane waves that propagate in direction $\boldsymbol{k}$ with the speed $c$. Since $\Phi(\boldsymbol{r},t)$ is real, we need

$$a(\boldsymbol{k},ck) = a(-\boldsymbol{k},-ck)^*\,, \tag{6.1.9}$$

and then this solution of the homogeneous wave equation

$$\left(\frac{1}{c^2}\frac{\partial^2}{\partial t^2} - \boldsymbol{\nabla}^2\right)\Phi(\boldsymbol{r},t) = 0 \tag{6.1.10}$$

reads

$$\mathrm{Re}\int \frac{(d\boldsymbol{k})}{(2\pi)^3} \frac{c}{2\pi k} a(\boldsymbol{k},ck)\, e^{i(\boldsymbol{k}\cdot\boldsymbol{r}-ckt)}\,, \tag{6.1.11}$$

where we can in fact drop the frequency argument in the amplitude factor, and simplify the notation by the replacement $\frac{c}{2\pi k}a(\boldsymbol{k},ck) \to a(\boldsymbol{k})$,

$$\mathrm{Re}\int \frac{(d\boldsymbol{k})}{(2\pi)^3} a(\boldsymbol{k})\, e^{i(\boldsymbol{k}\cdot\boldsymbol{r}-ckt)}\,. \tag{6.1.12}$$

This solution to the homogeneous wave equation describes a part of $\Phi(\boldsymbol{r},t)$ that is present irrespective of $\rho(\boldsymbol{r},t)$. Physically it could correspond to something originating in a distant star or being emitted from a not so distant radio station antenna.

We now focus on the contribution to $\Phi(\boldsymbol{r},t)$ from the charge distribution in question,

$$\Phi_\pm(\boldsymbol{r},t) = \int \frac{d\omega}{2\pi}\frac{(d\boldsymbol{k})}{(2\pi)^3} e^{-i\omega t} e^{i\boldsymbol{k}\cdot\boldsymbol{r}} \frac{4\pi}{k^2-\left(\frac{\omega}{c}\pm i\epsilon\right)^2}\rho(\boldsymbol{k},\omega)\Bigg|_{0<\epsilon\to 0}, \tag{6.1.13}$$

where

$$\rho(\boldsymbol{k},\omega) = \int \mathrm{d}t' \int (\mathrm{d}\boldsymbol{r}')\, \mathrm{e}^{\mathrm{i}\omega t'}\, \mathrm{e}^{-\mathrm{i}\boldsymbol{k}\cdot\boldsymbol{r}'} \rho(\boldsymbol{r}',t') \tag{6.1.14}$$

and, therefore,

$$\Phi_\pm(\boldsymbol{r},t) = \int \mathrm{d}t'(\mathrm{d}\boldsymbol{r}')\, G_\pm(\boldsymbol{r}-\boldsymbol{r}',t-t')\rho(\boldsymbol{r}',t') \tag{6.1.15}$$

with the *Green's* function*

$$G_\pm(\boldsymbol{r},t) = \int \frac{\mathrm{d}\omega}{2\pi}\frac{(\mathrm{d}\boldsymbol{k})}{(2\pi)^3} \frac{4\pi}{k^2-\left(\frac{\omega}{c}\pm\mathrm{i}\epsilon\right)^2}\, \mathrm{e}^{-\mathrm{i}\omega t}\, \mathrm{e}^{\mathrm{i}\boldsymbol{k}\cdot\boldsymbol{r}}\bigg|_{0<\epsilon\to 0} . \tag{6.1.16}$$

The defining property of the Green's function,

$$\left(\frac{1}{c^2}\frac{\partial^2}{\partial t^2} - \boldsymbol{\nabla}^2\right) G_\pm(\boldsymbol{r},t) = 4\pi\delta(\boldsymbol{r})\delta(t)\,, \tag{6.1.17}$$

is verified easily because the limit

$$\frac{k^2-\left(\frac{\omega}{c}\right)^2}{k^2-\left(\frac{\omega}{c}\pm\mathrm{i}\epsilon\right)^2}\Bigg|_{0<\epsilon\to 0} = 1 \tag{6.1.18}$$

is immediate.

We use spherical coordinates for the integration over $\boldsymbol{k}$ in (6.1.16), with the polar axis aligned with $\boldsymbol{r}$,

$$(\mathrm{d}\boldsymbol{k}) = \mathrm{d}k\, k^2 \mathrm{d}\theta \sin\theta\, \mathrm{d}\varphi\,, \qquad \boldsymbol{k}\cdot\boldsymbol{r} = kr\cos\theta\,, \tag{6.1.19}$$

so that the integration over the angular dependence gives

$$\int_0^\pi \mathrm{d}\theta\, \sin\theta \int_0^{2\pi} \mathrm{d}\varphi\, \mathrm{e}^{\mathrm{i}kr\cos\theta} = 4\pi\frac{\sin(kr)}{kr}\,, \tag{6.1.20}$$

and we arrive at

$$\begin{aligned} G_\pm(\boldsymbol{r},t) &= \frac{(4\pi)^2}{(2\pi)^3}\frac{1}{r}\int_{-\infty}^{\infty}\frac{\mathrm{d}\omega}{2\pi}\,\mathrm{e}^{-\mathrm{i}\omega t}\int_0^\infty \mathrm{d}k\, k\frac{\sin(kr)}{k^2-\left(\frac{\omega}{c}\pm\mathrm{i}\epsilon\right)^2}\Bigg|_{0<\epsilon\to 0} \\ &= \frac{2}{\pi}\frac{1}{r}\int_{-\infty}^{\infty}\frac{\mathrm{d}\omega}{2\pi}\,\mathrm{e}^{-\mathrm{i}\omega t}\frac{1}{2\mathrm{i}}\int_{-\infty}^{\infty}\mathrm{d}k\,\frac{k\,\mathrm{e}^{\mathrm{i}kr}}{k^2-\left(\frac{\omega}{c}\pm\mathrm{i}\epsilon\right)^2}\Bigg|_{0<\epsilon\to 0} . \end{aligned} \tag{6.1.21}$$

---

*George GREEN (1793–1841)

The $k$ integrand

$$\frac{k\,\mathrm{e}^{\mathrm{i}kr}}{k^2-\left(\frac{\omega}{c}\pm\mathrm{i}\epsilon\right)^2}=\frac{k\,\mathrm{e}^{\mathrm{i}kr}}{\left[k-\left(\frac{\omega}{c}\pm\mathrm{i}\epsilon\right)\right]\left[k+\left(\frac{\omega}{c}\pm\mathrm{i}\epsilon\right)\right]} \tag{6.1.22}$$

has poles at $k=\omega/c\pm\mathrm{i}\epsilon$ and $k=-\omega/c\mp\mathrm{i}\epsilon$. For each choice of sign, one pole is in the upper half-plane and the other pole is in the lower half-plane. The poles in the upper half-plane are at $k=\omega/c+\mathrm{i}\epsilon$ when the upper sign applies and at $k=-\omega/c+\mathrm{i}\epsilon$ for the lower sign. Since $r>0$ in $\mathrm{e}^{\mathrm{i}kr}$, we need to close the contour by a large semicircle in the upper half-plane, where the residue is

$$\frac{1}{2}\,\mathrm{e}^{\mathrm{i}(\pm\omega/c\,+\,\mathrm{i}\epsilon)r}=\frac{1}{2}\,\mathrm{e}^{\pm\mathrm{i}(\omega/c)r\,-\,\epsilon r}. \tag{6.1.23}$$

Accordingly,

$$\begin{aligned} G_\pm(\boldsymbol{r},t) &= \frac{2}{\pi}\frac{1}{r}\int_{-\infty}^{\infty}\frac{\mathrm{d}\omega}{2\pi}\,\mathrm{e}^{-\mathrm{i}\omega t}\frac{1}{2\mathrm{i}}2\pi\mathrm{i}\frac{1}{2}\,\mathrm{e}^{\pm\mathrm{i}(\omega/c)r\,-\,\epsilon r}\Big|_{0<\epsilon\to 0} \\ &= \frac{\delta(t\mp r/c)}{r}, \end{aligned} \tag{6.1.24}$$

after carrying out the now elementary limit $\epsilon\to 0$ and recognizing the Fourier representation of Dirac's delta function. So,

$$\begin{aligned} \Phi_\pm(\boldsymbol{r},t) &= \int(\mathrm{d}\boldsymbol{r}')\mathrm{d}t'\,\frac{\delta\left(t-t'\mp\frac{1}{c}|\boldsymbol{r}-\boldsymbol{r}'|\right)}{|\boldsymbol{r}-\boldsymbol{r}'|}\rho(\boldsymbol{r}',t') \\ &= \int(\mathrm{d}\boldsymbol{r}')\,\frac{\rho\left(\boldsymbol{r}',t\mp\frac{1}{c}|\boldsymbol{r}-\boldsymbol{r}'|\right)}{|\boldsymbol{r}-\boldsymbol{r}'|}, \end{aligned} \tag{6.1.25}$$

where the difference between the two choices of sign is now evident: for the upper sign we get the *retarded* solution in which $\rho(\boldsymbol{r}',t')$ at the *earlier* time $t'=t-|\boldsymbol{r}-\boldsymbol{r}'|/c$ contributes to $\Phi$ at time $t$; and for the lower sign we get the *advanced* solution in which the later time $t'=t+|\boldsymbol{r}-\boldsymbol{r}'|/c$ in $\rho(\boldsymbol{r}',t')$ is relevant for $\Phi(\boldsymbol{r},t)$.

Both are mathematically valid solutions, but physically the charge distribution is the source for the potential, and then the electromagnetic fields, which is to say that $\rho(\boldsymbol{r}',t')$ is the *cause* and $\Phi(\boldsymbol{r},t)$ is the *effect*. In the correct causal order, the cause precedes the effect in time, implying that $t'$ should be earlier than $t$. This selects the retarded solution (upper sign) on physical grounds, so that we will use the *retarded Green's function*

$$G_+(\boldsymbol{r}-\boldsymbol{r}',t-t') = \frac{1}{|\boldsymbol{r}-\boldsymbol{r}'|}\delta\Big(t-t'-\frac{1}{c}|\boldsymbol{r}-\boldsymbol{r}'|\Big). \tag{6.1.26}$$

Indeed, $t'$ is earlier than $t$ by $\frac{1}{c}|\boldsymbol{r}-\boldsymbol{r}'|$, the time it takes for light to travel from $\boldsymbol{r}'$ to $\boldsymbol{r}$. Very satisfactorily, then, we note that the effect at $\boldsymbol{r}$ occurs in the shortest time after the cause at $\boldsymbol{r}'$, as permitted by Einsteinian relativity, and not earlier.

## 6.2 Liénard–Wiechert potentials

The 4-potential in the Lorentz gauge is, therefore, given by

$$\begin{pmatrix}\Phi\\ \boldsymbol{A}\end{pmatrix}(\boldsymbol{r},t) = \int(\mathrm{d}\boldsymbol{r}')\,\frac{1}{|\boldsymbol{r}-\boldsymbol{r}'|}\begin{pmatrix}\rho\\ \frac{1}{c}\boldsymbol{j}\end{pmatrix}(\boldsymbol{r}',t-|\boldsymbol{r}-\boldsymbol{r}'|/c). \tag{6.2.1}$$

These are the *Liénard**–*Wiechert*† *potentials*, and the general solution is obtained by adding an arbitrary solution of the homogeneous wave equation to it, perhaps in the form (6.1.12),

$$\begin{pmatrix}\Phi(\boldsymbol{r},t)\\ \boldsymbol{A}(\boldsymbol{r},t)\end{pmatrix}_{\text{homogeneous}} = \mathrm{Re}\int\frac{(\mathrm{d}\boldsymbol{k})}{(2\pi)^3}\,\mathrm{e}^{\mathrm{i}(\boldsymbol{k}\cdot\boldsymbol{r}-ckt)}\begin{pmatrix}a_0(\boldsymbol{k})\\ \boldsymbol{a}(\boldsymbol{k})\end{pmatrix}, \tag{6.2.2}$$

where — if we wish to maintain the Lorentz gauge throughout — $a_0(\boldsymbol{k})$ and $\boldsymbol{a}(\boldsymbol{k})$ have to be chosen such that the Lorentz gauge condition (1.3.7) is obeyed. This translates immediately into the requirement

$$ka_0(\boldsymbol{k}) = \boldsymbol{k}\cdot\boldsymbol{a}(\boldsymbol{k}). \tag{6.2.3}$$

## 6.3 Retarded time

Let us look at the Liénard–Wiechert potentials for a single charge $e$ moving along the trajectory $\boldsymbol{R}(t)$ with velocity $\boldsymbol{V}(t) = \frac{\mathrm{d}}{\mathrm{d}t}\boldsymbol{R}(t)$, whereby $|\boldsymbol{V}(t)| < c$, of course. In this case, we have

$$\begin{aligned}\rho(\boldsymbol{r},t) &= e\delta\big(\boldsymbol{r}-\boldsymbol{R}(t)\big),\\ \boldsymbol{j}(\boldsymbol{r},t) &= e\boldsymbol{V}(t)\delta\big(\boldsymbol{r}-\boldsymbol{R}(t)\big)\end{aligned} \tag{6.3.1}$$

---

*Alfred-Marie LIÉNARD (1869–1958) †Emil Johann WIECHERT (1861–1928)

for the charge and current densities, as given by a single term in the sums in (1.4.4) and (1.4.5).

The scalar potential is then

$$\begin{aligned}\Phi(\boldsymbol{r},t) &= e\int(\mathrm{d}\boldsymbol{r}')\mathrm{d}t'\,\frac{\delta(\boldsymbol{r}'-\boldsymbol{R}(t'))\,\delta\left(t-t'-\frac{1}{c}|\boldsymbol{r}-\boldsymbol{r}'|\right)}{|\boldsymbol{r}-\boldsymbol{r}'|}\\ &= e\int\mathrm{d}t'\,\frac{\delta\left(t-t'-\frac{1}{c}|\boldsymbol{r}-\boldsymbol{R}(t')|\right)}{\left|\boldsymbol{r}-\boldsymbol{R}(t')\right|}\end{aligned} \tag{6.3.2}$$

after taking care of the integration over $\boldsymbol{r}'$. The contribution to the $t'$ integration is solely from the value $t' = t_{\mathrm{ret}}$, with $t_{\mathrm{ret}}$ solving

$$t_{\mathrm{ret}} + \frac{1}{c}|\boldsymbol{r}-\boldsymbol{R}(t_{\mathrm{ret}})| = t\,, \tag{6.3.3}$$

the so-called *retardation condition*, defining the *retarded time* $t_{\mathrm{ret}}$. One should think of $t_{\mathrm{ret}}$ as the *emission time*, the earlier time at which the charge emitted the electromagnetic field that is now observed at $\boldsymbol{r}$:

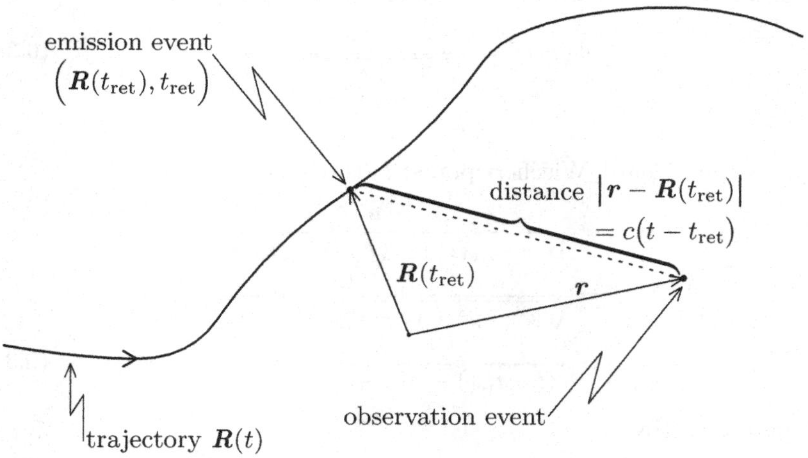


(6.3.4)

We then have $\delta\bigl(f(t')\bigr)$ with $f(t_{\mathrm{ret}}) = 0$ specifying the only $t'$ value for which the argument of the delta function vanishes, so that

$$
\begin{aligned}
\delta\Bigl(t - t' - \frac{1}{c}|\boldsymbol{r} - R(t')|\Bigr)
&= \delta(t' - t_{\mathrm{ret}}) \left| \frac{\partial}{\partial t'} \Bigl(t - t' - \frac{1}{c}|\boldsymbol{r} - \boldsymbol{R}(t')|\Bigr) \right|^{-1} \Bigg|_{t' = t_{\mathrm{ret}}} \\
&= \frac{\delta(t' - t_{\mathrm{ret}})}{1 - \dfrac{\boldsymbol{r} - \boldsymbol{R}(t_{\mathrm{ret}})}{|\boldsymbol{r} - \boldsymbol{R}(t_{\mathrm{ret}})|} \cdot \dfrac{\boldsymbol{V}(t_{\mathrm{ret}})}{c}}\,, \qquad (6.3.5)
\end{aligned}
$$

which takes into account that $|\boldsymbol{V}(t')| < c$ for all $t'$, including $t' = t_{\mathrm{ret}}$. We can now evaluate the $t'$ integral in (6.3.2) and obtain

$$
\Phi(\boldsymbol{r}, t) = \frac{e}{\bigl|\boldsymbol{r} - \boldsymbol{R}(t_{\mathrm{ret}})\bigr| - \bigl(\boldsymbol{r} - \boldsymbol{R}(t_{\mathrm{ret}})\bigr) \cdot \boldsymbol{V}(t_{\mathrm{ret}})/c}\,, \qquad (6.3.6)
$$

and likewise

$$
\boldsymbol{A}(\boldsymbol{r}, t) = \frac{1}{c} \boldsymbol{V}(t_{\mathrm{ret}}) \Phi(\boldsymbol{r}, t) \qquad (6.3.7)
$$

with the emission time $t_{\mathrm{ret}}$ determined by (6.3.3). These are the Liénard–Wiechert potentials for a single charge moving along an arbitrary trajectory.

As a consistency check, we verify that the two versions we have for a charge moving with constant velocity, $\boldsymbol{R}(t) = \boldsymbol{v}t = vt\boldsymbol{e}_z$, are the same. According to (3.5.7), with $\boldsymbol{r}', t'$ replaced by $\boldsymbol{r}, t$ now, we have

$$
\Phi(\boldsymbol{r}, t) = \frac{e}{\sqrt{\dfrac{x^2 + y^2}{\gamma^2} + (z - vt)^2}} \qquad (6.3.8)
$$

whereas the Liénard–Wiechert potential is ($\beta = v/c$)

$$
\begin{aligned}
\Phi(\boldsymbol{r}, t) &= \frac{e}{|\boldsymbol{r} - vt_{\mathrm{ret}}\boldsymbol{e}_z| - (\boldsymbol{r} - vt_{\mathrm{ret}}\boldsymbol{e}_z) \cdot \beta\boldsymbol{e}_z} \\
&= \frac{e}{\sqrt{x^2 + y^2 + (z - vt_{\mathrm{ret}})^2} - \beta(z - vt_{\mathrm{ret}})} \\
&= \frac{e}{c(t - t_{\mathrm{ret}}) - \beta(z - vt_{\mathrm{ret}})} \qquad (6.3.9)
\end{aligned}
$$

with $t_{\mathrm{ret}}$ given by

$$
\sqrt{x^2 + y^2 + (z - vt_{\mathrm{ret}})^2} = c(t - t_{\mathrm{ret}})\,. \qquad (6.3.10)
$$

Indeed, combining the various relations we have

$$\begin{aligned}\frac{x^2+y^2}{\gamma^2}+(z-vt)^2 &= (1-\beta^2)\left[c^2(t-t_{\text{ret}})^2-(z-vt_{\text{ret}})^2\right]\\ &\quad+\left[(z-vt_{\text{ret}})-v(t-t_{\text{ret}})\right]^2\\ &= c^2(t-t_{\text{ret}})^2+\beta^2(z-vt_{\text{ret}})^2\\ &\quad-2(z-vt_{\text{ret}})v(t-t_{\text{ret}})\\ &= \left[c(t-t_{\text{ret}})-\beta(z-vt_{\text{ret}})\right]^2 \qquad (6.3.11)\end{aligned}$$

and it follows that the two expressions for $\Phi(\boldsymbol{r},t)$ are identical, as they must be.

# Chapter 7

# Radiation Fields

## 7.1 Far fields

We return to the retarded 4-potential in the Lorentz gauge (6.2.1),

$$\begin{pmatrix} \Phi \\ \boldsymbol{A} \end{pmatrix}(\boldsymbol{r},t) = \int (\mathrm{d}\boldsymbol{r}') \frac{1}{|\boldsymbol{r}-\boldsymbol{r}'|} \begin{pmatrix} \rho \\ \frac{1}{c}\boldsymbol{j} \end{pmatrix} \left(\boldsymbol{r}', t - \frac{1}{c}|\boldsymbol{r}-\boldsymbol{r}'|\right), \tag{7.1.1}$$

and consider now the typical situation of a radiation problem: The charge and current densities are nonzero inside a finite volume of linear extension $a$,

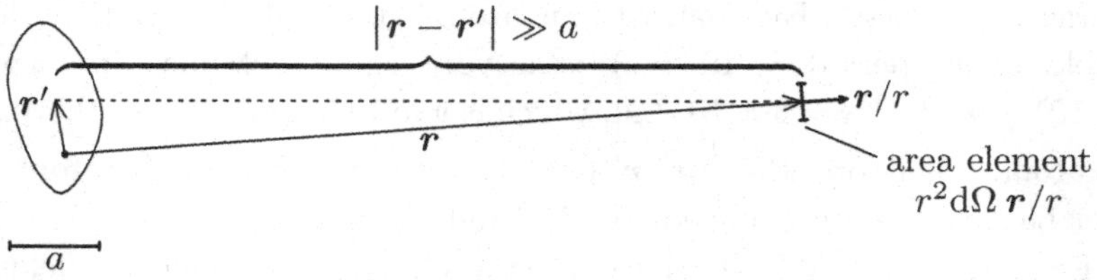


(7.1.2)

and the fields are observed *far away*, so that the radiation fields can be identified properly. In particular, we are interested in determining the energy flux through the area element $r^2 \mathrm{d}\Omega\, \boldsymbol{r}/r$ because this tells us about the energy that actually escapes from the source, the energy of the emitted electromagnetic radiation.

Since $r \gg a > r'$ (we choose the origin of the coordinate system inside the volume where $\rho$, $\boldsymbol{j}$ are nonvanishing), we have

$$\begin{aligned} |\boldsymbol{r}-\boldsymbol{r}'| &= \sqrt{r^2 - 2\boldsymbol{r}\cdot\boldsymbol{r}' + r'^2} \\ &= r - \frac{\boldsymbol{r}}{r}\cdot\boldsymbol{r}' + \cdots , \end{aligned} \tag{7.1.3}$$

where the second term is of the relative size $r'/r \ll 1$, and the ellipsis stands for terms of relative size $(r'/r)^2$ or smaller.

In $\dfrac{1}{|\boldsymbol{r}-\boldsymbol{r}'|}$ it is sufficient to just use the leading term $\dfrac{1}{|\boldsymbol{r}-\boldsymbol{r}'|} \cong \dfrac{1}{r}$, but we keep the next-to-leading term in the retarded time argument, so that

$$\begin{pmatrix}\Phi\\ \boldsymbol{A}\end{pmatrix}(\boldsymbol{r},t) = \frac{1}{r}\int(\mathrm{d}\boldsymbol{r}')\begin{pmatrix}\rho\\ \dfrac{1}{c}\boldsymbol{j}\end{pmatrix}\left(\boldsymbol{r}', t-\frac{r}{c}+\frac{1}{c}\boldsymbol{n}\cdot\boldsymbol{r}'\right)$$
$$+\ (\text{irrelevant terms}), \tag{7.1.4}$$

where $\boldsymbol{n} = \dfrac{\boldsymbol{r}}{r}$ is the unit vector for the direction of observation and we will see shortly why the "irrelevant terms" are indeed not relevant in the present context. The contribution $\dfrac{1}{c}\boldsymbol{n}\cdot\boldsymbol{r}'$ to the retarded time takes into account that there could be rapid changes of $\rho(\boldsymbol{r},t)$ or $\boldsymbol{j}(\boldsymbol{r},t)$ across the source, which will be the case whenever the charge and current distribution changes on a time scale that is short compared with the time that light needs to travel across the source.

For example, a FM radio antenna that is 3 m long and broadcasts at 100 MHz, is about the size set by the distance traveled by light during one cycle: $3\times10^8\,\frac{\mathrm{m}}{\mathrm{s}}/10^8\frac{1}{\mathrm{s}} = 3\,\mathrm{m}$, which says that we have a medium-size antenna that is neither big nor small on the scale set by the time-to-travel distance. By contrast, an atom (size $\sim 10^{-10}\,\mathrm{m}$) that emits visible light (period $\sim 10^{-15}\,\mathrm{s}$) is a very small antenna, in view of $3\times10^8\,\frac{\mathrm{m}}{\mathrm{s}}\times10^{-15}\mathrm{s} = 3\times10^{-7}\,\mathrm{m}$, which is a few thousand times the size of the atom. Therefore, the $\dfrac{1}{c}\boldsymbol{n}\cdot\boldsymbol{r}'$ term is not significant for the atom, but must be taken into account for the FM radio antenna.

From the potentials $\Phi$, $\boldsymbol{A}$ we get the electric and magnetic fields by differentiation,

$$\boldsymbol{E} = -\frac{1}{c}\frac{\partial}{\partial t}\boldsymbol{A} - \boldsymbol{\nabla}\Phi\,, \qquad \boldsymbol{B} = \boldsymbol{\nabla}\times\boldsymbol{A}\,. \tag{7.1.5}$$

Accordingly, the fields will be of the general form

$$\boldsymbol{E},\boldsymbol{B} = \left(\text{term} \propto \frac{1}{r}\right) + \left(\text{term} \propto \frac{1}{r^2}\right) + \cdots\,, \tag{7.1.6}$$

and in the energy current density we have

$$\boldsymbol{S} = \frac{c}{4\pi}\boldsymbol{E}\times\boldsymbol{B} = \left(\text{term} \propto \frac{1}{r^2}\right) + \left(\text{term} \propto \frac{1}{r^3}\right) + \cdots\,, \tag{7.1.7}$$

where only the leading term contributes to the outgoing energy flux

$$\boldsymbol{S}\cdot r^2 \mathrm{d}\Omega \frac{\boldsymbol{r}}{r} = \underbrace{r^2 \times \left(\text{term} \propto \frac{1}{r^2}\right)}_{\substack{\text{finite at large} \\ \text{distances } r}} + \underbrace{r^2 \times \left(\text{term} \propto \frac{1}{r^3}\right) + \cdots}_{\text{vanishing far away}} . \tag{7.1.8}$$

It follows that only the terms $\propto \frac{1}{r}$ in $\boldsymbol{E}$ and $\boldsymbol{B}$ contribute to the energy flux away from the sources composed of the charge density $\rho(\boldsymbol{r},t)$ and the current density $\boldsymbol{j}(\boldsymbol{r},t)$, so that these $\frac{1}{r}$ contributions in $\boldsymbol{E}$ and $\boldsymbol{B}$ make up the *radiation field*, and all other terms are irrelevant for the radiation. This includes in particular the "irrelevant terms" of (7.1.4).

Now, to actually determine the radiation field we need to differentiate the potentials, where the $\boldsymbol{r}$ dependence is in the $\frac{1}{r}$ prefactor and in the time argument

$$t_r = t - \frac{r}{c} + \frac{1}{c}\boldsymbol{n}\cdot\boldsymbol{r}' \quad \text{with} \quad \boldsymbol{n} = \frac{\boldsymbol{r}}{r} . \tag{7.1.9}$$

We take a look at the gradient of a function of the product form $\frac{1}{r}f(t_r)$,

$$\boldsymbol{\nabla}\left(\frac{1}{r}f(t_r)\right) = \left(\boldsymbol{\nabla}\frac{1}{r}\right)f(t_r) + \frac{1}{r}\frac{\partial}{\partial t_r}f(t_r)\boldsymbol{\nabla}t_r , \tag{7.1.10}$$

where

$$\boldsymbol{\nabla}\frac{1}{r} = -\frac{\boldsymbol{r}}{r^3} = -\frac{\boldsymbol{n}}{r^2} \tag{7.1.11}$$

and

$$\begin{aligned} \boldsymbol{\nabla}t_r &= -\frac{1}{c}\frac{\boldsymbol{r}}{r} + \frac{1}{c}\boldsymbol{\nabla}\left(\frac{\boldsymbol{r}}{r}\cdot\boldsymbol{r}'\right) \\ &= -\frac{1}{c}\frac{\boldsymbol{r}}{r} + \frac{1}{c}\left(\frac{\boldsymbol{r}'}{r} - \frac{\boldsymbol{r}\,\boldsymbol{r}}{r^3}\cdot\boldsymbol{r}'\right) \\ &= -\left[\frac{\boldsymbol{n}}{c} + \frac{\boldsymbol{n}}{c}\times\left(\boldsymbol{n}\times\frac{\boldsymbol{r}'}{r}\right)\right] . \end{aligned} \tag{7.1.12}$$

In view of $\dfrac{\partial}{\partial t_r}f(t_r) = \dfrac{\partial}{\partial t}f(t_r)$, we thus have

$$\begin{aligned}\nabla\Big(\frac{1}{r}f(t_r)\Big) &= -\frac{\boldsymbol{n}}{c}\frac{\partial}{\partial t}\Big(\frac{1}{r}f(t_r)\Big)\\ &\quad -\frac{\boldsymbol{n}}{r^2}f(t_r) - \frac{\boldsymbol{n}\times(\boldsymbol{n}\times\boldsymbol{r}')}{r^2}\frac{1}{c}\frac{\partial}{\partial t}f(t_r)\\ &= -\frac{\boldsymbol{n}}{c}\frac{\partial}{\partial t}\Big(\frac{1}{r}f(t_r)\Big) + O\Big(\frac{1}{r^2}\Big). \end{aligned} \tag{7.1.13}$$

The terms of order $\dfrac{1}{r^2}$ are irrelevant for the radiation field, as we have just discussed, and therefore

$$\nabla\Big(\frac{1}{r}f(t_r)\Big) = -\frac{\boldsymbol{n}}{c}\frac{\partial}{\partial t}\Big(\frac{1}{r}f(t_r)\Big) \tag{7.1.14}$$

applies in the present context of getting the radiation field from the potentials.

For the magnetic field, this yields

$$\begin{aligned}\boldsymbol{B} = \nabla\times\boldsymbol{A} &= \nabla\times\frac{1}{r}\int(\mathrm{d}\boldsymbol{r}')\,\frac{1}{c}\boldsymbol{j}(\boldsymbol{r}',t_r)\\ &= -\frac{\boldsymbol{n}}{c}\times\frac{1}{r}\int(\mathrm{d}\boldsymbol{r}')\,\frac{1}{c}\frac{\partial}{\partial t}\boldsymbol{j}(\boldsymbol{r}',t_r)\,, \end{aligned} \tag{7.1.15}$$

where we write "=" rather than "≅", remembering that these are statements about the radiation field, not about $\boldsymbol{B}(\boldsymbol{r},t)$ everywhere and in all aspects.

In the expression for the electric field, we have two terms

$$\begin{aligned}\boldsymbol{E} &= -\frac{1}{c}\frac{\partial}{\partial t}\boldsymbol{A} - \nabla\Phi\\ &= -\frac{1}{c}\frac{1}{r}\int(\mathrm{d}\boldsymbol{r}')\,\frac{1}{c}\frac{\partial}{\partial t}\boldsymbol{j}(\boldsymbol{r}',t_r) + \frac{\boldsymbol{n}}{c}\frac{1}{r}\int(\mathrm{d}\boldsymbol{r}')\,\frac{\partial}{\partial t}\rho(\boldsymbol{r}',t_r)\,, \end{aligned} \tag{7.1.16}$$

and we exploit the continuity equation

$$\frac{\partial}{\partial t'}\rho(\boldsymbol{r}',t') = -\nabla'\cdot\boldsymbol{j}(\boldsymbol{r}',t') \tag{7.1.17}$$

to relate the time derivation of $\rho$ to a derivative of $\boldsymbol{j}$,

$$\begin{aligned}\frac{\partial}{\partial t}\rho(\boldsymbol{r}',t_r) &= \frac{\partial}{\partial t_r}\rho(\boldsymbol{r}',t_r)\\ &= -\underbrace{\boldsymbol{\nabla}'\cdot\boldsymbol{j}(\boldsymbol{r}',t_r)}_{\text{only}}\\ &= -\boldsymbol{\nabla}'\cdot\boldsymbol{j}(\boldsymbol{r}',t_r) + \underbrace{\boldsymbol{\nabla}'\cdot\boldsymbol{j}(\boldsymbol{r}',t_r)}_{\text{only}}\,, \qquad (7.1.18)\end{aligned}$$

where we must carefully note that $t_r$ contains an $\boldsymbol{r}'$ dependence as well,

$$\boldsymbol{\nabla}' t_r = \frac{\boldsymbol{n}}{c}\,, \qquad (7.1.19)$$

and we get

$$\frac{\partial}{\partial t}\rho(\boldsymbol{r}',t_r) = -\boldsymbol{\nabla}'\cdot\boldsymbol{j}(\boldsymbol{r}',t_r) + \frac{\boldsymbol{n}}{c}\cdot\frac{\partial}{\partial t}\boldsymbol{j}(\boldsymbol{r}',t_r)\,. \qquad (7.1.20)$$

The integral of $\boldsymbol{\nabla}'\cdot\boldsymbol{j}(\boldsymbol{r}',t_r)$ over all of $\boldsymbol{r}'$ space equals a surface integral over an infinitely remote surface and, therefore, it vanishes. Then,

$$\begin{aligned}\boldsymbol{E} &= -\frac{1}{c}\frac{1}{r}\int(\mathrm{d}\boldsymbol{r}')\,\frac{1}{c}\frac{\partial}{\partial t}\boldsymbol{j}(\boldsymbol{r}',t_r) + \frac{\boldsymbol{n}}{c}\frac{\boldsymbol{n}}{c}\cdot\frac{1}{r}\int(\mathrm{d}\boldsymbol{r}')\,\frac{\partial}{\partial t}\boldsymbol{j}(\boldsymbol{r}',t_r)\\ &= \boldsymbol{n}\times\left(\frac{\boldsymbol{n}}{c}\times\frac{1}{r}\int(\mathrm{d}\boldsymbol{r}')\,\frac{1}{c}\frac{\partial}{\partial t}\boldsymbol{j}(\boldsymbol{r}',t_r)\right)\\ &= -\boldsymbol{n}\times\boldsymbol{B} \qquad (7.1.21)\end{aligned}$$

is obtained, upon recognizing the expression for $\boldsymbol{B}$ in (7.1.15). Since $\boldsymbol{B}\perp\boldsymbol{n}$, we have also

$$\boldsymbol{B} = \boldsymbol{n}\times\boldsymbol{E}\,, \qquad (7.1.22)$$

so that $\boldsymbol{E}$, $\boldsymbol{B}$, $\boldsymbol{n}$ constitute a right-handed set of pairwise perpendicular vectors, just as we established in Section 2.3 for $\boldsymbol{E}$, $\boldsymbol{B}$, $\boldsymbol{v}$ of a unidirectional electromagnetic pulse; see (2.3.7). The direction of propagation $\boldsymbol{v}$ is, of course, the direction of the unit vector $\boldsymbol{n}$ in the present context.

## 7.2 Emitted power

The energy current density of the radiation field is then

$$\boldsymbol{S} = \frac{c}{4\pi}\boldsymbol{E}\times\boldsymbol{B} = \frac{c}{4\pi}\Big(-\boldsymbol{n}\times\boldsymbol{B}\Big)\times\boldsymbol{B} = \frac{c}{4\pi}\boldsymbol{B}^2\,\boldsymbol{n}\,, \qquad (7.2.1)$$

and the total energy flux into solid angle $\mathrm{d}\Omega$ in the direction $\boldsymbol{n}$ is, therefore, given by

$$\boldsymbol{S}\cdot\mathrm{d}\Omega\, r^2\boldsymbol{n} = \frac{c}{4\pi}\big(r\boldsymbol{B}\big)^2\mathrm{d}\Omega\,. \tag{7.2.2}$$

This is the energy per unit time flowing through the area element $r^2\mathrm{d}\Omega$, the *power* radiated into solid angle $\mathrm{d}\Omega$,

$$\frac{\mathrm{d}P}{\mathrm{d}\Omega}\mathrm{d}\Omega = \frac{c}{4\pi}\big(r\boldsymbol{B}\big)^2\mathrm{d}\Omega\,, \tag{7.2.3}$$

so that

$$\frac{\mathrm{d}P}{\mathrm{d}\Omega} = \frac{c}{4\pi}\big(r\boldsymbol{B}\big)^2 = \frac{1}{4\pi c^3}\left|\boldsymbol{n}\times\int(\mathrm{d}\boldsymbol{r}')\,\frac{\partial}{\partial t}\boldsymbol{j}(\boldsymbol{r}',t_r)\right|^2 \tag{7.2.4}$$

after inserting the expression for $\boldsymbol{B}$ in (7.1.15). The total radiated power is obtained by an integration over all solid angle,

$$P = \int \mathrm{d}\Omega\,\frac{\mathrm{d}P}{\mathrm{d}\Omega}\,. \tag{7.2.5}$$

These expressions for $\dfrac{\mathrm{d}P}{\mathrm{d}\Omega}$ and $P$ are associated with the name of Larmor,* but what is called the "Larmor formula" is a specialized application to dipole radiation; see below.

As a first, simple yet important, application we consider a single charge $e$ moving along the trajectory $\boldsymbol{R}(t)$ such that its velocity $\boldsymbol{V}(t) = \dfrac{\mathrm{d}}{\mathrm{d}t}\boldsymbol{R}(t)$ is small compared with the speed of light at all times, $|\,\boldsymbol{V}(t)| \ll c$. Then the values of $\boldsymbol{r}'$ in $t_r = t - r/c + \boldsymbol{n}\cdot\boldsymbol{r}'/c$, which equal $\boldsymbol{R}(t_{\mathrm{ret}})$, will be bounded by some characteristic length, and $\boldsymbol{n}\cdot\boldsymbol{r}'/c$ is not larger than the corresponding characteristic short time. We can therefore, ignore this *internal retardation* for slowly moving charges, so that, with $\rho(\boldsymbol{r},t) = e\delta\big(\boldsymbol{r}-\boldsymbol{R}(t)\big)$ and $\boldsymbol{j}(\boldsymbol{r},t) = e\,\boldsymbol{V}(t)\delta\big(\boldsymbol{r}-\boldsymbol{R}(t)\big)$, the charge and current densities of (6.3.1),

$$\int(\mathrm{d}\boldsymbol{r}')\,\frac{\partial}{\partial t}\boldsymbol{j}(\boldsymbol{r}',t_r) \to \int(\mathrm{d}\boldsymbol{r}')\,\frac{\partial}{\partial t}\boldsymbol{j}(\boldsymbol{r}',t_{\mathrm{e}}) \tag{7.2.6}$$

where the replacement

$$t_r = t - \frac{r}{c} + \frac{1}{c}\boldsymbol{n}\cdot\boldsymbol{r}' \to t - \frac{r}{c} = t_{\mathrm{e}} \tag{7.2.7}$$

*Sir Joseph LARMOR (1857–1942)

is carried out, or

$$\int (\mathrm{d}\boldsymbol{r}')\, \frac{\partial}{\partial t}\boldsymbol{j}(\boldsymbol{r}', t_r) \to \frac{\mathrm{d}}{\mathrm{d}t_\mathrm{e}} \int (\mathrm{d}\boldsymbol{r}')\, \boldsymbol{j}(\boldsymbol{r}', t_\mathrm{e}) = e\frac{\mathrm{d}\boldsymbol{V}}{\mathrm{d}t}(t_\mathrm{e}) = e\frac{\mathrm{d}^2\boldsymbol{R}}{\mathrm{d}t^2}(t_\mathrm{e})\,. \tag{7.2.8}$$

The radiated power per solid angle is then

$$\begin{aligned}\frac{\mathrm{d}P}{\mathrm{d}\Omega} &= \frac{1}{4\pi c^3}\left|\boldsymbol{n} \times e\frac{\mathrm{d}^2\boldsymbol{R}}{\mathrm{d}t^2}(t_\mathrm{e})\right|^2 \\ &= \frac{e^2}{4\pi c^3}\left|\frac{\mathrm{d}^2\boldsymbol{R}}{\mathrm{d}t^2}(t_\mathrm{e})\right|^2 (\sin\theta)^2\,,\end{aligned} \tag{7.2.9}$$

where $\theta$ is the angle between the direction of observation $\boldsymbol{n}$ and the acceleration $\dfrac{\mathrm{d}^2\boldsymbol{R}}{\mathrm{d}t^2} = \dfrac{\mathrm{d}\boldsymbol{V}}{\mathrm{d}t}$ at the *emission time* $t_\mathrm{e}$:

(7.2.10)

For the total radiated power we need the integral of $\dfrac{\mathrm{d}P}{\mathrm{d}\Omega}$ over all of $4\pi$ solid angle, where $\displaystyle\int \frac{\mathrm{d}\Omega}{4\pi}(\sin\theta)^2 = \frac{2}{3}$ since the $(\sin\theta)^2$ term refers to two of the three spatial directions. This yields

$$P = \frac{2}{3}\frac{e^2}{c^3}\left|\frac{\mathrm{d}^2\boldsymbol{R}}{\mathrm{d}t^2}(t_\mathrm{e})\right|^2\,, \tag{7.2.11}$$

which is *Larmor's formula* for the radiation of a moving charge. The $(\sin\theta)^2$ dependence of $\dfrac{\mathrm{d}P}{\mathrm{d}\Omega}$ is characteristic of the so-called *dipole radiation*, about which we shall say more in a short while. Its main feature is that there is no radiation emitted in the forward and backward direction (as specified by the acceleration at the emission time), radiation is to transverse directions, with maximal intensity in the plane perpendicular to the acceleration.

We note that the general formula (7.2.4) seems to suggest that there is radiation whenever the electric current is time-dependent, because its time derivative $\dfrac{\partial}{\partial t}\boldsymbol{j}(\boldsymbol{r}', t_r)$ appears in the expression for $\dfrac{\mathrm{d}P}{\mathrm{d}\Omega}$, but in fact this is not enough. A charge in uniform motion has a time-dependent current, too, but it does not radiate, because the *acceleration* vanishes. More generally, one can conceive situations with truly time-dependent $\rho(\boldsymbol{r},t)$ and $\boldsymbol{j}(\boldsymbol{r},t)$ but without *any* radiation at all.

## 7.3 Larmor formula

Now consider a collection of charges, all moving so slowly that a common emission time applies to all of them. This means that we have $t_{\mathrm{e}}$ in

$$\frac{\mathrm{d}P}{\mathrm{d}\Omega}=\frac{1}{4\pi c^3}\left|\boldsymbol{n}\times\int(\mathrm{d}\boldsymbol{r}')\frac{\partial}{\partial t}\boldsymbol{j}(\boldsymbol{r}',t_{\mathrm{e}})\right|^2 \tag{7.3.1}$$

rather than $t_r=t_{\mathrm{e}}+\boldsymbol{n}\cdot\boldsymbol{r}'/c$ as in (7.2.4). In view of

$$\begin{aligned}\int(\mathrm{d}\boldsymbol{r}')\,\boldsymbol{j}(\boldsymbol{r}',t_{\mathrm{e}}) &= \int(\mathrm{d}\boldsymbol{r}')\,\boldsymbol{j}'(\boldsymbol{r}',t_{\mathrm{e}})\cdot\underbrace{(\boldsymbol{\nabla}'\boldsymbol{r}')}_{=\mathbf{1}}\\ &= \underbrace{\int(\mathrm{d}\boldsymbol{r}')\,\boldsymbol{\nabla}'\cdot\Bigl(\boldsymbol{j}(\boldsymbol{r}',t_{\mathrm{e}})\boldsymbol{r}'\Bigr)}_{\substack{\text{vanishing}\\ \text{surface integral}}}-\int(\mathrm{d}\boldsymbol{r}')\,\boldsymbol{r}'\underbrace{\boldsymbol{\nabla}'\cdot\boldsymbol{j}}_{=-\frac{\partial}{\partial t}\rho}\\ &= \frac{\mathrm{d}}{\mathrm{d}t_{\mathrm{e}}}\int(\mathrm{d}\boldsymbol{r}')\,\boldsymbol{r}'\rho(\boldsymbol{r}',t_{\mathrm{e}})=\frac{\mathrm{d}}{\mathrm{d}t_{\mathrm{e}}}\boldsymbol{d}(t_{\mathrm{e}})\,,\end{aligned} \tag{7.3.2}$$

with the *electric dipole moment*

$$\boldsymbol{d}(t)=\int(\mathrm{d}\boldsymbol{r})\,\boldsymbol{r}\rho(\boldsymbol{r},t)\,, \tag{7.3.3}$$

we get another *Larmor formula*,

$$\frac{\mathrm{d}P}{\mathrm{d}\Omega}=\frac{1}{4\pi c^3}\left|\boldsymbol{n}\times\left(\frac{\mathrm{d}}{\mathrm{d}t}\right)^2\boldsymbol{d}(t)\right|^2\Biggr|_{t\,=\,t_{\mathrm{e}}\,=\,t-r/c}\,. \tag{7.3.4}$$

Upon denoting the angle between $\boldsymbol{n}$ and $\dfrac{\mathrm{d}^2}{\mathrm{d}t^2}\boldsymbol{d}(t_{\mathrm{e}})$ by $\theta$, this reads

$$\frac{\mathrm{d}P}{\mathrm{d}\Omega}=\frac{1}{4\pi c^3}\left|\ddot{\boldsymbol{d}}(t_{\mathrm{e}})\right|^2(\sin\theta)^2\,, \tag{7.3.5}$$

wherein $\ddot{\ }$ stands for the second derivative with respect to the argument. We use the overdot notation only for total time derivatives, usually of integrated quantities.

The total power radiated is here given by the square of $\ddot{\boldsymbol{d}}$,

$$P = \frac{2}{3c^3}\left|\ddot{\boldsymbol{d}}(t_{\mathrm{e}})\right|^2, \tag{7.3.6}$$

where clearly the instant of emission is the relevant one (time $t_{\mathrm{e}}$), not the instant of observation (time $t = t_{\mathrm{e}} + r/c$). This result, for $P$, and the previous one, for $\frac{\mathrm{d}P}{\mathrm{d}\Omega}$ in (7.3.5), are also referred to as Larmor formulas, for electric dipole radiation to be specific.

The replacement $t_r \to t_{\mathrm{e}}$ is not so well justified for an extended charge and current distribution as it was for the single point charge. Let us, therefore, consider the leading correction that we obtain by taking the difference between $t_r$ and $t_{\mathrm{e}}$ into account to first order,

$$\begin{aligned} \boldsymbol{j}(\boldsymbol{r}', t_r) &= \boldsymbol{j}(\boldsymbol{r}', t_{\mathrm{e}} + \boldsymbol{n}\cdot\boldsymbol{r}'/c) \\ &\cong \boldsymbol{j}(\boldsymbol{r}', t_{\mathrm{e}}) + \boldsymbol{n}\cdot\boldsymbol{r}'\frac{1}{c}\frac{\partial}{\partial t_{\mathrm{e}}}\boldsymbol{j}(\boldsymbol{r}', t_{\mathrm{e}}), \end{aligned} \tag{7.3.7}$$

neglecting contributions proportional to $\frac{1}{c^2}$ or higher powers of $\frac{1}{c}$. Then we have

$$\int (\mathrm{d}\boldsymbol{r}')\,\frac{\partial}{\partial t}\boldsymbol{j}(\boldsymbol{r}', t_r) = \underbrace{\ddot{\boldsymbol{d}}(t_{\mathrm{e}})}_{\substack{\text{as}\\ \text{before}}} + \underbrace{\frac{1}{c}\frac{\mathrm{d}^2}{\mathrm{d}t_{\mathrm{e}}^{\,2}}\boldsymbol{n}\cdot\int(\mathrm{d}\boldsymbol{r}')\,\boldsymbol{r}'\boldsymbol{j}(\boldsymbol{r}', t_{\mathrm{e}})}_{\text{additional term} \,\propto\, \frac{1}{c}}, \tag{7.3.8}$$

and we break up the additional term into two contributions in accordance with

$$\begin{aligned} &\boldsymbol{n}\cdot\frac{\mathrm{d}^2}{\mathrm{d}t_{\mathrm{e}}^{\,2}}\int(\mathrm{d}\boldsymbol{r}')\left[\frac{1}{2c}(\boldsymbol{r}'\boldsymbol{j} - \boldsymbol{j}\,\boldsymbol{r}') + \frac{1}{2c}(\boldsymbol{r}'\boldsymbol{j} + \boldsymbol{j}\,\boldsymbol{r}')\right] \\ &= -\boldsymbol{n}\times\frac{\mathrm{d}^2}{\mathrm{d}t_{\mathrm{e}}^{\,2}}\int(\mathrm{d}\boldsymbol{r}')\,\frac{1}{2c}\boldsymbol{r}'\times\boldsymbol{j}(\boldsymbol{r}', t_{\mathrm{e}}) \\ &\quad + \boldsymbol{n}\cdot\frac{1}{2c}\frac{\mathrm{d}^2}{\mathrm{d}t_{\mathrm{e}}^{\,2}}\int(\mathrm{d}\boldsymbol{r}')\,[\boldsymbol{r}'\boldsymbol{j}(\boldsymbol{r}', t_{\mathrm{e}}) + \boldsymbol{j}(\boldsymbol{r}', t_{\mathrm{e}})\,\boldsymbol{r}']\,. \end{aligned} \tag{7.3.9}$$

We recognize the *magnetic dipole moment*

$$\boldsymbol{\mu}(t) = \int(\mathrm{d}\boldsymbol{r})\,\frac{1}{2c}\boldsymbol{r}\times\boldsymbol{j}(\boldsymbol{r}, t) \tag{7.3.10}$$

in the first of these terms, which combines with the electric dipole moment to give

$$\frac{\mathrm{d}P}{\mathrm{d}\Omega} = \frac{1}{4\pi c^3}\left|\boldsymbol{n} \times \left[\ddot{\boldsymbol{d}}(t_\mathrm{e}) - \boldsymbol{n} \times \ddot{\boldsymbol{\mu}}(t_\mathrm{e})\right]\right|^2, \qquad (7.3.11)$$

the Larmor formula for both electric and magnetic dipole radiation.

Before dealing with the second term in (7.3.9), let us note that

$$\frac{\mathrm{d}P}{\mathrm{d}\Omega} = \frac{1}{4\pi c^3}\left[\left(\boldsymbol{n} \times \ddot{\boldsymbol{d}}\right)^2 + \left(\boldsymbol{n} \times \ddot{\boldsymbol{\mu}}\right)^2 + 2\boldsymbol{n} \cdot \left(\ddot{\boldsymbol{d}} \times \ddot{\boldsymbol{\mu}}\right)\right] \qquad (7.3.12)$$

has one term for electric dipole radiation, one for magnetic dipole radiation, and a third term for the interference between electric and magnetic dipole radiation. Whereas the interference term is important for the spatial distribution of the radiation, giving constructive interference for directions $\boldsymbol{n}$ for which $\boldsymbol{n} \cdot \left(\ddot{\boldsymbol{d}} \times \ddot{\boldsymbol{\mu}}\right) > 0$ and destructive interference in the opposite directions, it does not contribute to the total radiated power

$$P = \frac{2}{3c^3}\left(\ddot{\boldsymbol{d}}^{\,2} + \ddot{\boldsymbol{\mu}}^2\right), \qquad (7.3.13)$$

because $\int \mathrm{d}\Omega\, \boldsymbol{n} = 0$.

As a consequence of the factor $\frac{1}{c}$ in the expression for $\boldsymbol{\mu}(t)$, the radiated power for the magnetic dipole radiation tends to be smaller by a factor $\left(\frac{v}{c}\right)^2$ than the power of electric dipole radiation, whereby $v$ is a characteristic velocity of the moving charges. For special geometric circumstances, it can however happen that $\boldsymbol{d}(t) = 0$ while $\boldsymbol{\mu}(t) \neq 0$, and then the magnetic dipole radiation dominates.

The remaining last term in (7.3.9),

$$\begin{aligned}
&\int (\mathrm{d}\boldsymbol{r}')\left(\boldsymbol{r}'\boldsymbol{j}(\boldsymbol{r}',t_\mathrm{e}) + \boldsymbol{j}(\boldsymbol{r}',t_\mathrm{e})\,\boldsymbol{r}'\right) \\
&= \int (\mathrm{d}\boldsymbol{r}')\left[\underbrace{\boldsymbol{\nabla}'\left(\boldsymbol{r}'\boldsymbol{j}(\boldsymbol{r}',t_\mathrm{e})\,\boldsymbol{r}'\right)}_{} - \boldsymbol{r}'\underbrace{\left(\boldsymbol{\nabla}'\cdot\boldsymbol{j}(\boldsymbol{r}',t_\mathrm{e})\right)}_{=-\frac{\partial}{\partial t_\mathrm{e}}\rho(\boldsymbol{r}',t_\mathrm{e})}\boldsymbol{r}'\right] \\
&= \frac{\mathrm{d}}{\mathrm{d}t_\mathrm{e}}\int (\mathrm{d}\boldsymbol{r}')\left(\boldsymbol{r}'\boldsymbol{r}' - \frac{1}{3}r'^2\mathbf{1}\right)\rho(\boldsymbol{r}',t_\mathrm{e}) + \frac{1}{3}\mathbf{1}\frac{\mathrm{d}}{\mathrm{d}t_\mathrm{e}}\int (\mathrm{d}\boldsymbol{r}')\,r'^2\rho(\boldsymbol{r}',t_\mathrm{e}) \\
&= \frac{\mathrm{d}}{\mathrm{d}t_\mathrm{e}}\left[\frac{1}{3}\mathbf{Q}(t_\mathrm{e}) + \frac{1}{3}\mathbf{1}\int (\mathrm{d}\boldsymbol{r}')\,r'^2\rho(\boldsymbol{r}',t_\mathrm{e})\right],
\end{aligned} \qquad (7.3.14)$$

involves the electric quadrupole moment dyadic,

$$\mathsf{Q}(t) = \int (\mathrm{d}\boldsymbol{r})\,(3\boldsymbol{r}\,\boldsymbol{r} - \mathbf{1}r^2)\rho(\boldsymbol{r},t)\,, \tag{7.3.15}$$

and a contribution $\propto \mathbf{1}$, which does not contribute to the radiated power,

$$\frac{\mathrm{d}P}{\mathrm{d}\Omega} = \frac{1}{4\pi c^3}\left|\boldsymbol{n}\times\left(\ddot{\boldsymbol{d}}(t_\mathrm{e}) - \boldsymbol{n}\times\ddot{\boldsymbol{\mu}}(t_\mathrm{e}) + \frac{1}{6c}\dddot{\mathsf{Q}}(t_\mathrm{e})\cdot\boldsymbol{n}\right)\right|^2. \tag{7.3.16}$$

Here we have included electric dipole radiation, magnetic dipole radiation, and electric quadrupole radiation.

## 7.4 Longitudinal and transverse components of a field

Let us take a brief look at the radiation fields as obtained from the *radiation gauge*, specified in (1.3.8) by $\boldsymbol{\nabla}\cdot\boldsymbol{A} = 0$, for which [see (1.3.13)]

$$-\boldsymbol{\nabla}^2\Phi = 4\pi\rho \tag{7.4.1}$$

is an instantaneous Poisson equation, solved by

$$\Phi(\boldsymbol{r},t) = \int (\mathrm{d}\boldsymbol{r}')\,\frac{\rho(\boldsymbol{r}',t)}{|\boldsymbol{r}-\boldsymbol{r}'|} \equiv (-\boldsymbol{\nabla}^2)^{-1}4\pi\rho(\boldsymbol{r},t)\,, \tag{7.4.2}$$

where the Coulomb potential, the Green's function for the Laplace* differential operator, is symbolically written as the inverse Laplace operator. The vector potential obeys the inhomogeneous wave equation (1.3.14),

$$\left(\frac{1}{c^2}\frac{\partial^2}{\partial t^2} - \boldsymbol{\nabla}^2\right)\boldsymbol{A} = \frac{4\pi}{c}\boldsymbol{j}_\mathrm{eff}\,, \tag{7.4.3}$$

where the effective current density

$$\boldsymbol{j}_\mathrm{eff} = \boldsymbol{j} - \frac{1}{4\pi}\frac{\partial}{\partial t}\boldsymbol{\nabla}\Phi \tag{7.4.4}$$

is divergenceless,

$$\begin{aligned}\boldsymbol{\nabla}\cdot\boldsymbol{j}_\mathrm{eff} &= \boldsymbol{\nabla}\cdot\boldsymbol{j} - \frac{1}{4\pi}\frac{\partial}{\partial t}\underbrace{\boldsymbol{\nabla}^2\Phi}_{=-4\pi\rho} \\ &= \boldsymbol{\nabla}\cdot\boldsymbol{j} + \frac{\partial}{\partial t}\rho = 0\,,\end{aligned} \tag{7.4.5}$$

---

*Pierre Simon, Marquis de LAPLACE (1749–1827)

a necessary property that ensures the consistency of the radiation-gauge condition. For, the divergence of the left-hand side of (7.4.3) vanishes, as a consequence of $\boldsymbol{\nabla} \cdot \boldsymbol{A} = 0$, and therefore the right-hand side must be divergenceless as well.

Here is a bit of useful terminology. A vector field $\boldsymbol{F}(\boldsymbol{r})$ with a vanishing curl, $\boldsymbol{\nabla} \times \boldsymbol{F} = 0$ is called a *longitudinal field*, and when the divergence vanishes, $\boldsymbol{\nabla} \cdot \boldsymbol{F} = 0$, we call it a *transverse field*. This refers to the direction of the gradient, in the sense that a longitudinal field is parallel to the gradient, and a transverse field is perpendicular to the gradient. Upon expressing $\boldsymbol{F}(\boldsymbol{r})$ in terms of its spatial Fourier transform $\boldsymbol{F}(\boldsymbol{k})$,

$$\boldsymbol{F}(\boldsymbol{r}) = \int \frac{(\mathrm{d}\boldsymbol{k})}{(2\pi)^3} \, \mathrm{e}^{\mathrm{i}\boldsymbol{k} \cdot \boldsymbol{r}} \, \boldsymbol{F}(\boldsymbol{k}) \,, \tag{7.4.6}$$

we note that $\boldsymbol{\nabla} \to \mathrm{i}\boldsymbol{k}$ and therefore,

$$\begin{aligned} \boldsymbol{k} \times \boldsymbol{F}(\boldsymbol{k}) &= 0 \quad \text{for a longitudinal field,} \\ \text{and} \quad \boldsymbol{k} \cdot \boldsymbol{F}(\boldsymbol{k}) &= 0 \quad \text{for a transverse field.} \end{aligned} \tag{7.4.7}$$

The radiation field is transverse because $\boldsymbol{\nabla} \cdot \boldsymbol{E} = 0$ and $\boldsymbol{\nabla} \cdot \boldsymbol{B} = 0$ far away from the sources $\rho$ and $\boldsymbol{j}$. In fact, the magnetic field is always transverse, whereas the electric field has a longitudinal part where the charge density is nonzero: $\boldsymbol{\nabla} \cdot \boldsymbol{E} = 4\pi\rho$, but this longitudinal component of $\boldsymbol{E}$ has nothing to do with radiation.

What about the vector potential $\boldsymbol{A}$? The fundamental observation is that a gauge transformation $\boldsymbol{A} \to \boldsymbol{A} + \boldsymbol{\nabla}\lambda$ effects solely the longitudinal component of $\boldsymbol{A}$, whereas the transverse component of $\boldsymbol{A}$ is gauge-invariant. [We exploited this fact earlier, in Section 5.4, when we eliminated the gauge-dependent terms from the Lagrange function for the electromagnetic field in interaction with charge-carrying massive particles. The effective current density of (7.4.4) is the transverse current density of (5.4.15).] The radiation gauge reduces the vector potential to its gauge-invariant transverse part by eliminating the longitudinal part altogether. Accordingly, we can find the vector potential in the radiation gauge by picking out the transverse part of $\boldsymbol{A}$ as given in any other gauge, such as the Lorentz gauge.

The decomposition of an arbitrary vector field $\boldsymbol{F}(\boldsymbol{r})$ into its longitudinal and transverse parts is immediate in terms of the Fourier transform,

$$\boldsymbol{F}(\boldsymbol{k}) = \underbrace{\frac{\boldsymbol{k}\,\boldsymbol{k}}{k^2} \cdot \boldsymbol{F}(\boldsymbol{k})}_{\text{longitudinal}} + \underbrace{\left(\boldsymbol{1} - \frac{\boldsymbol{k}\,\boldsymbol{k}}{k^2}\right) \cdot \boldsymbol{F}(\boldsymbol{k})}_{\text{transverse}} \,, \tag{7.4.8}$$

so that

$$\boldsymbol{F}(\boldsymbol{r}) = \boldsymbol{F}_{\|}(\boldsymbol{r}) + \boldsymbol{F}_{\perp}(\boldsymbol{r}) \tag{7.4.9}$$

with the longitudinal component

$$\boldsymbol{F}_{\|}(\boldsymbol{r}) = \int \frac{(\mathrm{d}\boldsymbol{k})}{(2\pi)^3} \frac{\boldsymbol{k}\,\boldsymbol{k}}{k^2} \cdot \boldsymbol{F}(\boldsymbol{k})\, \mathrm{e}^{\mathrm{i}\boldsymbol{k}\cdot\boldsymbol{r}} = \frac{\boldsymbol{\nabla}\boldsymbol{\nabla}}{\boldsymbol{\nabla}^2} \cdot \boldsymbol{F}(\boldsymbol{r}) \tag{7.4.10}$$

and the transverse component

$$\begin{aligned} \boldsymbol{F}_{\perp}(\boldsymbol{r}) &= \int \frac{(\mathrm{d}\boldsymbol{k})}{(2\pi)^3} \left(\boldsymbol{1} - \frac{\boldsymbol{k}\,\boldsymbol{k}}{k^2}\right) \cdot \boldsymbol{F}(\boldsymbol{k})\, \mathrm{e}^{\mathrm{i}\boldsymbol{k}\cdot\boldsymbol{r}} \\ &= \left(\boldsymbol{1} - \frac{\boldsymbol{\nabla}\,\boldsymbol{\nabla}}{\boldsymbol{\nabla}^2}\right) \cdot \boldsymbol{F}(\boldsymbol{r}), \end{aligned} \tag{7.4.11}$$

where $\dfrac{1}{\boldsymbol{\nabla}^2} = (\boldsymbol{\nabla}^2)^{-1}$ is to be understood as explained at (7.4.2).

Caveat: Do not confuse these longitudinal and transverse components of a vector field with the parallel and perpendicular components of Section 3.5, although we use the subscripts $_{\|}$ and $_{\perp}$ for both.

With the vector potential in the Lorentz gauge as given in (6.2.1) and (7.1.1), we conclude that the retarded vector potential in the radiation gauge is

$$\boldsymbol{A}(\boldsymbol{r},t) = \left(\boldsymbol{1} - \frac{\boldsymbol{\nabla}\,\boldsymbol{\nabla}}{\boldsymbol{\nabla}^2}\right) \cdot \int (\mathrm{d}\boldsymbol{r}') \frac{\frac{1}{c}\boldsymbol{j}\left(\boldsymbol{r}', t - \frac{1}{c}|\boldsymbol{r}-\boldsymbol{r}'|\right)}{|\boldsymbol{r}-\boldsymbol{r}'|}, \tag{7.4.12}$$

and for the radiation field we have, as usual,

$$\begin{aligned} \frac{1}{|\boldsymbol{r}-\boldsymbol{r}'|} &\to \frac{1}{r}, \\ t - \frac{1}{c}|\boldsymbol{r}-\boldsymbol{r}'| &\to t - \frac{r}{c} + \boldsymbol{n}\cdot\boldsymbol{r}'/c = t_r, \\ \boldsymbol{\nabla} &\to -\boldsymbol{n}\frac{1}{c}\frac{\partial}{\partial t}, \end{aligned} \tag{7.4.13}$$

so that

$$\begin{aligned} \boldsymbol{A}(\boldsymbol{r},t) &= (\boldsymbol{1} - \boldsymbol{n}\boldsymbol{n}) \cdot \frac{1}{r} \int (\mathrm{d}\boldsymbol{r}')\, \frac{1}{c}\boldsymbol{j}(\boldsymbol{r}', t_r) \\ &= -\boldsymbol{n} \times \left[\boldsymbol{n} \times \frac{1}{r} \int (\mathrm{d}\boldsymbol{r}')\, \frac{1}{c}\boldsymbol{j}(\boldsymbol{r}', t_r)\right] \end{aligned} \tag{7.4.14}$$

for the radiation field.

The magnetic field

$$\begin{aligned}\boldsymbol{B} = \boldsymbol{\nabla} \times \boldsymbol{A} &\to -\boldsymbol{n} \times \frac{1}{c}\frac{\partial}{\partial t}\boldsymbol{A} \\ &= -\boldsymbol{n} \times \frac{1}{cr}\int (\mathrm{d}\boldsymbol{r}')\,\frac{1}{c}\frac{\partial}{\partial t}\boldsymbol{j}(\boldsymbol{r}', t_r)\end{aligned} \tag{7.4.15}$$

is transverse by its nature, and we recover the expression in (7.1.15). The two terms contributing to the electric field,

$$\boldsymbol{E} = -\frac{1}{c}\frac{\partial}{\partial t}\boldsymbol{A} - \boldsymbol{\nabla}\Phi\,, \tag{7.4.16}$$

naturally split $\boldsymbol{E}$ into its transverse (radiation) part and its longitudinal (electrostatic) part when the radiation gauge is used — because then $\boldsymbol{A}$ and its time derivative are transverse, and the gradient of $\Phi$ is always longitudinal — and we get

$$\boldsymbol{E} \to -\frac{1}{c}\frac{\partial}{\partial t}\boldsymbol{A} = -\boldsymbol{n} \times \boldsymbol{B} \tag{7.4.17}$$

for the radiation field. Of course, this is the same expression for $\boldsymbol{E}$ in terms of $\boldsymbol{j}$ that we found earlier in the Lorentz gauge, see (7.1.21).

## 7.5 Charge point of view

We integrate the continuity equation for energy (1.4.12),

$$\frac{\partial}{\partial t}U + \boldsymbol{\nabla} \cdot \boldsymbol{S} = -\boldsymbol{j} \cdot \boldsymbol{E}\,, \tag{7.5.1}$$

over a large but finite volume $V$ that encloses all electric charges and currents, so that $\rho$, $\boldsymbol{j}$ vanish on the surface $S$ of $V$ and everywhere outside of $V$. This integration gives

$$\frac{\mathrm{d}}{\mathrm{d}t}E + P = \dot{W}\,, \tag{7.5.2}$$

where

$$E(t) = \int_V (\mathrm{d}\boldsymbol{r})\, U(\boldsymbol{r}, t) \tag{7.5.3}$$

is the energy stored in the electromagnetic field inside volume $V$,

$$P(t) = \int_S \mathrm{d}\boldsymbol{S} \cdot \boldsymbol{S}(\boldsymbol{r}, t) \tag{7.5.4}$$

is the power of the radiation leaving volume $V$ through its surface $S$, and

$$\dot{W}(t) = \int (\mathrm{d}\boldsymbol{r})\big(-\boldsymbol{j}(\boldsymbol{r},t)\cdot \boldsymbol{E}(\boldsymbol{r},t)\big) \qquad (7.5.5)$$

is the power associated with the work done by the moving charges on the electromagnetic field. This work is partly converted into energy stored inside $V$, and partly into energy radiated into the space surrounding volume $V$.

When evaluating $\dot{W}$ explicitly, we need the electric field $\boldsymbol{E}(\boldsymbol{r},t)$ in the region where $\boldsymbol{j}(\boldsymbol{r},t) \neq 0$, that is: we need the *near field*, not the *far field* that we considered in Sections 7.1–7.4 when studying the radiation field. Let us be content with balancing the energy for electric dipole radiation, for which $P = \frac{2}{3c^3}\ddot{\boldsymbol{d}}^{\,2}$ is proportional to $\left(\frac{1}{c}\right)^3$. Therefore, we need the scalar potential to order $\left(\frac{1}{c}\right)^3$ and the vector potential to order $\left(\frac{1}{c}\right)^2$ in

$$\boldsymbol{E} = -\frac{1}{c}\frac{\partial}{\partial t}\boldsymbol{A} - \boldsymbol{\nabla}\Phi\,. \qquad (7.5.6)$$

We use the potentials in the Lorentz gauge for this purpose. Thus

$$\begin{aligned}\Phi(\boldsymbol{r},t) &= \int (\mathrm{d}\boldsymbol{r}')\,\frac{1}{|\boldsymbol{r}-\boldsymbol{r}'|}\rho\Big(\boldsymbol{r}',t-\frac{1}{c}|\boldsymbol{r}-\boldsymbol{r}'|\Big)\\ &= \int (\mathrm{d}\boldsymbol{r}')\,\frac{1}{|\boldsymbol{r}-\boldsymbol{r}'|}\bigg[\rho(\boldsymbol{r}',t) - \frac{1}{c}|\boldsymbol{r}-\boldsymbol{r}'|\frac{\partial}{\partial t}\rho(\boldsymbol{r}',t)\\ &\qquad + \frac{1}{2}\Big(\frac{1}{c}|\boldsymbol{r}-\boldsymbol{r}'|\Big)^2\frac{\partial^2}{\partial t^2}\rho(\boldsymbol{r}',t)\\ &\qquad - \frac{1}{6}\Big(\frac{1}{c}|\boldsymbol{r}-\boldsymbol{r}'|\Big)^3\frac{\partial^3}{\partial t^3}\rho(\boldsymbol{r}',t)\\ &\qquad + \cdots\bigg],\end{aligned} \qquad (7.5.7)$$

where the ellipsis indicates terms of order $\left(\frac{1}{c}\right)^4$ or higher powers of $\frac{1}{c}$, which we shall ignore in the present context. This gives

$$\begin{aligned}-\boldsymbol{\nabla}\Phi(\boldsymbol{r},t) &= -\boldsymbol{\nabla}\bigg[\int (\mathrm{d}\boldsymbol{r}')\,\frac{\rho(\boldsymbol{r}',t)}{|\boldsymbol{r}-\boldsymbol{r}'|} + \frac{1}{2c^2}\int (\mathrm{d}\boldsymbol{r}')\,|\boldsymbol{r}-\boldsymbol{r}'|\frac{\partial^2}{\partial t^2}\rho(\boldsymbol{r}',t)\bigg]\\ &\quad + \frac{1}{3c^3}\int (\mathrm{d}\boldsymbol{r}')\,(\boldsymbol{r}-\boldsymbol{r}')\frac{\partial^3}{\partial t^3}\rho(\boldsymbol{r}',t)\\ &= -\boldsymbol{\nabla}[\cdots] - \frac{1}{3c^3}\Big(\frac{\mathrm{d}}{\mathrm{d}t}\Big)^3\boldsymbol{d}(t)\,,\end{aligned} \qquad (7.5.8)$$

where we took into account that the total charge is conserved,

$$\int (\mathrm{d}\boldsymbol{r}') \frac{\partial}{\partial t}\rho(\boldsymbol{r}',t) = \frac{\mathrm{d}}{\mathrm{d}t}\int (\mathrm{d}\boldsymbol{r}')\,\rho(\boldsymbol{r}',t) = 0\,. \tag{7.5.9}$$

The contribution to $\dot{W}$ from the scalar potential is, therefore,

$$\begin{aligned}
&\int (\mathrm{d}\boldsymbol{r})\Big(-\boldsymbol{j}(\boldsymbol{r},t)\cdot(-\boldsymbol{\nabla}[\cdots])\Big) + \frac{1}{3c^3}\dddot{\boldsymbol{d}}(t)\cdot \underbrace{\int (\mathrm{d}\boldsymbol{r})\,\boldsymbol{j}(\boldsymbol{r},t)}_{=\dot{\boldsymbol{d}}(t)}\\
&= \int (\mathrm{d}\boldsymbol{r})\,[\cdots]\,\underbrace{(-\boldsymbol{\nabla}\cdot\boldsymbol{j})}_{=\frac{\partial}{\partial t}\rho} + \frac{1}{3c^3}\dot{\boldsymbol{d}}(t)\cdot\dddot{\boldsymbol{d}}(t)\\
&= \frac{\mathrm{d}}{\mathrm{d}t}\left\{\frac{1}{2}\int (\mathrm{d}\boldsymbol{r})(\mathrm{d}\boldsymbol{r}')\,\frac{\rho(\boldsymbol{r},t)\rho(\boldsymbol{r}',t)}{|\boldsymbol{r}-\boldsymbol{r}'|}\right.\\
&\qquad\left. + \frac{1}{4c^2}\int (\mathrm{d}\boldsymbol{r})(\mathrm{d}\boldsymbol{r}')\,\frac{\partial\rho(\boldsymbol{r},t)}{\partial t}\,|\boldsymbol{r}-\boldsymbol{r}'|\,\frac{\partial\rho(\boldsymbol{r}',t)}{\partial t}\right\}\\
&\quad + \frac{1}{3c^3}\dot{\boldsymbol{d}}(t)\cdot\dddot{\boldsymbol{d}}(t)\,,
\end{aligned} \tag{7.5.10}$$

after recognizing a total time derivative upon inserting the expression for $[\cdots]$ from (7.5.8). This is to be supplemented by the contribution from the $-\frac{1}{c}\frac{\partial}{\partial t}\boldsymbol{A}$ term in $\boldsymbol{E}$, for which we need to use

$$\begin{aligned}
\boldsymbol{A}(\boldsymbol{r},t) &= \int (\mathrm{d}\boldsymbol{r}')\,\frac{1}{|\boldsymbol{r}-\boldsymbol{r}'|}\frac{1}{c}\boldsymbol{j}\Big(\boldsymbol{r}',t-\frac{1}{c}|\boldsymbol{r}-\boldsymbol{r}'|\Big)\\
&= \frac{1}{c}\int (\mathrm{d}\boldsymbol{r}')\,\frac{\boldsymbol{j}(\boldsymbol{r}',t)}{|\boldsymbol{r}-\boldsymbol{r}'|} - \frac{1}{c^2}\int (\mathrm{d}\boldsymbol{r}')\,\frac{\partial}{\partial t}\boldsymbol{j}(\boldsymbol{r}',t)\,,
\end{aligned} \tag{7.5.11}$$

consistently discarding all terms of higher order in $\frac{1}{c}$. This gives

$$\begin{aligned}
-\frac{1}{c}\frac{\partial}{\partial t}\boldsymbol{A}(\boldsymbol{r},t) &= -\frac{1}{c^2}\int (\mathrm{d}\boldsymbol{r}')\,\frac{1}{|\boldsymbol{r}-\boldsymbol{r}'|}\frac{\partial}{\partial t}\boldsymbol{j}(\boldsymbol{r}',t)\\
&\quad + \frac{1}{c^3}\frac{\mathrm{d}^2}{\mathrm{d}t^2}\underbrace{\int (\mathrm{d}\boldsymbol{r}')\,\boldsymbol{j}(\boldsymbol{r}',t)}_{=\frac{\mathrm{d}}{\mathrm{d}t}\boldsymbol{d}(t)}\,,
\end{aligned} \tag{7.5.12}$$

so that the $-\frac{1}{c}\frac{\partial}{\partial t}\boldsymbol{A}$ contribution to the power $\dot{W}$ is

$$\int (\mathrm{d}\boldsymbol{r}) \left( -\boldsymbol{j} \cdot \left( -\frac{1}{c}\frac{\partial}{\partial t}\boldsymbol{A} \right) \right) = \frac{\mathrm{d}}{\mathrm{d}t} \left\{ \frac{1}{2c^2} \int (\mathrm{d}\boldsymbol{r})(\mathrm{d}\boldsymbol{r}') \frac{\boldsymbol{j}(\boldsymbol{r},t) \cdot \boldsymbol{j}(\boldsymbol{r}',t)}{|\boldsymbol{r} - \boldsymbol{r}'|} \right\} - \frac{1}{c^3} \dot{\boldsymbol{d}}(t) \cdot \dddot{\boldsymbol{d}}(t) \,. \tag{7.5.13}$$

Collecting all terms, we have — up to order $\left(\frac{1}{c}\right)^3$ —

$$\begin{aligned} \dot{W} &= \frac{\mathrm{d}}{\mathrm{d}t} \left\{ \frac{1}{2} \int (\mathrm{d}\boldsymbol{r})(\mathrm{d}\boldsymbol{r}') \frac{\rho(\boldsymbol{r},t)\rho(\boldsymbol{r}',t) + \frac{1}{c}\boldsymbol{j}(\boldsymbol{r},t) \cdot \frac{1}{c}\boldsymbol{j}(\boldsymbol{r}',t)}{|\boldsymbol{r} - \boldsymbol{r}'|} \right. \\ &\qquad \left. + \frac{1}{4c^2} \int (\mathrm{d}\boldsymbol{r})(\mathrm{d}\boldsymbol{r}') \frac{\partial \rho(\boldsymbol{r},t)}{\partial t} |\boldsymbol{r} - \boldsymbol{r}'| \frac{\partial \rho(\boldsymbol{r},t)}{\partial t} \right\} \\ &\qquad - \frac{2}{3c^3} \dot{\boldsymbol{d}}(t) \cdot \dddot{\boldsymbol{d}}(t) \\ &= \frac{\mathrm{d}}{\mathrm{d}t} \left[ \{\cdots\} - \frac{2}{3c^3} \dot{\boldsymbol{d}}(t) \cdot \ddot{\boldsymbol{d}}(t) \right] + \frac{2}{3c^3} \ddot{\boldsymbol{d}}(t)^2 \,, \end{aligned} \tag{7.5.14}$$

and the comparison with $\dot{W} = \frac{\mathrm{d}}{\mathrm{d}t}E + P$ tells us that

$$E(t) = \{\cdots\} - \frac{1}{3c^3} \frac{\mathrm{d}}{\mathrm{d}t} \dot{\boldsymbol{d}}(t)^2 \tag{7.5.15}$$

with $\{\cdots\}$ from (7.5.14), and

$$P = \frac{2}{3c^3} \ddot{\boldsymbol{d}}(t)^2 \,, \tag{7.5.16}$$

consistent — as one would expect — with the Larmor formula for the total radiated power in the case of electric dipole radiation, which is (7.3.6) above.

Accordingly, whether we adopt the earlier field point of view to calculate the radiated power, or the present charge point of view, we obtain the same result for the total radiated power. This is, of course, as it should be. Depending on the finer details that we may be interested in, either approach may be preferable. The field point of view gave us the spatial distribution of the radiation, in addition to the total power, whereas here we obtained an expression for the energy stored in the volume, now expressed in terms of the charge and current density, rather than the fields, as we had it before.

## 7.6 Simple model antenna

We observed above, in the context of (7.2.9) and (7.3.5), that a simple dipole radiates a pattern $\propto (\sin\theta)^2$, which is not very selective of the direction. This is typical of a *small* antenna, whereas large antennas tend to have much better spatial selection for the radiation. For an illustration of this point, let us consider a simple model of an antenna, a model that is a bit over-idealized but reasonably realistic nevertheless.

We assume a harmonic current of strength $I$ along the $z$ axis, and limited to the finite stretch $-\frac{1}{2}L < z < \frac{1}{2}L$,

$$\boldsymbol{j}(\boldsymbol{r},t) = \boldsymbol{e}_z I\delta(x)\delta(y)\eta(L^2 - 4z^2)\cos(\omega t)\,, \tag{7.6.1}$$

where $\eta(\ )$ denotes Heaviside's* unit step function. The continuity equation (1.2.1) implies that

$$\begin{aligned}\rho(\boldsymbol{r},t) &= -\frac{I}{\omega}\delta(x)\delta(y)\frac{\partial}{\partial z}\eta(L^2 - 4z^2)\sin(\omega t)\\ &= \frac{I}{\omega}\delta(x)\delta(y)\big[\delta(z - L/2) - \delta(z + L/2)\big]\sin(\omega t)\end{aligned} \tag{7.6.2}$$

is the corresponding time-dependent charge density, where the build-up of charge at the ends of the antenna at $z = \pm L/2$ is the feature that is least realistic. This build-up would not occur in a centrally fed real-life antenna, but let us not be bothered by this detail; a more realistic antenna model is the subject matter of Exercise 67.

In addition to the length $L$ of the antenna, there is the wavelength $\lambda$ of the emitted radiation,

$$\lambda = \frac{2\pi}{k} = \frac{2\pi c}{\omega}\,, \tag{7.6.3}$$

and we expect that a small antenna ($L \ll \lambda$) shows the familiar dipole pattern, as internal retardation does not matter then, whereas a large antenna ($L \gg \lambda$) should exhibit quite another pattern because internal retardation could matter a lot and give rise to interference effects between the radiation emitted by different parts of the antenna.

The general formula (7.2.4) for $\dfrac{\mathrm{d}P}{\mathrm{d}\Omega}$ applies, so that we need to evaluate

$$\int(\mathrm{d}\boldsymbol{r}')\,\frac{\partial}{\partial t}\boldsymbol{j}(\boldsymbol{r}',t_r) = \int(\mathrm{d}\boldsymbol{r}')\,\boldsymbol{e}_z I\delta(x')\delta(y')\eta\big(L^2 - 4z'^2\big)\frac{\partial}{\partial t}\cos(\omega t_r) \tag{7.6.4}$$

*Oliver HEAVISIDE (1850–1925)

with $t_r = t - r/c + \boldsymbol{n}\cdot\boldsymbol{r}'/c$. We choose the coordinates as indicated in this figure:

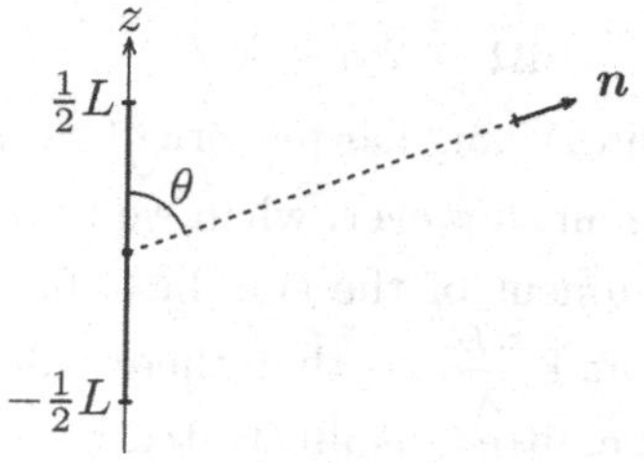


(7.6.5)

so that $\boldsymbol{n}\cdot\boldsymbol{r}' = z'\cos\theta$ for $x' = y' = 0$, and

$$\int (\mathrm{d}\boldsymbol{r}')\,\frac{\partial}{\partial t}\boldsymbol{j}(\boldsymbol{r}',t_r) = \boldsymbol{e}_z I \frac{\partial}{\partial t_\mathrm{e}} \int_{-L/2}^{L/2} \mathrm{d}z'\,\cos\Big(\omega t_\mathrm{e} + \frac{\omega}{c}z'\cos\theta\Big)\,, \tag{7.6.6}$$

where $t_\mathrm{e} = t - r/c$ is the usual emission time. When breaking up the cosine of the sum into a sum of cos( ) cos( ) and sin( ) sin( ), the product of sine factors does not contribute to the integral because $\sin(\frac{\omega}{c}z'\cos\theta)$ is odd in $z'$. Accordingly,

$$\begin{aligned}\int (\mathrm{d}\boldsymbol{r}')\,\frac{\partial}{\partial t}\boldsymbol{j}(\boldsymbol{r}',t_r) &= \boldsymbol{e}_z I \frac{\partial}{\partial t_\mathrm{e}} \int_{-L/2}^{L/2} \mathrm{d}z'\,\cos(\omega t_\mathrm{e})\cos\Big(\frac{\omega}{c}z'\cos\theta\Big)\\ &= -\boldsymbol{e}_z I\omega\sin(\omega t_\mathrm{e})\frac{\sin\Big(\dfrac{\omega L}{2c}\cos\theta\Big)}{\dfrac{\omega}{2c}\cos\theta}\,,\end{aligned} \tag{7.6.7}$$

and with $|\boldsymbol{n}\times\boldsymbol{e}_z|^2 = (\sin\theta)^2$ we obtain

$$\frac{\mathrm{d}P}{\mathrm{d}\Omega} = \frac{1}{4\pi c^3}(2cI)^2\sin(\omega t_\mathrm{e})^2\left(\tan\theta\sin\Big(\frac{\omega L}{2c}\cos\theta\Big)\right)^2. \tag{7.6.8}$$

We average this over one period of the oscillation, $\sin(\omega t_\mathrm{e})^2 \to \frac{1}{2}$, and recognize that $\dfrac{\omega L}{2c} = \dfrac{\pi L}{\lambda}$, and thus arrive at

$$\frac{\mathrm{d}P}{\mathrm{d}\Omega} = \frac{I^2}{2\pi c}(\tan\theta)^2\sin\Big(\frac{\pi L}{\lambda}\cos\theta\Big)^2 \tag{7.6.9}$$

for the time-averaged radiated power.

Indeed, in the case of a small antenna, $\frac{L}{\lambda} \ll 1$, this turns into

$$\frac{\mathrm{d}P}{\mathrm{d}\Omega} = \frac{I^2}{2\pi c}\left(\frac{\pi L}{\lambda}\right)^2 (\sin\theta)^2 \tag{7.6.10}$$

with the familiar $\propto (\sin\theta)^2$ angular pattern of electric dipole radiation. The situation is very different, however, when we have a large antenna, $\frac{L}{\lambda} \gg 1$, because then the argument of the sine in (7.6.9) has values between the large numbers $-\frac{\pi L}{\lambda}$ and $\frac{\pi L}{\lambda}$, so that there will be many directions into which no radiation is emitted, as illustrated here for $\frac{L}{\lambda} = 3.8$:

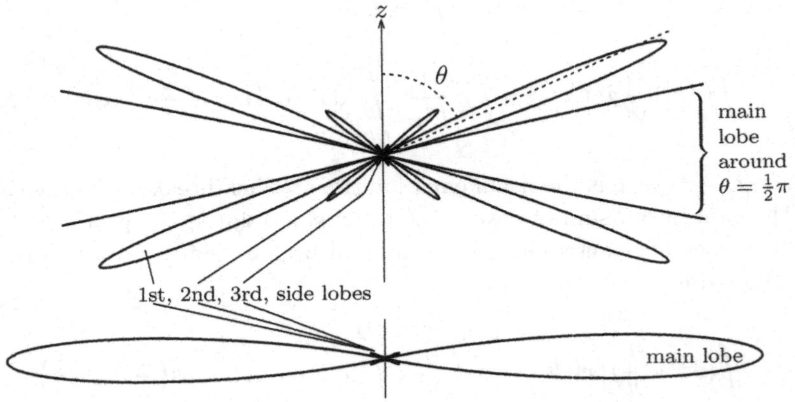


(7.6.11)

More specifically, for $n < \frac{L}{\lambda} < n+1$, with integer $n$, there are $n$ side lobes each above and below the $xy$ plane (where $\theta = \pi/2$), and for a large antenna, there will thus be many side lobes. For the example shown in the figure, we have $n = 3$ and thus three side lobes which, as is clearly visible, are much smaller than the main lobe. In the bottom plot where all of the main lobe is in sight, the side lobes are barely noticeable; they can be seen in the top plot which is scaled up by a factor of 20.

The total radiated power is

$$\begin{aligned} P &= \int \mathrm{d}\Omega\,\frac{\mathrm{d}P}{\mathrm{d}\Omega} = 2\pi\int_0^\pi \mathrm{d}\theta\,\sin\theta\,\frac{\mathrm{d}P}{\mathrm{d}\Omega} \\ &= \frac{I^2}{c}\int_0^\pi \mathrm{d}\theta\,\sin\theta\,(\tan\theta)^2\sin\left(\frac{\pi L}{\lambda}\cos\theta\right)^2, \end{aligned} \tag{7.6.12}$$

where we substitute

$$x = \frac{\pi L}{\lambda}\cos\theta\,, \quad \mathrm{d}x = -\frac{\pi L}{\lambda}\mathrm{d}\theta\sin\theta\,, \quad (\tan\theta)^2 = \frac{1-\left(\frac{\lambda}{\pi L}x\right)^2}{\left(\frac{\lambda}{\pi L}x\right)^2}\,, \tag{7.6.13}$$

to express $P$ in the form

$$P = \frac{I^2}{c}\frac{\pi L}{\lambda}\int\limits_{-\pi L/\lambda}^{\pi L/\lambda}\mathrm{d}x\left(\frac{\sin x}{x}\right)^2\left[1-\left(\frac{\lambda}{\pi L}x\right)^2\right]. \tag{7.6.14}$$

For a large antenna, $L \gg \lambda$, this simplifies to

$$P = \frac{I^2}{c}\frac{\pi L}{\lambda}\underbrace{\int\limits_{-\infty}^{\infty}\mathrm{d}x\left(\frac{\sin x}{x}\right)^2}_{=\pi} = \frac{\pi^2 I^2}{c}\frac{L}{\lambda}\,, \tag{7.6.15}$$

which is proportional to the length $L$ of the antenna.

By contrast, the power radiated into the $xy$ plane, where $\theta = \frac{\pi}{2}$ in $\frac{\mathrm{d}P}{\mathrm{d}\Omega}$, is

$$\begin{aligned}\left.\frac{\mathrm{d}P}{\mathrm{d}\Omega}\right|_{\theta=\frac{1}{2}\pi} &= \frac{I^2}{2\pi c}(\sin\theta)^2\left.\left(\frac{\sin\left(\frac{\pi L}{\lambda}\cos\theta\right)}{\cos\theta}\right)^2\right|_{\substack{\cos\theta=0\\ \sin\theta=1}} \\ &= \frac{I^2}{2\pi c}\left(\frac{\pi L}{\lambda}\right)^2,\end{aligned} \tag{7.6.16}$$

which is proportional to $L^2$. This suggests that a large fraction of the radiation is emitted into the plane perpendicular to the extension of the antenna. Let us be more quantitative about this matter and determine the total power emitted into the main lobe,

$$\begin{aligned}P_{\mathrm{main}} &= \frac{I^2}{c}\frac{\pi L}{\lambda}\int\limits_{-\pi}^{\pi}\mathrm{d}x\left(\frac{\sin x}{x}\right)^2\left[1-\left(\frac{\lambda}{\pi L}x\right)^2\right] \\ &\overset{(L\gg\lambda)}{=} \frac{I^2}{c}\frac{\pi L}{\lambda}\int\limits_{-\pi}^{\pi}\mathrm{d}x\left(\frac{\sin x}{x}\right)^2,\end{aligned} \tag{7.6.17}$$

so that the fraction of the power emitted into the main lobe is

$$\frac{P_{\text{main}}}{P} = \frac{\displaystyle\int_{-\pi}^{\pi} \mathrm{d}x \left(\frac{\sin x}{x}\right)^2 \left[1 - \left(\frac{\lambda}{\pi L}x\right)^2\right]}{\displaystyle\int_{-\pi L/\lambda}^{\pi L/\lambda} \mathrm{d}x \left(\frac{\sin x}{x}\right)^2 \left[1 - \left(\frac{\lambda}{\pi L}x\right)^2\right]}$$

$$\overset{(L \gg \lambda)}{=} \frac{2}{\pi}\int_0^{\pi} \mathrm{d}x \left(\frac{\sin x}{x}\right)^2 . \tag{7.6.18}$$

The exact numerical value — if you are curious: to 4 significant digits, it is 0.9028 — is not of great interest, but we can easily estimate it with successive integrations by part,

$$\begin{aligned}
\frac{2}{\pi}\int_0^{\pi} \mathrm{d}x \left(\frac{\sin x}{x}\right)^2 &= \frac{2}{\pi}\int_0^{2\pi} \mathrm{d}x\, \frac{\sin x}{x} \\
&= 1 - \frac{2}{\pi}\int_{2\pi}^{\infty} \mathrm{d}x\, \frac{\sin x}{x} \\
&= 1 - \frac{2}{\pi}\left(-\frac{\cos x}{x}\right)\bigg|_{x\,=\,2\pi}^{\infty} + \frac{2}{\pi}\int_{2\pi}^{\infty} \mathrm{d}x\, \frac{\cos x}{x^2} \\
&= 1 - \frac{1}{\pi^2} + \frac{4}{\pi}\int_{2\pi}^{\infty} \mathrm{d}x\, \frac{\sin x}{x^3} \\
&= 1 - \frac{1}{\pi^2} + \frac{1}{2\pi^4} + \cdots \\
&= 1 - 0.1013 + 0.0051 + \cdots ,
\end{aligned} \tag{7.6.19}$$

which tells us that about 90% of the radiation is emitted into the main lobe by a large antenna, that is: into the angular range

$$\frac{\pi}{2} - \frac{\lambda}{L} \lesssim \theta \lesssim \frac{\pi}{2} + \frac{\lambda}{L} \tag{7.6.20}$$

around the $\theta = \frac{\pi}{2}$ plane perpendicular to the antenna.

# Chapter 8

# Spectral Properties of Radiation

The energy considerations of Sections 7.2, 7.3, and 7.5 established the spatial distribution of the power emitted by a radiating source, and also the total power, but other than that we did not gain any detailed knowledge of the properties of the electromagnetic radiation. We now turn our attention to the spectral properties of the radiation and develop a point of view that is, in a sense, complementary to the temporal picture of Chapter 7.

## 8.1 Fourier-transformed fields

We return once more to (6.2.1), the retarded potentials in the Lorentz gauge,

$$\begin{pmatrix} \Phi(\boldsymbol{r},t) \\ \boldsymbol{A}(\boldsymbol{r},t) \end{pmatrix} = \int (\mathrm{d}\boldsymbol{r}')\mathrm{d}t' \, \frac{\delta\Big(t - t' - \frac{1}{c}|\boldsymbol{r} - \boldsymbol{r}'|\Big)}{|\boldsymbol{r} - \boldsymbol{r}'|} \begin{pmatrix} \rho(\boldsymbol{r}',t') \\ \frac{1}{c}\boldsymbol{j}(\boldsymbol{r}',t') \end{pmatrix}, \qquad (8.1.1)$$

and replace the Dirac delta function by its Fourier representation

$$\delta\Big(t - t' - \frac{1}{c}|\boldsymbol{r} - \boldsymbol{r}'|\Big) = \int \frac{\mathrm{d}\omega}{2\pi} \, \mathrm{e}^{-\mathrm{i}\omega(t - t')} \, \mathrm{e}^{\mathrm{i}\omega|\boldsymbol{r} - \boldsymbol{r}'|/c} \qquad (8.1.2)$$

to arrive at

$$\begin{pmatrix} \Phi(\boldsymbol{r},t) \\ \boldsymbol{A}(\boldsymbol{r},t) \end{pmatrix} = \int \frac{\mathrm{d}\omega}{2\pi} \, \mathrm{e}^{-\mathrm{i}\omega t} \int (\mathrm{d}\boldsymbol{r}') \, \frac{\mathrm{e}^{\mathrm{i}\omega|\boldsymbol{r} - \boldsymbol{r}'|/c}}{|\boldsymbol{r} - \boldsymbol{r}'|} \begin{pmatrix} \rho(\boldsymbol{r}',\omega) \\ \frac{1}{c}\boldsymbol{j}(\boldsymbol{r}',\omega) \end{pmatrix}, \qquad (8.1.3)$$

where $\rho(\boldsymbol{r},\omega)$ and $\boldsymbol{j}(\boldsymbol{r},\omega)$ are the Fourier transforms of the respective function of time. You should have a feeling of *déjà vu* here as you should be reminded of (6.1.24) and its context.

For Fourier transformation between time and frequency, the familiar basic relations are

$$f(t) = \int \frac{\mathrm{d}\omega}{2\pi}\, \mathrm{e}^{-\mathrm{i}\omega t} f(\omega)\,, \tag{8.1.4}$$

which expresses the time dependence in terms of the frequency dependence, and

$$f(\omega) = \int \mathrm{d}t\, \mathrm{e}^{\mathrm{i}\omega t} f(t)\,, \tag{8.1.5}$$

which expresses the frequency-dependent function in terms of its time-dependent partner. Accordingly, we have

$$\begin{pmatrix} \Phi(\boldsymbol{r},t) \\ \boldsymbol{A}(\boldsymbol{r},t) \end{pmatrix} = \int \frac{\mathrm{d}\omega}{2\pi}\, \mathrm{e}^{-\mathrm{i}\omega t} \begin{pmatrix} \Phi(\boldsymbol{r},\omega) \\ \boldsymbol{A}(\boldsymbol{r},\omega) \end{pmatrix} \tag{8.1.6}$$

with

$$\begin{pmatrix} \Phi(\boldsymbol{r},\omega) \\ \boldsymbol{A}(\boldsymbol{r},\omega) \end{pmatrix} = \int (\mathrm{d}\boldsymbol{r}') \frac{\mathrm{e}^{\mathrm{i}\omega|\boldsymbol{r}-\boldsymbol{r}'|/c}}{|\boldsymbol{r}-\boldsymbol{r}'|} \begin{pmatrix} \rho(\boldsymbol{r}',\omega) \\ \frac{1}{c}\boldsymbol{j}(\boldsymbol{r}',\omega) \end{pmatrix}. \tag{8.1.7}$$

We note that the time-dependent quantities are real, $f(t) = f(t)^*$, so that

$$f(\omega)^* = f(-\omega) \tag{8.1.8}$$

and

$$|f(\omega)|^2 = f(\omega)^* f(\omega) = f(-\omega) f(\omega)\,, \tag{8.1.9}$$

reminding us that the sign of $\omega$ is not relevant: the $\omega < 0$ part of $f(\omega)$ contains no information that is not already available in the $\omega > 0$ part.

We now turn our attention to the radiation fields. The analog of the replacement

$$\frac{\delta\Big(t - t' - \frac{1}{c}|\boldsymbol{r}-\boldsymbol{r}'|\Big)}{|\boldsymbol{r}-\boldsymbol{r}'|} \longrightarrow \frac{\delta\Big(t - t' - \frac{r}{c} + \frac{1}{c}\frac{\boldsymbol{r}}{r}\cdot\boldsymbol{r}'\Big)}{r} \tag{8.1.10}$$

in Section 7.1 is here

$$\frac{\mathrm{e}^{\mathrm{i}\omega|\boldsymbol{r}-\boldsymbol{r}'|/c}}{|\boldsymbol{r}-\boldsymbol{r}'|} \longrightarrow \frac{\mathrm{e}^{\mathrm{i}\frac{\omega}{c}r}\, \mathrm{e}^{-\mathrm{i}\frac{\omega}{c}\frac{\boldsymbol{r}}{r}\cdot\boldsymbol{r}'}}{r} = \frac{1}{r}\,\mathrm{e}^{\mathrm{i}kr}\,\mathrm{e}^{-\mathrm{i}\boldsymbol{k}\cdot\boldsymbol{r}'} \tag{8.1.11}$$

where $\boldsymbol{k} = \frac{\omega}{c}\frac{\boldsymbol{r}}{r} = \frac{\omega}{c}\boldsymbol{n}$ is the wave vector of the emitted radiation — with propagation direction $\boldsymbol{n} = \frac{\boldsymbol{r}}{r}$ and circular frequency $\omega$ — and $k = |\boldsymbol{k}| = \frac{\omega}{c}$ is its length, the wave number.

For the potentials of the *radiation field* we thus have

$$\begin{pmatrix} \Phi(\boldsymbol{r},\omega) \\ \boldsymbol{A}(\boldsymbol{r},\omega) \end{pmatrix} = \frac{\mathrm{e}^{\mathrm{i}kr}}{r} \int (\mathrm{d}\boldsymbol{r}')\,\mathrm{e}^{-\mathrm{i}\boldsymbol{k}\cdot\boldsymbol{r}'} \begin{pmatrix} \rho(\boldsymbol{r}',\omega) \\ \frac{1}{c}\boldsymbol{j}(\boldsymbol{r}',\omega) \end{pmatrix}$$
$$= \frac{\mathrm{e}^{\mathrm{i}kr}}{r} \begin{pmatrix} \rho(\boldsymbol{k},\omega) \\ \frac{1}{c}\boldsymbol{j}(\boldsymbol{k},\omega) \end{pmatrix}, \tag{8.1.12}$$

where we meet the fully Fourier-transformed charge and current densities,

$$\begin{pmatrix} \rho(\boldsymbol{k},\omega) \\ \frac{1}{c}\boldsymbol{j}(\boldsymbol{k},\omega) \end{pmatrix} = \int (\mathrm{d}\boldsymbol{r})\mathrm{d}t\;\mathrm{e}^{-\mathrm{i}\boldsymbol{k}\cdot\boldsymbol{r}+\mathrm{i}\omega t} \begin{pmatrix} \rho(\boldsymbol{r},t) \\ \frac{1}{c}\boldsymbol{j}(\boldsymbol{r},t) \end{pmatrix}. \tag{8.1.13}$$

In this expression, $\boldsymbol{k}$ and $\omega$ are not related to each other, but they are on the right-hand side of (8.1.12), the equations for $\Phi(\boldsymbol{r},\omega)$ and $\boldsymbol{A}(\boldsymbol{r},\omega)$, because there $\boldsymbol{k}^2 = \left(\frac{\omega}{c}\right)^2$, sometimes referred to as being "on shell" — namely, the so-called mass shell, here for the massless light quanta.

The potentials, Fourier transformed from time to frequency, have thus a very simple form: the outgoing spherical wave factor $\frac{1}{r}\mathrm{e}^{\mathrm{i}kr}$ is multiplied by the on-shell fully Fourier-transformed charge or current densities. For the latter, the continuity equation (1.2.1),

$$\frac{\partial}{\partial t}\rho(\boldsymbol{r},t) + \boldsymbol{\nabla}\cdot\boldsymbol{j}(\boldsymbol{r},t) = 0\,, \tag{8.1.14}$$

turns into (recall $\frac{\partial}{\partial t} \to -\mathrm{i}\omega$, $\boldsymbol{\nabla} \to \mathrm{i}\boldsymbol{k}$)

$$\omega\rho(\boldsymbol{k},\omega) = \boldsymbol{k}\cdot\boldsymbol{j}(\boldsymbol{k},\omega)\,, \tag{8.1.15}$$

and as a consequence we note that

$$\frac{\omega}{c}\Phi(\boldsymbol{r},\omega) = \boldsymbol{k}\cdot\boldsymbol{A}(\boldsymbol{r},\omega) \tag{8.1.16}$$

for the radiation fields, Fourier transformed from time to frequency. Here we recognize, of course, the Lorentz gauge condition (1.3.7), in conjunction with $\boldsymbol{\nabla} \to \mathrm{i}\frac{\omega}{c}\boldsymbol{n} = \mathrm{i}\boldsymbol{k}$, as applicable to the radiation fields; see (7.1.14): $\boldsymbol{\nabla} \to -\frac{\boldsymbol{n}}{c}\frac{\partial}{\partial t}$ for time dependent fields, combined with $\frac{\partial}{\partial t} \to -\mathrm{i}\omega$ upon Fourier transformation.

In view of this observation, we get

$$\boldsymbol{B}(\boldsymbol{r},t) = \boldsymbol{\nabla} \times \boldsymbol{A}(\boldsymbol{r},t) \to \boldsymbol{B}(\boldsymbol{r},\omega) = \mathrm{i}\boldsymbol{k} \times \boldsymbol{A}(\boldsymbol{r},\omega) \tag{8.1.17}$$

for the magnetic radiation field, and

$$\boldsymbol{E}(\boldsymbol{r},t) = -\frac{1}{c}\frac{\partial}{\partial t}\boldsymbol{A}(\boldsymbol{r},t) - \boldsymbol{\nabla}\Phi(\boldsymbol{r},t) \tag{8.1.18}$$

implies

$$\begin{aligned}
\boldsymbol{E}(\boldsymbol{r},\omega) &= \mathrm{i}\frac{\omega}{c}\boldsymbol{A}(\boldsymbol{r},\omega) - \mathrm{i}\boldsymbol{k}\Phi(\boldsymbol{r},\omega) \\
&= \mathrm{i}\frac{\omega}{c}\boldsymbol{A}(\boldsymbol{r},\omega) - \mathrm{i}\boldsymbol{k}\frac{\boldsymbol{k}}{\omega/c}\cdot\boldsymbol{A}(\boldsymbol{r},\omega) \\
&= \mathrm{i}\frac{\omega}{c}\Big(\boldsymbol{1} - \boldsymbol{n}\,\boldsymbol{n}\Big)\cdot\boldsymbol{A}(\boldsymbol{r},\omega) \\
&= -\mathrm{i}\frac{\omega}{c}\boldsymbol{n}\times\Big(\boldsymbol{n}\times\boldsymbol{A}(\boldsymbol{r},\omega)\Big) \\
&= -\boldsymbol{n}\times\Big(\mathrm{i}\boldsymbol{k}\times\boldsymbol{A}(\boldsymbol{r},\omega)\Big)\,,
\end{aligned} \tag{8.1.19}$$

or

$$\boldsymbol{E}(\boldsymbol{r},\omega) = -\boldsymbol{n}\times\boldsymbol{B}(\boldsymbol{r},\omega) \tag{8.1.20}$$

for the electric radiation field with

$$\boldsymbol{B}(\boldsymbol{r},\omega) = \frac{\mathrm{e}^{\mathrm{i}kr}}{r}\mathrm{i}\boldsymbol{k}\times\frac{1}{c}\boldsymbol{j}(\boldsymbol{k},\omega)\,. \tag{8.1.21}$$

It follows, not unexpectedly, that $\boldsymbol{n}\cdot\boldsymbol{B}(\boldsymbol{r},\omega) = 0$ and $\boldsymbol{n}\cdot\boldsymbol{E}(\boldsymbol{r},\omega) = 0$, as well as

$$\boldsymbol{B}(\boldsymbol{r},\omega) = \boldsymbol{n}\times\boldsymbol{E}(\boldsymbol{r},\omega)\,, \tag{8.1.22}$$

so that $\boldsymbol{n}$, $\boldsymbol{E}(\boldsymbol{r},\omega)$, $\boldsymbol{B}(\boldsymbol{r},\omega)$ are a right-handed trio of mutually perpendicular vectors, for all frequencies $\omega$, and everywhere in the radiation field, far away from where the sources are nonzero. This is once more the observation we made at the end of Section 7.1.

## 8.2 Spectral distribution

Next, we determine the total energy radiated into the direction specified by unit vector $\boldsymbol{n}$, for which we consider the energy current flowing through an area element $r^2 \mathrm{d}\Omega$ in the radiation field:

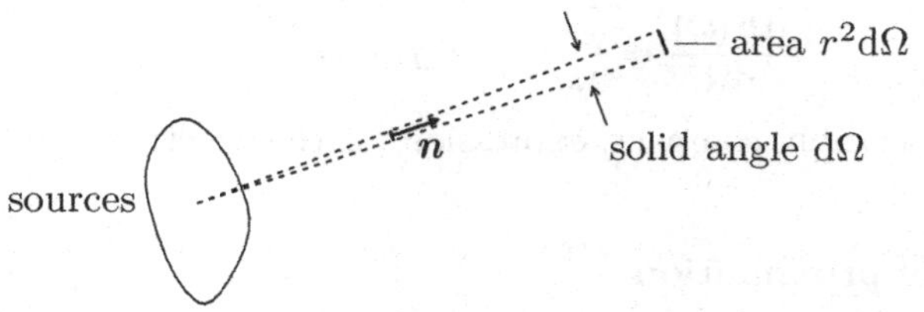


(8.2.1)

This total radiated energy is

$$\int_{-\infty}^{\infty} \mathrm{d}t\, \boldsymbol{S}(\boldsymbol{r},t)\cdot\boldsymbol{n}\, r^2\mathrm{d}\Omega = \mathrm{d}\Omega \int_0^{\infty} \mathrm{d}\omega\, \frac{\mathrm{d}E(\omega)}{\mathrm{d}\Omega}, \tag{8.2.2}$$

where $\boldsymbol{S}(\boldsymbol{r},t) = \frac{c}{4\pi}\boldsymbol{E}(\boldsymbol{r},t)\times\boldsymbol{B}(\boldsymbol{r},t)$ is the familiar energy current density (Poynting vector), and the right-hand side will identify the spectral distribution $\frac{\mathrm{d}E(\omega)}{\mathrm{d}\Omega}$, the energy radiated per unit frequency and per unit solid angle.

Now,

$$\begin{aligned}
\int_{-\infty}^{\infty} \mathrm{d}t\, \boldsymbol{S}(\boldsymbol{r},t) &= \frac{c}{4\pi}\int_{-\infty}^{\infty} \mathrm{d}t \int_{-\infty}^{\infty} \frac{\mathrm{d}\omega}{2\pi}\, \boldsymbol{E}(\boldsymbol{r},\omega)^* \mathrm{e}^{\mathrm{i}\omega t} \times \boldsymbol{B}(\boldsymbol{r},t) \\
&= \frac{c}{4\pi}\int_{-\infty}^{\infty} \frac{\mathrm{d}\omega}{2\pi}\, \boldsymbol{E}(\boldsymbol{r},\omega)^* \times \boldsymbol{B}(\boldsymbol{r},\omega) \\
&= \frac{c}{4\pi}\int_0^{\infty} \frac{\mathrm{d}\omega}{\pi}\, [-\boldsymbol{n}\times\boldsymbol{B}(\boldsymbol{r},\omega)^*]\times\boldsymbol{B}(\boldsymbol{r},\omega) \\
&= \boldsymbol{n}\frac{c}{4\pi^2}\int_0^{\infty} \mathrm{d}\omega\, |\boldsymbol{B}(\boldsymbol{r},\omega)|^2,
\end{aligned} \tag{8.2.3}$$

where we take into account the observation of (8.1.8) and (8.1.9) that the negative frequencies contribute exactly the same amount to the $\omega$ integral as the positive frequencies. It follows that

$$\frac{\mathrm{d}E(\omega)}{\mathrm{d}\Omega} = \frac{c}{4\pi^2}|r\boldsymbol{B}(\boldsymbol{r},\omega)|^2, \tag{8.2.4}$$

so that, with $r\boldsymbol{B}(\boldsymbol{r},\omega) = \mathrm{e}^{ikr}\, \mathrm{i}\boldsymbol{k} \times \frac{1}{c}\boldsymbol{j}(\boldsymbol{k},\omega)$ from (8.1.21),

$$\frac{\mathrm{d}E(\omega)}{\mathrm{d}\Omega} = \frac{1}{4\pi^2 c}\left|\boldsymbol{k} \times \boldsymbol{j}(\boldsymbol{k},\omega)\right|^2, \tag{8.2.5}$$

is the final, remarkably compact, expression for the spectral distribution.

## 8.3 Dipole approximation

In

$$\boldsymbol{j}(\boldsymbol{k},\omega) = \int (\mathrm{d}\boldsymbol{r}')\, \mathrm{e}^{-\mathrm{i}\boldsymbol{k}\cdot\boldsymbol{r}'} \boldsymbol{j}(\boldsymbol{r}',\omega), \tag{8.3.1}$$

the exponential factor accounts for the internal retardation and is, therefore, of great importance except for very small sources. The simplifying approximation $\mathrm{e}^{-\mathrm{i}\boldsymbol{k}\cdot\boldsymbol{r}'} \cong 1$ is only justified when *electric dipole radiation* dominates and the internal retardation is not significant. Then

$$\begin{aligned}\frac{\mathrm{d}E(\omega)}{\mathrm{d}\Omega} &= \frac{\omega^2}{4\pi^2 c^3}\left|\boldsymbol{n} \times \int (\mathrm{d}\boldsymbol{r})\, \boldsymbol{j}(\boldsymbol{r},\omega)\right|^2 \\ &= \frac{\omega^2}{4\pi^2 c^3}(\sin\theta)^2 \left|\int (\mathrm{d}\boldsymbol{r})\, \boldsymbol{j}(\boldsymbol{r},\omega)\right|^2\end{aligned} \tag{8.3.2}$$

exhibits the familiar $(\sin\theta)^2$ pattern of electric dipole radiation, whereby here $\theta$ is the angle between the direction $\boldsymbol{n}$ of propagation of the radiation and the direction of $\int (\mathrm{d}\boldsymbol{r})\, \boldsymbol{j}(\boldsymbol{r},\omega)$. Upon integrating over $\mathrm{d}\Omega$, this gives

$$E(\omega) = \frac{2}{3}\frac{\omega^2}{\pi c^3}\left|\int (\mathrm{d}\boldsymbol{r})\, \boldsymbol{j}(\boldsymbol{r},\omega)\right|^2 \tag{8.3.3}$$

for the spectral density of the emitted radiation: $E(\omega)\mathrm{d}\omega$ is the energy emitted into the frequency interval $\omega \cdots \omega + \mathrm{d}\omega$.

As an example, we consider once more a slowly moving point charge, for which

$$\boldsymbol{j}(\boldsymbol{r}',t) = e\boldsymbol{v}(t)\delta\big(\boldsymbol{r}' - \boldsymbol{r}(t)\big), \tag{8.3.4}$$

provided that $\boldsymbol{r}(t)$ is the trajectory and $\boldsymbol{v}(t) = \frac{\mathrm{d}}{\mathrm{d}t}\boldsymbol{r}(t)$ is the velocity at time $t$. Here,

$$\int (\mathrm{d}\boldsymbol{r}')\, \boldsymbol{j}(\boldsymbol{r}',t) = e\boldsymbol{v}(t) \tag{8.3.5}$$

and

$$\int (\mathrm{d}\boldsymbol{r}')\,\boldsymbol{j}(\boldsymbol{r}',\omega) = e\boldsymbol{v}(\omega)\,, \tag{8.3.6}$$

and noting that $\boldsymbol{v}(t) \to \boldsymbol{v}(\omega)$ is accompanied by $\frac{\mathrm{d}}{\mathrm{d}t}\boldsymbol{v}(t) \to -\mathrm{i}\omega\boldsymbol{v}(\omega) = \dot{\boldsymbol{v}}(\omega)$, we arrive at

$$E(\omega) = \frac{2}{3}\frac{e^2}{\pi c^3}\,|\,\dot{\boldsymbol{v}}(\omega)\,|^2. \tag{8.3.7}$$

The total radiated energy

$$E_{\mathrm{rad}} = \int_0^\infty \mathrm{d}\omega\, E(\omega) = \frac{2}{3}\frac{e^2}{\pi c^3}\int_0^\infty \mathrm{d}\omega\,|\,\dot{\boldsymbol{v}}(\omega)\,|^2 \tag{8.3.8}$$

should agree with what we get from the time-dependent radiated power, see (7.2.11),

$$E_{\mathrm{rad}} = \int_{-\infty}^{\infty} \mathrm{d}t\, P(t) = \frac{2e^2}{3c^3}\int_{-\infty}^{\infty} \mathrm{d}t\,|\,\dot{\boldsymbol{v}}(t)\,|^2. \tag{8.3.9}$$

We check that: For real $f(t)$, we have

$$\begin{aligned}
\frac{1}{\pi}\int_0^\infty \mathrm{d}\omega\,|\,f(\omega)\,|^2 &= \int_{-\infty}^{\infty}\frac{\mathrm{d}\omega}{2\pi}\,|\,f(\omega)\,|^2 \\
&= \int_{-\infty}^{\infty}\frac{\mathrm{d}\omega}{2\pi}\,f(\omega)^*\int_{-\infty}^{\infty}\mathrm{d}t\,\mathrm{e}^{\mathrm{i}\omega t}f(t) \\
&= \int_{-\infty}^{\infty}\mathrm{d}t\,f(t)\left[\int_{-\infty}^{\infty}\frac{\mathrm{d}\omega}{2\pi}\,\mathrm{e}^{-\mathrm{i}\omega t}f(\omega)\right]^* \\
&= \int_{-\infty}^{\infty}\mathrm{d}t\,f(t)f(t)^* = \int_{-\infty}^{\infty}\mathrm{d}t\,f(t)^2\,,
\end{aligned} \tag{8.3.10}$$

and it follows that the two expressions for $E_{\mathrm{rad}}$ in (8.3.8) and (8.3.9) give the same value, indeed.

## 8.4 Impulsive scattering

A second, more complicated example is that of impulsive scattering that changes the velocity of a moving charge $e$ abruptly, with uniform motion before and after the scattering event:

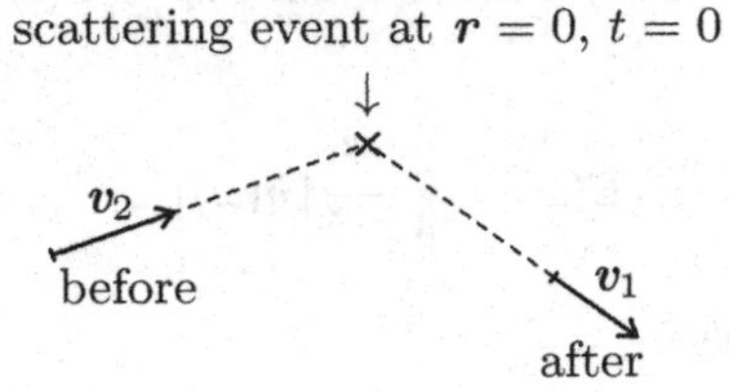


(8.4.1)

The corresponding electric current is

$$\boldsymbol{j}(\boldsymbol{r},t) = \begin{cases} e\boldsymbol{v}_2\,\delta(\boldsymbol{r}-\boldsymbol{v}_2 t) & \text{for} \quad t<0\,, \\ e\boldsymbol{v}_1\,\delta(\boldsymbol{r}-\boldsymbol{v}_1 t) & \text{for} \quad t>0\,. \end{cases} \tag{8.4.2}$$

We have here

$$\begin{aligned} \boldsymbol{j}(\boldsymbol{k},\omega) &= \int (\mathrm{d}\boldsymbol{r})\,\mathrm{e}^{-\mathrm{i}\boldsymbol{k}\cdot\boldsymbol{r}} \int \mathrm{d}t\;\mathrm{e}^{\mathrm{i}\omega t}\,\boldsymbol{j}(\boldsymbol{r},t) \\ &= \int_{-\infty}^{0} \mathrm{d}t\;\mathrm{e}^{\mathrm{i}\omega t} e\boldsymbol{v}_2\,\mathrm{e}^{-\mathrm{i}\boldsymbol{k}\cdot\boldsymbol{v}_2 t} \\ &\quad + \int_{0}^{\infty} \mathrm{d}t\;\mathrm{e}^{\mathrm{i}\omega t} e\boldsymbol{v}_1\,\mathrm{e}^{-\mathrm{i}\boldsymbol{k}\cdot\boldsymbol{v}_1 t} \\ &= e\boldsymbol{v}_2 \frac{-\mathrm{i}}{\omega-\boldsymbol{k}\cdot\boldsymbol{v}_2} + e\boldsymbol{v}_1 \frac{\mathrm{i}}{\omega-\boldsymbol{k}\cdot\boldsymbol{v}_1}\,, \end{aligned} \tag{8.4.3}$$

where we make use of

$$\begin{aligned} \int_0^\infty \mathrm{d}t\;\mathrm{e}^{\mathrm{i}\lambda t} &= \int_0^\infty \mathrm{d}t\,\mathrm{e}^{\mathrm{i}\lambda t}\,\mathrm{e}^{-\epsilon t}\Big|_{0\,<\,\epsilon\,\to\,0} \\ &= \frac{1}{\epsilon-\mathrm{i}\lambda}\Big|_{0\,<\,\epsilon\,\to\,0} = \frac{\mathrm{i}}{\lambda} \end{aligned} \tag{8.4.4}$$

and likewise

$$\int_{-\infty}^{0} \mathrm{d}t\;\mathrm{e}^{\mathrm{i}\lambda t} = \frac{-\mathrm{i}}{\lambda}\,, \tag{8.4.5}$$

thereby ignoring all mathematical subtleties, such as

$$\lim_{0<\epsilon\to 0} \frac{1}{\epsilon - \mathrm{i}\lambda} = \mathcal{P}\frac{\mathrm{i}}{\lambda} + \pi\delta(\lambda) \tag{8.4.6}$$

with $\mathcal{P}$ denoting the principal value, on the ground that nothing happening in the very distant past ($t = -\infty$) or the very distant future ($t = \infty$) can possibly be relevant in a physical context. And so,

$$\boldsymbol{j}(\boldsymbol{k},\omega) = \mathrm{i}\frac{e}{\omega}\left(\frac{\boldsymbol{v}_1}{1-\boldsymbol{n}\cdot\boldsymbol{v}_1/c} - \frac{\boldsymbol{v}_2}{1-\boldsymbol{n}\cdot\boldsymbol{v}_2/c}\right), \tag{8.4.7}$$

implying

$$\frac{\mathrm{d}E(\omega)}{\mathrm{d}\Omega} = \frac{e^2}{4\pi^2c^3}\left|\boldsymbol{n}\times\left(\frac{\boldsymbol{v}_1}{1-\boldsymbol{n}\cdot\boldsymbol{v}_1/c} - \frac{\boldsymbol{v}_2}{1-\boldsymbol{n}\cdot\boldsymbol{v}_2/c}\right)\right|^2. \tag{8.4.8}$$

Most remarkably, this does not depend on the frequency $\omega$ of the emitted radiation. This unphysical feature is the price we pay for the over-idealization that treats the scattering as an instantaneous process of no duration. Real scattering events take time — they are of duration $T$, say — and we should regard our result as an approximation that is valid for frequencies $\omega$ that are small on the scale set by the duration $T$, $\omega \ll \frac{1}{T}$.

If both velocities are small, $|\boldsymbol{v}_1| \ll c$ and $|\boldsymbol{v}_2| \ll c$, then the denominators can be replaced by unity, and we get

$$\begin{aligned}\frac{\mathrm{d}E(\omega)}{\mathrm{d}\Omega} &= \frac{e^2}{4\pi^2c^3}\left|\boldsymbol{n}\times(\boldsymbol{v}_1-\boldsymbol{v}_2)\right|^2 \\ &= \frac{e^2}{4\pi^2c^3}\left|\boldsymbol{v}_1-\boldsymbol{v}_2\right|^2(\sin\theta)^2,\end{aligned} \tag{8.4.9}$$

which exhibits the familiar characteristics of electric dipole radiation, with $\theta$ denoting the angle between $\boldsymbol{n}$ and $\boldsymbol{v}_1 - \boldsymbol{v}_2$. Under these nonrelativistic circumstances the angle-integrated spectral density is

$$E(\omega) = \frac{2}{3}\frac{e^2}{\pi c^3}\left|\boldsymbol{v}_1-\boldsymbol{v}_2\right|^2, \tag{8.4.10}$$

which is of the dipole form of (8.3.3),

$$E(\omega) = \frac{2}{3}\frac{e^2}{\pi c^3}\left|\dot{\boldsymbol{v}}(\omega)\right|^2, \tag{8.4.11}$$

because

$$\dot{\boldsymbol{v}}(t) = (\boldsymbol{v}_1-\boldsymbol{v}_2)\,\delta(t) \tag{8.4.12}$$

in the present context of point scattering, and then

$$\dot{\boldsymbol{v}}(\omega) = \int \mathrm{d}t\ \mathrm{e}^{\mathrm{i}\omega t}\, \dot{\boldsymbol{v}}(t) = \boldsymbol{v}_1 - \boldsymbol{v}_2\,, \tag{8.4.13}$$

indeed.

The other extreme situation is when both velocities are close to the speed of light, $|\boldsymbol{v}_1|, |\boldsymbol{v}_2| \lesssim c$, and then the denominators in (8.4.8) are quite important. Let us look at one of the terms,

$$\frac{e^2}{4\pi^2c^3}\left|\frac{\boldsymbol{n}\times\boldsymbol{v}}{1-\boldsymbol{n}\cdot\boldsymbol{v}/c}\right|^2 = \frac{e^2}{4\pi^2 c}\left(\frac{v}{c}\right)^2\left(\frac{\sin\theta}{1-\dfrac{v}{c}\cos\theta}\right)^2, \tag{8.4.14}$$

where $\theta$ is the angle between $\boldsymbol{n}$ and $\boldsymbol{v}$. The denominator is particularly small for $\theta \gtrsim 0$, but for $\theta = 0$ the numerator vanishes, so that radiation is predominantly emitted into the forward direction, yet not into the exact forward direction of $\theta = 0$. For small angles $\theta$, we have

$$\begin{aligned} \sin\theta \cong \theta \quad \text{and} \quad 1-\frac{v}{c}\cos\theta &\cong 1-\frac{v}{c}+\frac{1}{2}\frac{v}{c}\theta^2 \\ &\cong \frac{1}{2}\left(1-\left(\frac{v}{c}\right)^2\right)+\frac{1}{2}\theta^2 \end{aligned} \tag{8.4.15}$$

where $1-\frac{v}{c} \ll 1$ is used twice, once in putting

$$1-\frac{v}{c} = \frac{1-(v/c)^2}{1+v/c} \cong \frac{1}{2}\left(1-\left(\frac{v}{c}\right)^2\right) = \frac{1}{2\gamma^2}\,, \tag{8.4.16}$$

and then in $\frac{v}{c}\theta^2 \cong \theta^2$. In summary, the right-hand side of (8.4.14) is well approximated by

$$\frac{e^2}{\pi^2 c}\left(\frac{v}{c}\right)^2\left(\frac{\theta}{\gamma^{-2}+\theta^2}\right)^2 \cong \frac{e^2}{\pi^2 c}\left(\frac{\theta}{\gamma^{-2}+\theta^2}\right)^2 \tag{8.4.17}$$

for the small $\theta$ values that are important. Here,

$$\frac{\theta}{\dfrac{1}{\gamma^2}+\theta^2} = \begin{cases} 0 & \text{for} \quad \theta = 0\,, \\ \dfrac{1}{2}\gamma & \text{for} \quad \theta = \dfrac{1}{\gamma}\,, \\ \dfrac{1}{\theta} & \text{for} \quad \dfrac{1}{\gamma} \ll \theta \ll 1\,, \end{cases} \tag{8.4.18}$$

which peaks at $\theta = \frac{1}{\gamma}$:

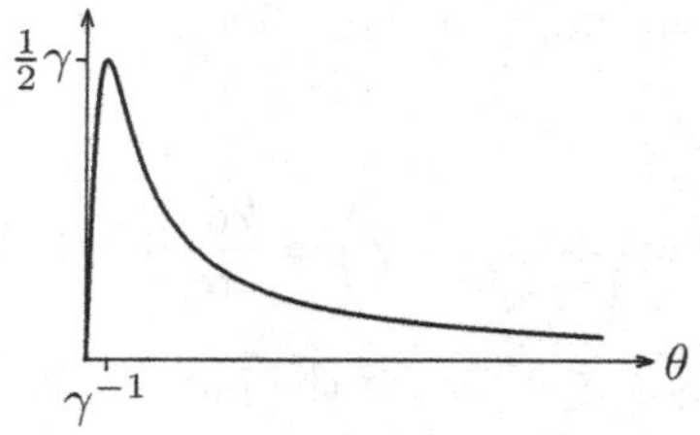


(8.4.19)

Accordingly, the maximal intensity is emitted under angle $\theta = \frac{1}{\gamma}$, and this maximum is $\frac{e^2}{\pi^2}\frac{\gamma^2}{4c}$, and we have this picture:

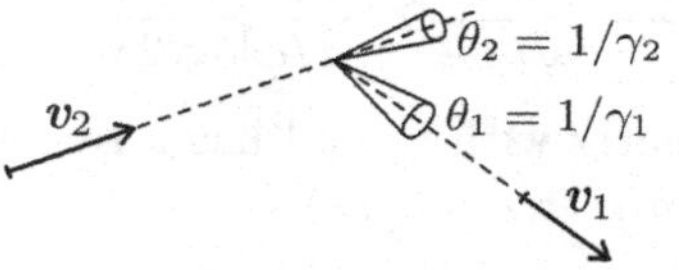


(8.4.20)

whereby we assume that the directions of $\boldsymbol{v}_1$ and $\boldsymbol{v}_2$ are sufficiently different to ensure a good separation of the two cones into which most of the radiation is emitted.

## 8.5 Bremsstrahlung

For a process of sudden stopping, $\boldsymbol{v}_2 = \boldsymbol{v}$, $\boldsymbol{v}_1 = 0$, there is only the single cone associated with the "before" velocity $\boldsymbol{v}_2$, and we have

$$\frac{\mathrm{d}E(\omega)}{\mathrm{d}\Omega} = \frac{e^2}{4\pi^2 c}\left(\frac{v}{c}\right)^2 \frac{(\sin\theta)^2}{\left(1 - \frac{v}{c}\cos\theta\right)^2} \tag{8.5.1}$$

for this impulsive version of *bremsstrahlung* (German for *radiation of braking*). The angular integration gives the spectral density ($\zeta = \cos\theta$)

$$\begin{aligned} E(\omega) = \int \mathrm{d}\Omega \, \frac{\mathrm{d}E(\omega)}{\mathrm{d}\Omega} &= \frac{e^2}{2\pi c}\left(\frac{v}{c}\right)^2 \int_{-1}^{1} \mathrm{d}\zeta \, \frac{1-\zeta^2}{\left(1 - \frac{v}{c}\zeta\right)^2} \\ &= \frac{e^2}{\pi c}\left[\frac{c}{v}\log\frac{1+v/c}{1-v/c} - 2\right], \end{aligned} \tag{8.5.2}$$

which has the correct value in the $v \ll c$ limit,

$$E(\omega) \cong \frac{2}{3}\frac{e^2}{\pi c}\left(\frac{v}{c}\right)^2 \quad \text{for} \quad \frac{v}{c} \ll 1\,, \tag{8.5.3}$$

and is approximated by

$$E(\omega) \cong \frac{e^2}{\pi c}\left(\log\frac{2}{1-v/c} - 2\right) = \frac{2e^2}{\pi c}\left[\log(2\gamma) - 1\right] \tag{8.5.4}$$

for relativistic velocities $\frac{v}{c} \lesssim 1$. So, if the particle is moving at relativistic speed before being stopped suddenly — with mechanical energy $= \gamma mc^2$ and kinetic energy $= (\gamma - 1)mc^2$ — then the range of emitted frequencies, roughly given by $\omega \lesssim (\gamma - 1)\frac{mc^2}{E(\omega)}$, is determined by

$$\frac{(\gamma - 1)mc^2}{E(\omega)} \cong \frac{mc^2}{e^2/c}\frac{\gamma}{\log(2\gamma) - 1}\,, \tag{8.5.5}$$

which grows about linearly with $\gamma$, and has a frequency pre-factor that is, for an electron (mass $m_{\text{el}}$, charge $-e_0$),

$$\frac{m_{\text{el}}c^2}{e_0^2/c} = \frac{m_{\text{el}}c^2}{\hbar}\Big/\frac{e_0^2}{\hbar c} = \frac{511\ \text{keV}}{0.658\ \text{eV fs}}\Big/\frac{1}{137} \cong 10^{23}/\text{s}\,, \tag{8.5.6}$$

a very large frequency, indeed.

# Chapter 9

# Time-Dependent Spectral Distribution

We shall now address the question of time-dependent spectral densities, which is not a contradiction in terms, but rather something quite familiar. Just think of switching a light source on and off: You have radiation with certain spectral properties at some time, and no radiation at other times. Or think of a chunk of metal heated up: First it is dark, then it glows red, later it shines brightly in yellow, finally it is almost white. Clearly, you have different spectral distributions at different times, whereby the epochs are long compared with the period of the oscillations of the radiation in question. For visible light with its period of the order of $10^{-15}$ s, there is good sense in speaking about having this frequency now, and another some milliseconds later, but it would be utterly meaningless to speak of a change of color within a femtosecond.

## 9.1 Time-dependent power spectrum

We return to the expression for $\dfrac{\mathrm{d}E(\omega)}{\mathrm{d}\Omega}$ in (8.2.5),

$$\begin{aligned}\frac{\mathrm{d}E(\omega)}{\mathrm{d}\Omega} &= \frac{1}{4\pi^2 c}\left|\boldsymbol{k}\times\boldsymbol{j}(\boldsymbol{k},\omega)\right|^2 \\ &= \frac{\omega^2}{4\pi^2 c^3}\left|\boldsymbol{n}\times\int \mathrm{d}t\, \mathrm{e}^{\mathrm{i}\omega t}\,\boldsymbol{j}(\boldsymbol{k},t)\right|^2 \\ &= \frac{\omega^2}{4\pi^2 c^3}\left[\boldsymbol{n}\times\int \mathrm{d}t\, \mathrm{e}^{-\mathrm{i}\omega t}\,\boldsymbol{j}(\boldsymbol{k},t)^*\right]\cdot\left[\boldsymbol{n}\times\int \mathrm{d}t'\, \mathrm{e}^{\mathrm{i}\omega t'}\,\boldsymbol{j}(\boldsymbol{k},t')\right],\end{aligned} \tag{9.1.1}$$

and focus on the two time integrations,

$$\int \mathrm{d}t \int \mathrm{d}t'\, \mathrm{e}^{-\mathrm{i}\omega(t-t')}\, \boldsymbol{j}(\boldsymbol{k},t)^* \, \boldsymbol{j}(\boldsymbol{k},t')$$
$$= \int \mathrm{d}T \int \mathrm{d}\tau\, \mathrm{e}^{-\mathrm{i}\omega\tau}\, \boldsymbol{j}\big(\boldsymbol{k},T+\tfrac{1}{2}\tau\big)^* \, \boldsymbol{j}\big(\boldsymbol{k},T-\tfrac{1}{2}\tau\big)\,, \qquad (9.1.2)$$

where

$$T = \frac{1}{2}(t+t') \qquad (9.1.3)$$

is the time of the epoch, and

$$\tau = t - t' \qquad (9.1.4)$$

is the relative time. The important range for $\tau$ is roughly $-\frac{1}{\omega}\cdots\frac{1}{\omega}$, and the epoch time $T$ is only defined physically on a scale that is long compared with the period $\frac{1}{\omega}$ of the radiation. Keeping this in mind, we thus introduce the *time-dependent power spectrum* $\frac{\mathrm{d}P(\omega,T)}{\mathrm{d}\Omega}$ in accordance with

$$\frac{\mathrm{d}E(\omega)}{\mathrm{d}\Omega} = \int \mathrm{d}T\, \frac{\mathrm{d}P(\omega,T)}{\mathrm{d}\Omega}\,, \qquad (9.1.5)$$

and conclude that it is given by

$$\begin{aligned}\frac{\mathrm{d}P(\omega,T)}{\mathrm{d}\Omega} &= \frac{\omega^2}{4\pi^2c^3}\int_{-\infty}^{\infty}\mathrm{d}\tau\, \mathrm{e}^{-\mathrm{i}\omega\tau}\, \Big[\boldsymbol{n}\times\boldsymbol{j}\big(\boldsymbol{k},T+\tfrac{1}{2}\tau\big)^*\Big] \\ &\qquad\qquad \cdot\Big[\boldsymbol{n}\times\boldsymbol{j}\big(\boldsymbol{k},T-\tfrac{1}{2}\tau\big)\Big] \\ &= \frac{1}{4\pi^2c}\int_{-\infty}^{\infty}\mathrm{d}\tau\, \mathrm{e}^{-\mathrm{i}\omega\tau}\, \Big[\boldsymbol{k}\times\boldsymbol{j}\big(\boldsymbol{k},T+\tfrac{1}{2}\tau\big)^*\Big] \\ &\qquad\qquad \cdot\Big[\boldsymbol{k}\times\boldsymbol{j}\big(\boldsymbol{k},T-\tfrac{1}{2}\tau\big)\Big]\,. \end{aligned} \qquad (9.1.6)$$

## 9.2 Constant acceleration

As a first application, we consider a charge $e$ that is uniformly accelerated, $\boldsymbol{v}(t) = \boldsymbol{a}t$ with constant acceleration $\boldsymbol{a}$, for which

$$\boldsymbol{j}(\boldsymbol{r},t) = e\boldsymbol{a}t\,\delta\big(\boldsymbol{r}-\tfrac{1}{2}\boldsymbol{a}t^2\big) \qquad (9.2.1)$$

is the electric current density, thereby assuming that the motion is nonrelativistic during the relevant period of time. Here, then,

$$\boldsymbol{j}(\boldsymbol{k},t) = \int (\mathrm{d}\boldsymbol{r})\, \mathrm{e}^{-\mathrm{i}\boldsymbol{k}\cdot\boldsymbol{r}}\, \boldsymbol{j}(\boldsymbol{r},t) = e\boldsymbol{a}t\, \mathrm{e}^{-\mathrm{i}\boldsymbol{k}\cdot\boldsymbol{a}t^2/2} \tag{9.2.2}$$

so that

$$\begin{aligned}&\left[\boldsymbol{n}\times\boldsymbol{j}(\boldsymbol{k},T+\tfrac{1}{2}\tau)^*\right]\cdot\left[\boldsymbol{n}\times\boldsymbol{j}(\boldsymbol{k},T-\tfrac{1}{2}\tau)\right]\\ &= e^2(\boldsymbol{n}\times\boldsymbol{a})^2\Big(T^2-\frac{1}{4}\tau^2\Big)\,\mathrm{e}^{\mathrm{i}\boldsymbol{k}\cdot\boldsymbol{v}(T)\,\tau}\end{aligned} \tag{9.2.3}$$

and

$$\begin{aligned}\frac{\mathrm{d}P(\omega,T)}{\mathrm{d}\Omega} &= \frac{\omega^2}{4\pi^2c^3}e^2(\boldsymbol{n}\times\boldsymbol{a})^2\int_{-\infty}^{\infty}\mathrm{d}\tau\;\mathrm{e}^{-\mathrm{i}\omega\tau}\Big(T^2-\frac{1}{4}\tau^2\Big)\,\mathrm{e}^{\mathrm{i}\boldsymbol{k}\cdot\boldsymbol{v}(T)\tau}\\ &= \frac{\omega^2e^2}{2\pi c^3}(\boldsymbol{n}\times\boldsymbol{a})^2\left[T^2\delta(\omega-\boldsymbol{k}\cdot\boldsymbol{v}(T))+\frac{1}{4}\delta''(\omega-\boldsymbol{k}\cdot\boldsymbol{v}(T))\right]\end{aligned} \tag{9.2.4}$$

or, with $\delta(\lambda x) = \dfrac{1}{|\lambda|}\delta(x)$ and $\delta''(\lambda x) = \dfrac{1}{|\lambda|^3}\delta''(x)$,

$$\frac{\mathrm{d}P(\omega,T)}{\mathrm{d}\Omega} = \frac{e^2(\boldsymbol{n}\times\boldsymbol{a})^2}{2\pi c^3}\left[\omega T^2\delta(1-\boldsymbol{n}\cdot\boldsymbol{v}(T)/c)+\frac{1}{4\omega}\delta''(1-\boldsymbol{n}\cdot\boldsymbol{v}(T)/c)\right]. \tag{9.2.5}$$

Since the velocity of the moving charge cannot exceed the speed of light, $|\boldsymbol{v}(T)| < c$, the arguments of the Dirac delta functions never vanishes, and it follows that there is no radiated power,

$$\frac{\mathrm{d}P(\omega,T)}{\mathrm{d}\Omega} = 0\,. \tag{9.2.6}$$

This states that the uniformly accelerated charge does not radiate at all.

But, the Larmor formula (7.2.11) says that the total radiated power is

$$P = \frac{2}{3}\frac{e^2}{c^3}|\boldsymbol{a}|^2 \neq 0\,, \tag{9.2.7}$$

which is assuredly nonzero. There appears to be a contradiction, which in fact is really only apparent. For, the Larmor result (9.2.7) states a time-independent radiated power, and if there is no time dependence, Fourier transformation from time to frequency gives a $\delta(\omega)$ dependence: $\omega = 0$ only, which, however, is not radiation but rather a static situation. Now, the delta functions in (9.2.4) have $\omega(1-\boldsymbol{n}\cdot\boldsymbol{v}(T)/c)$ as their common argument, so that the $\omega = 0$ contribution is literally there.

Nevertheless, we have encountered a weird situation, and it should be clear that we are just facing a consequence of the unphysical assumption that the acceleration is uniform for *all* times (and the motion remains nonrelativistic despite the eternal acceleration). We must be more realistic and take into account that the period of constant acceleration has a finite duration,

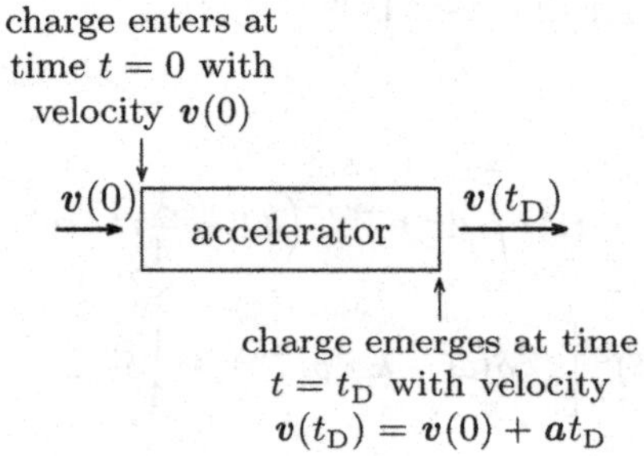


(9.2.8)

so that we have

$$\boldsymbol{v}(t) = \begin{cases} \boldsymbol{v}(0) & \text{for} \quad t < 0 \\ \boldsymbol{v}(0) + \boldsymbol{a}t & \text{for} \quad 0 < t < t_{\mathrm{D}} \\ \boldsymbol{v}(0) + \boldsymbol{a}t_{\mathrm{D}} & \text{for} \quad t_{\mathrm{D}} < t \end{cases} \tag{9.2.9}$$

for the velocity and

$$\dot{\boldsymbol{v}}(t) = \begin{cases} 0 & \text{for} \quad t < 0 \\ \boldsymbol{a} & \text{for} \quad 0 < t < t_{\mathrm{D}} \\ 0 & \text{for} \quad t_{\mathrm{D}} < t \end{cases} \tag{9.2.10}$$

for the acceleration. Then, recalling (8.4.11),

$$E(\omega) = \frac{2}{3}\frac{e^2}{\pi c^3}\left|\dot{\boldsymbol{v}}(\omega)\right|^2, \tag{9.2.11}$$

here with

$$\begin{aligned} \dot{\boldsymbol{v}}(\omega) &= \int \mathrm{d}t\, \mathrm{e}^{\mathrm{i}\omega t}\, \dot{\boldsymbol{v}}(t) = \boldsymbol{a}\int_0^{t_{\mathrm{D}}} \mathrm{d}t\, \mathrm{e}^{\mathrm{i}\omega t} \\ &= \boldsymbol{a}\, \mathrm{e}^{\mathrm{i}\omega t_{\mathrm{D}}/2}\, \frac{2}{\omega} \sin\frac{\omega t_{\mathrm{D}}}{2}\,, \end{aligned} \tag{9.2.12}$$

we have

$$E(\omega) = \frac{2}{3}\frac{e^2}{c^3}\frac{(at_{\rm D})^2}{\pi}\left(\frac{\sin\frac{\omega t_{\rm D}}{2}}{\frac{\omega t_{\rm D}}{2}}\right)^2 \tag{9.2.13}$$

for the spectral density of the radiation emitted by a charge that experiences constant acceleration for a finite period. This spectral density is maximal for $\omega = 0$,

$$E(\omega = 0) = \frac{2}{3\pi}\frac{e^2}{c^3}\left(at_{\rm D}\right)^2 , \tag{9.2.14}$$

and has 90% of the radiation emitted into frequencies below $\omega = 2\pi/t_{\rm D}$:

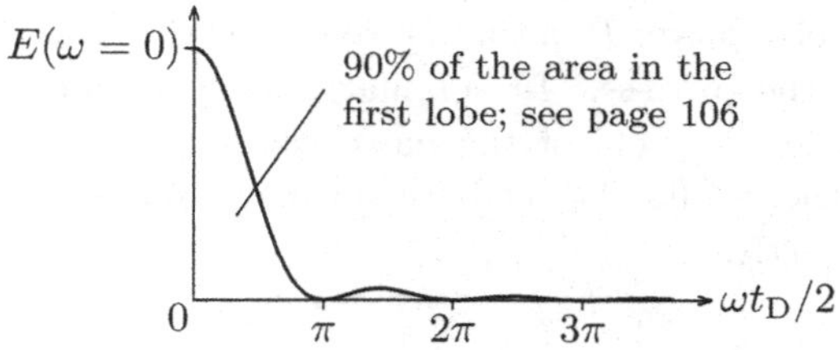


(9.2.15)

We recognize that the emission process is a coherent process that involves all times $t$ in the interval $0 < t < t_{\rm D}$, that is: the whole period of acceleration.

Yet another aspect is illuminated by returning to (7.5.14) and noticing that we had

$$P(t) = -\frac{2}{3c^3}\dot{\boldsymbol{d}}(t)\cdot\dddot{\boldsymbol{d}}(t) \tag{9.2.16}$$

for the radiated power before putting aside a total time derivative. In the present context,

$$\begin{aligned} \dot{\boldsymbol{d}}(t) &= e\boldsymbol{v}(t) \\ \text{and}\quad \dddot{\boldsymbol{d}}(t) &= e\ddot{\boldsymbol{v}}(t) = e\boldsymbol{a}\,\delta(t) - e\boldsymbol{a}\,\delta(t-t_{\rm D}) , \end{aligned} \tag{9.2.17}$$

so that

$$P(t) = -\frac{2}{3c^3}e^2\boldsymbol{a}\cdot\left[\boldsymbol{v}(0)\,\delta(t) - \boldsymbol{v}(t_{\rm D})\,\delta(t-t_{\rm D})\right] , \tag{9.2.18}$$

which gives

$$\int_{-\infty}^{\infty} \mathrm{d}t\, P(t) = -\frac{2}{3c^3} e^2 \boldsymbol{a} \cdot \underbrace{\left[\boldsymbol{v}(0) - \boldsymbol{v}(t_\mathrm{D})\right]}_{= -\boldsymbol{a} t_\mathrm{D}} = \frac{2}{3c^3} e^2 |\boldsymbol{a}|^2 t_\mathrm{D} \tag{9.2.19}$$

for the total energy emitted into radiation during the acceleration period, consistent with what we get from (9.2.13),

$$\int_0^{\infty} \mathrm{d}\omega\, E(\omega) = \frac{2}{3c^3} e^2 |\boldsymbol{a}|^2 t_\mathrm{D} \,. \tag{9.2.20}$$

We observe that

$$\frac{1}{t_\mathrm{D}} \int \mathrm{d}t\, P(t) = \frac{1}{t_\mathrm{D}} \int \mathrm{d}\omega\, E(\omega) = \frac{2}{3c^3} e^2 |\boldsymbol{a}|^2 \tag{9.2.21}$$

is the value of the total power $P$ in (9.2.7), telling us that this is the correct *average* power for the process. In summary, we thus note that despite the various unphysical aspects of the naive constant-acceleration-for-all-times calculation, the results obtained are not meaningless, we just have to interpret them correctly.

## 9.3 Cherenkov radiation

How about a charge in uniform motion — constant velocity, that is? Then

$$\boldsymbol{j}(\boldsymbol{r},t) = e\boldsymbol{v}\, \delta(\boldsymbol{r} - \boldsymbol{v}t) \quad \text{with constant } \boldsymbol{v} \tag{9.3.1}$$

and

$$\boldsymbol{j}(\boldsymbol{k},t) = e\boldsymbol{v}\, \mathrm{e}^{-\mathrm{i}\boldsymbol{k}\cdot\boldsymbol{v}t} \,, \tag{9.3.2}$$

so that

$$\left[\boldsymbol{n} \times \boldsymbol{j}(\boldsymbol{k}, T + \tfrac{1}{2}\tau)^*\right] \cdot \left[\boldsymbol{n} \times \boldsymbol{j}(\boldsymbol{k}, T - \tfrac{1}{2}\tau)\right] = e^2 (\boldsymbol{n} \times \boldsymbol{v})^2 \, \mathrm{e}^{\mathrm{i}\boldsymbol{k}\cdot\boldsymbol{v}\tau} \tag{9.3.3}$$

in (9.1.6). It follows that

$$\begin{aligned} \frac{\mathrm{d}P(\omega,T)}{\mathrm{d}\Omega} &= \frac{\omega^2}{4\pi^2 c^3} e^2 |\boldsymbol{n} \times \boldsymbol{v}|^2 \int \mathrm{d}\tau\, \mathrm{e}^{-\mathrm{i}\omega\tau}\, \mathrm{e}^{\mathrm{i}\boldsymbol{k}\cdot\boldsymbol{v}\tau} \\ &= \frac{\omega^2}{4\pi^2 c^3} e^2 |\boldsymbol{n} \times \boldsymbol{v}|^2\, 2\pi\, \delta(\omega - \boldsymbol{k}\cdot\boldsymbol{v}) \end{aligned} \tag{9.3.4}$$

or, with $\boldsymbol{k} = \omega c \boldsymbol{n}$, $\boldsymbol{n} \cdot \boldsymbol{v} = v \cos\theta$, and $|\boldsymbol{n} \times \boldsymbol{v}|^2 = v^2 (\sin\theta)^2$,

$$\begin{aligned}\frac{\mathrm{d}P(\omega,T)}{\mathrm{d}\Omega} &= \frac{\omega}{2\pi c} e^2 \left(\frac{v}{c}\right)^2 (\sin\theta)^2 \, \delta\Big(1 - \frac{v}{c}\cos\theta\Big) \\ &= \frac{\omega}{2\pi c} e^2 \left[\left(\frac{v}{c}\right)^2 - 1\right] \delta\Big(1 - \frac{v}{c}\cos\theta\Big). \end{aligned} \tag{9.3.5}$$

With $\frac{v}{c} < 1$, we have $1 - \frac{v}{c}\cos\theta > 0$ for all angles $\theta$, and conclude that a charge in uniform motion does not radiate — hardly a surprising result if one remembers that there is another frame of reference in which the charge is just standing still and nothing depends on time.

But, what if we could have $v > c$? This, by itself, can of course not happen. What is possible, however, is a charge moving in a medium (water, say) in which light propagates at a reduced speed: $\frac{v}{n}$ with index of refraction $n$, which is 1.33 for water at a wavelength of 600 nm (yellow light). There is then a range of velocities

$$\frac{c}{n} < v < c \tag{9.3.6}$$

which are above the speed of light in the medium and, of course, below the speed of light in vacuum.

We deal with this situation in a simplified but not over-idealized way, by assuming that the medium is homogeneous in the volume of relevance and that its index of refraction does not depend much on the frequency for the relevant frequency range. In other words, in the macroscopic Maxwell's equations [see also the context of (3.4.22)]

$$\begin{aligned} \boldsymbol{\nabla} \cdot \boldsymbol{D} &= 4\pi\rho\,, & \boldsymbol{\nabla} \cdot \boldsymbol{B} &= 0\,, \\ \boldsymbol{\nabla} \times \boldsymbol{H} - \frac{1}{c}\frac{\partial}{\partial t}\boldsymbol{D} &= \frac{4\pi}{c}\boldsymbol{j}\,, & \boldsymbol{\nabla} \times \boldsymbol{E} + \frac{1}{c}\frac{\partial}{\partial t}\boldsymbol{B} &= 0\,, \end{aligned} \tag{9.3.7}$$

with $\boldsymbol{D} = \epsilon \boldsymbol{E}$ and $\boldsymbol{B} = \mu \boldsymbol{H}$, we regard the electric permittivity $\epsilon$ (*vulgo* "dielectric constant") and the magnetic permeability $\mu$ as constants; and then the index of refraction $n = \sqrt{\epsilon\mu}$ is constant as well. For the present purpose, this is a well justified simplifying assumption.

So, with $\epsilon$ and $\mu$ constant, we can cast these Maxwell's equations in the medium in the form

$$\begin{aligned}
&\boldsymbol{\nabla}\cdot\left(\sqrt{\epsilon}\boldsymbol{E}\right) = 4\pi\frac{\rho}{\sqrt{\epsilon}}\,, \qquad \boldsymbol{\nabla}\cdot\left(\frac{1}{\sqrt{\mu}}\boldsymbol{B}\right) = 0\,,\\
&\boldsymbol{\nabla}\times\left(\frac{1}{\sqrt{\mu}}\boldsymbol{B}\right) - \frac{1}{c/\sqrt{\epsilon\mu}}\frac{\partial}{\partial t}\left(\sqrt{\epsilon}\boldsymbol{E}\right) = \frac{4\pi}{c/\sqrt{\epsilon\mu}}\frac{\boldsymbol{j}}{\sqrt{\epsilon}}\,,\\
&\boldsymbol{\nabla}\times\left(\sqrt{\epsilon}\boldsymbol{E}\right) + \frac{1}{c/\sqrt{\epsilon\mu}}\frac{\partial}{\partial t}\left(\frac{1}{\sqrt{\mu}}\boldsymbol{B}\right) = 0\,,
\end{aligned} \tag{9.3.8}$$

where we simply divided or multiplied by $\sqrt{\epsilon}$ or $\sqrt{\mu}$. The resulting set of equations is obtained from the microscopic equations (1.1.1) by the replacements

$$\rho \to \frac{\rho}{\sqrt{\epsilon}} \quad \text{and} \quad \boldsymbol{j} \to \frac{\boldsymbol{j}}{\sqrt{\epsilon}} \tag{9.3.9}$$

for the charge and current densities,

$$\boldsymbol{E} \to \sqrt{\epsilon}\boldsymbol{E} \quad \text{and} \quad \boldsymbol{B} \to \frac{1}{\sqrt{\mu}}\boldsymbol{B} \tag{9.3.10}$$

for the electric and magnetic fields, and

$$c \to \frac{c}{\sqrt{\epsilon\mu}} = \frac{c}{n} \tag{9.3.11}$$

for the speed of light. Therefore, if we now consider a charge $e$ moving through a dielectric medium with an index of refraction $n$ — that is: $\sqrt{\epsilon} = n$, $\mu = 1$ — then we need to perform the replacements

$$e \to \frac{e}{\sqrt{\epsilon}} = \frac{e}{n} \quad \text{and} \quad c \to \frac{c}{n} \tag{9.3.12}$$

on the right-hand side of (9.3.5). This takes us to

$$\frac{\mathrm{d}P(\omega,T)}{\mathrm{d}\Omega} = \frac{\omega}{2\pi}\frac{e^2}{nc}\left(\left(\frac{nv}{c}\right)^2 - 1\right)\delta\left(1 - \frac{nv}{c}\cos\theta\right), \tag{9.3.13}$$

and radiation occurs for $\frac{c}{n} < v < c$, emitted into the direction specified by

$$\cos\theta = \frac{c}{nv} < 1\,. \tag{9.3.14}$$

This is the so-called Cherenkov* radiation, which one can see as blue light in the water basins of nuclear reactors. It results from the fast motion of highly energetic electrons through the water, which are products of nuclear reactions or beta decay processes.

*Pavel Alekseyevich CHERENKOV (1904–1990)

In reality, the index of refraction will depend on the frequency $\omega$, that is: $n = n(\omega)$, and at sufficiently high frequencies it will be so close to $n \cong 1$ that $v > \frac{c}{n}$ is no longer true. Therefore, extremely high frequencies will not be emitted, and the integral for the total radiated energy,

$$-\frac{\mathrm{d}E}{\mathrm{d}T} = \int_0^\infty \mathrm{d}\omega \int \mathrm{d}\Omega \, \frac{\mathrm{d}P(\omega, T)}{\mathrm{d}\Omega}, \tag{9.3.15}$$

will have no contributions above a certain critical frequency.

In this expression, the angular integration is immediate,

$$\begin{aligned}\int \mathrm{d}\Omega \, \frac{\mathrm{d}P(\omega,T)}{\mathrm{d}\Omega} &= \omega \frac{e^2}{nc}\left(\left(\frac{nv}{c}\right)^2 - 1\right)\int_0^\pi \mathrm{d}\theta \, \sin\theta \, \delta\Big(1 - \frac{nv}{c}\cos\theta\Big) \\ &= \omega \frac{e^2}{n^2 v}\left(\left(\frac{nv}{c}\right)^2 - 1\right)\eta(nv - c)\end{aligned} \tag{9.3.16}$$

where the Heaviside unit step function $\eta(\ )$ permits only these frequencies for which $n(\omega) > \frac{c}{v}$. We have

$$-\frac{\mathrm{d}E}{\mathrm{d}T} = \int_0^\infty \mathrm{d}\omega \, \omega \frac{e^2 v}{c^2}\left(1 - \left(\frac{c}{nv}\right)^2\right)\eta\big(n(\omega)v - c\big) \tag{9.3.17}$$

for the energy loss per unit time of the moving charge resulting from the Cherenkov radiation — per unit epoch time $T$, that is. Since this is a mild process for a charge moving at ultrarelativistic speeds — kinetic energy of the order of MeV, compared with the few-eV energy of the emitted photons — we can regard the velocity $v$ as constant in time (or very slowly changing). It is then easy to switch from energy loss per time to energy loss per covered distance, $\mathrm{d}s = v\mathrm{d}T$,

$$-\frac{\mathrm{d}E}{\mathrm{d}s} = -\frac{1}{v}\frac{\mathrm{d}E}{\mathrm{d}T} = \int_0^\infty \mathrm{d}\omega \, \omega \frac{e^2}{c^2}\left(1 - \left(\frac{c}{nv}\right)^2\right)\eta\big(n(\omega)v - c\big), \tag{9.3.18}$$

and it is even more revealing to note that the energy per photon is $\hbar\omega$, and thus observe that the number of photons emitted per unit distance is ($e \to -e_0$ for electrons now)

$$\frac{\mathrm{d}N}{\mathrm{d}s} = \frac{e_0^2}{\hbar c}\frac{1}{c}\int_0^\infty \mathrm{d}\omega \left(1 - \left(\frac{c}{n(\omega)v}\right)^2\right)\eta\big(n(\omega)v - c\big), \tag{9.3.19}$$

where once again we come across the fine structure constant $\alpha = \frac{e_0^2}{\hbar c} \cong \frac{1}{137}$. To get a rough estimate, we regard the $\omega$ integral as defining a typical frequency $\overline{\omega} = 2\pi \frac{c}{\overline{\lambda}}$, and get

$$\frac{dN}{ds} = \frac{2\pi\alpha}{\overline{\lambda}} \tag{9.3.20}$$

where $\overline{\lambda}$ is the wavelength associated with $\overline{\omega}$. If $\overline{\lambda} \cong 5 \times 10^{-5}$cm (visible light), then

$$\frac{dN}{ds} \cong 10^3/\text{cm}, \tag{9.3.21}$$

which overestimates the actual number of photons somewhat — a better number is $10^2$/cm.

A common picture of Cherenkov radiation is analogous to the shock wave of supersonic aircraft:

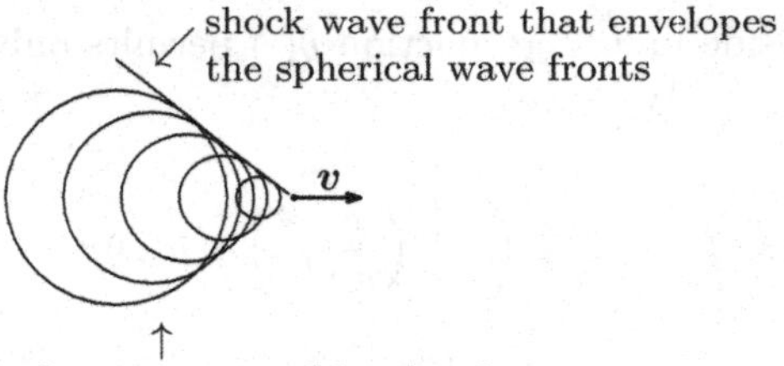


(9.3.22)

whereby the angle $\theta$ of (9.3.14) is identified by

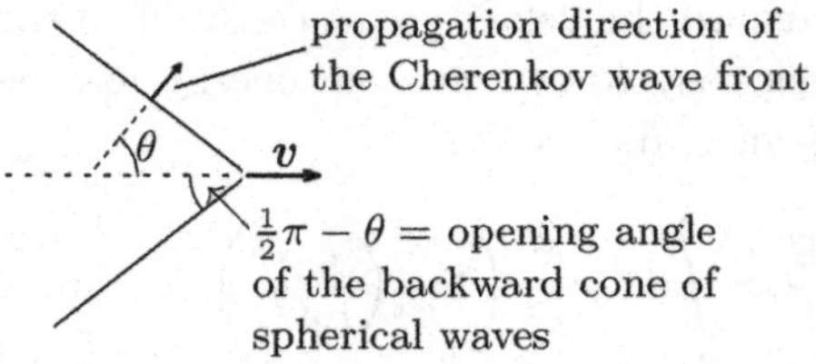


(9.3.23)

Note that the Cherenkov radiation is emitted into the forward direction, $\theta < \frac{\pi}{2}$.

# Chapter 10

# Synchrotron Radiation

## 10.1 Kinematics

We now turn to the situation of *synchrotron radiation*, which is emitted by a charge in a homogeneous magnetic field $\boldsymbol{B}$, where the Lorentz force acts such that the charge $e$ is moving with constant speed $v$ on a circle with radius $R$. More specifically, we have

$$\frac{\mathrm{d}}{\mathrm{d}t}\boldsymbol{p} = \frac{e}{c}\boldsymbol{v} \times \boldsymbol{B} \tag{10.1.1}$$

for the time derivative of the momentum, and

$$\frac{\mathrm{d}}{\mathrm{d}t}E = 0 \tag{10.1.2}$$

for the time derivative of the energy. The momentum is related to the velocity by

$$\boldsymbol{p} = \frac{m\boldsymbol{v}}{\sqrt{1-(v/c)^2}} = \gamma m\boldsymbol{v} = \frac{E}{c^2}\boldsymbol{v}\,, \tag{10.1.3}$$

where the relativistic relation

$$E = \gamma mc^2 \tag{10.1.4}$$

is taken into account. Since $E$ is constant in time, the time derivative of $\boldsymbol{p}$ appears as

$$\frac{E}{c^2}\frac{\mathrm{d}}{\mathrm{d}t}\boldsymbol{v} = \frac{e}{c}\boldsymbol{v} \times \boldsymbol{B}\,, \tag{10.1.5}$$

so that

$$\frac{\mathrm{d}}{\mathrm{d}t}\boldsymbol{v} = \boldsymbol{\omega}_0 \times \boldsymbol{v} \tag{10.1.6}$$

with

$$\boldsymbol{\omega}_0 = -\frac{ec\boldsymbol{B}}{E}\,. \tag{10.1.7}$$

These equations state that $\boldsymbol{v}$ precesses around the direction of $\boldsymbol{B}$ with the angular velocity

$$\omega_0 = \left|\frac{ec\boldsymbol{B}}{E}\right|, \tag{10.1.8}$$

the so-called Larmor frequency. In the nonrelativistic limit ($\frac{v}{c} \ll 1$, $\gamma \cong 1$), it turns into the cyclotron frequency

$$\omega_0 \cong \left|\frac{e\boldsymbol{B}}{mc}\right|. \tag{10.1.9}$$

In practice, the motion of the charge is confined to a plane, for which we take the $xy$ plane,

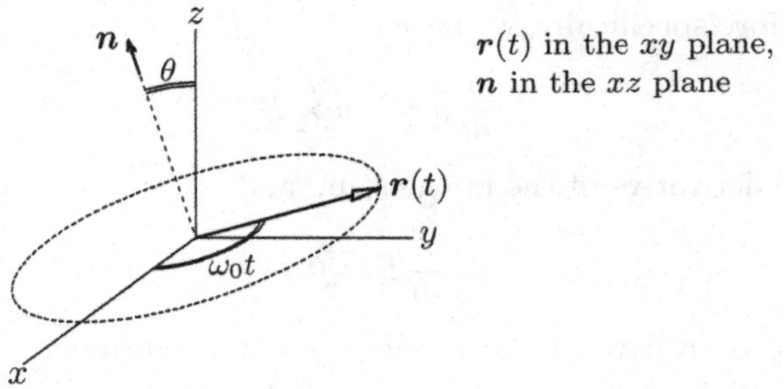


(10.1.10)

with the trajectory

$$\boldsymbol{r}(t) = R\begin{pmatrix}\cos(\omega_0 t)\\ \sin(\omega_0 t)\\ 0\end{pmatrix} \tag{10.1.11}$$

and the velocity

$$\boldsymbol{v}(t) = \omega_0 R\begin{pmatrix}-\sin(\omega_0 t)\\ \cos(\omega_0 t)\\ 0\end{pmatrix} \tag{10.1.12}$$

and the acceleration $\dot{\boldsymbol{v}}(t) = -{\omega_0}^2\boldsymbol{r}(t)$. We thus have

$$v = \omega_0 R \tag{10.1.13}$$

for the speed and

$$|\dot{\boldsymbol{v}}(t)| = {\omega_0}^2 R = \frac{v^2}{R} \tag{10.1.14}$$

for the centripetal acceleration. And if the charge were moving slowly, so that the standard nonrelativistic Larmor formula (7.2.11) would apply, the total radiated power would be given by

$$P_{\text{nonrel}} = \frac{2}{3}\frac{e^2}{c^3}|\dot{\boldsymbol{v}}|^2 = \frac{2}{3}\frac{e^2}{R}\omega_0\left(\frac{v}{c}\right)^3. \tag{10.1.15}$$

As we will see, in (10.3.15) below, the expression for arbitrary velocity consists of this nonrelativistic radiated power times a relativistic correction factor.

## 10.2 Time-dependent spectrum

It will be slightly more convenient to use the alternative form

$$\begin{aligned}\frac{\mathrm{d}P(\omega,T)}{\mathrm{d}\Omega} = \frac{\omega^2}{4\pi^2 c}\int \mathrm{d}\tau\, \mathrm{e}^{-\mathrm{i}\omega\tau}\Bigl[\frac{1}{c}\boldsymbol{j}\bigl(\boldsymbol{k},T+\tfrac{1}{2}\tau\bigr)^* \cdot \frac{1}{c}\boldsymbol{j}\bigl(\boldsymbol{k},T-\tfrac{1}{2}\tau\bigr) \\ - \rho\bigl(\boldsymbol{k},T+\tfrac{1}{2}\tau\bigr)^*\rho\bigl(\boldsymbol{k},T-\tfrac{1}{2}\tau\bigr)\Bigr]\end{aligned} \tag{10.2.1}$$

for the time-dependent power spectrum than the equivalent expression in (9.1.6). We obtain (10.2.1) when using

$$\begin{aligned}|\boldsymbol{k}\times\boldsymbol{j}(\boldsymbol{k},\omega)|^2 &= \bigl[\boldsymbol{k}\times\boldsymbol{j}(\boldsymbol{k},\omega)^*\bigr]\cdot\bigl[\boldsymbol{k}\times\boldsymbol{j}(\boldsymbol{k},\omega)\bigr] \\ &= \boldsymbol{k}^2\boldsymbol{j}(\boldsymbol{k},\omega)^*\cdot\boldsymbol{j}(\boldsymbol{k},\omega) - \boldsymbol{k}\cdot\boldsymbol{j}(\boldsymbol{k},\omega)^*\,\underbrace{\boldsymbol{k}\cdot\boldsymbol{j}(\boldsymbol{k},\omega)}_{=\omega\rho(\boldsymbol{k},\omega)} \\ &= \omega^2\Bigl[\frac{1}{c}\boldsymbol{j}(\boldsymbol{k},\omega)^*\cdot\frac{1}{c}\boldsymbol{j}(\boldsymbol{k},\omega) - \rho(\boldsymbol{k},\omega)^*\rho(\boldsymbol{k},\omega)\Bigr]\end{aligned} \tag{10.2.2}$$

in (9.1.1) before carrying out the steps that take us to (9.1.6).

In the present context we have the electric charge and current densities

$$\begin{aligned}\rho(\boldsymbol{r}',t) &= e\,\delta\bigl(\boldsymbol{r}'-\boldsymbol{r}(t)\bigr)\,, \\ \boldsymbol{j}(\boldsymbol{r}',t) &= e\boldsymbol{v}(t)\,\delta\bigl(\boldsymbol{r}'-\boldsymbol{r}(t)\bigr)\,,\end{aligned} \tag{10.2.3}$$

with $\boldsymbol{r}(t)$ and $\boldsymbol{v}(t)$ from (10.1.11) and (10.1.12) and we choose $\boldsymbol{k} = \dfrac{\omega}{c}\boldsymbol{n}$ to lie in the $xz$ plane,

$$\boldsymbol{k} = \frac{\omega}{c}\begin{pmatrix} \sin\theta \\ 0 \\ \cos\theta \end{pmatrix}, \tag{10.2.4}$$

as indicated in (10.1.10). Then

$$\rho(\boldsymbol{k},t) = \int (\mathrm{d}\boldsymbol{r}')\, \mathrm{e}^{-\mathrm{i}\boldsymbol{k}\cdot\boldsymbol{r}'}\rho(\boldsymbol{r}',t) = e\,\mathrm{e}^{-\mathrm{i}\boldsymbol{k}\cdot\boldsymbol{r}(t)} \tag{10.2.5}$$

and

$$\boldsymbol{j}(\boldsymbol{k},t) = e\boldsymbol{v}(t)\,\mathrm{e}^{-\mathrm{i}\boldsymbol{k}\cdot\boldsymbol{r}(t)}, \tag{10.2.6}$$

and we get

$$\begin{aligned} \frac{\mathrm{d}P(\omega,T)}{\mathrm{d}\Omega} = \frac{e^2\omega^2}{4\pi^2 c}\int \mathrm{d}\tau\, \mathrm{e}^{-\mathrm{i}\omega\tau}\, \mathrm{e}^{\mathrm{i}\boldsymbol{k}\cdot[\boldsymbol{r}(T+\frac{1}{2}\tau)-\boldsymbol{r}(T-\frac{1}{2}\tau)]} \\ \times\left[\frac{1}{c^2}\boldsymbol{v}\left(T+\tfrac{1}{2}\tau\right)\cdot\boldsymbol{v}\left(T-\tfrac{1}{2}\tau\right)-1\right]. \end{aligned} \tag{10.2.7}$$

With $\boldsymbol{k}$, $\boldsymbol{r}(t)$, $\boldsymbol{v}(t)$ as parameterized above, we have

$$\begin{aligned} \boldsymbol{r}(T+\tfrac{1}{2}\tau)-\boldsymbol{r}(T-\tfrac{1}{2}\tau) &= R\begin{pmatrix} \cos(\omega_0 T+\frac{1}{2}\omega_0\tau)-\cos(\omega_0 T-\frac{1}{2}\omega_0\tau) \\ \sin(\omega_0 T+\frac{1}{2}\omega_0\tau)-\sin(\omega_0 T-\frac{1}{2}\omega_0\tau) \\ 0 \end{pmatrix} \\ &= R\begin{pmatrix} -2\sin(\omega_0 T)\sin(\frac{1}{2}\omega_0\tau) \\ 2\cos(\omega_0 T)\sin(\frac{1}{2}\omega_0\tau) \\ 0 \end{pmatrix} \\ &= \frac{2}{\omega_0}\sin(\tfrac{1}{2}\omega_0\tau)\boldsymbol{v}(T) \end{aligned} \tag{10.2.8}$$

with its obvious geometrical meaning, so that

$$\boldsymbol{k}\cdot\left[\boldsymbol{r}(T+\tfrac{1}{2}\tau)-\boldsymbol{r}(T-\tfrac{1}{2}\tau)\right] = -2\frac{\omega}{c}R\sin(\omega_0 T)\sin(\tfrac{1}{2}\omega_0\tau)\sin\theta, \tag{10.2.9}$$

and

$$\begin{aligned} \boldsymbol{v}(T+\tfrac{1}{2}\tau)\cdot\boldsymbol{v}(T-\tfrac{1}{2}\tau) &= v^2\left[\cos(\omega_0 T+\tfrac{1}{2}\omega_0\tau)\cos(\omega_0 T-\tfrac{1}{2}\omega_0\tau)\right. \\ &\quad \left.+\sin(\omega_0 T+\tfrac{1}{2}\omega_0\tau)\sin(\omega_0 T-\tfrac{1}{2}\omega_0\tau)\right] \\ &= v^2\cos(\omega_0\tau), \end{aligned} \tag{10.2.10}$$

and arrive at

$$\frac{dP(\omega,T)}{d\Omega} = \frac{e^2\omega^2}{4\pi^2 c}\int d\tau\, e^{-i\omega\tau}\left[\left(\frac{v}{c}\right)^2\cos(\omega_0\tau) - 1\right] \times e^{-2i\frac{\omega}{c}R\sin(\omega_0 T)\sin(\frac{1}{2}\omega_0\tau)\sin\theta}. \tag{10.2.11}$$

This expression for the time-dependent power spectrum of synchrotron radiation is the take-off point for much of what follows.

## 10.3 Total radiated power

For the integration over all directions, the specific choice $\boldsymbol{n} = \begin{pmatrix}\sin\theta\\ 0\\ \cos\theta\end{pmatrix}$ of (10.1.10) and (10.2.4) is not appropriate, and we just return to the form (10.2.7) with general $\boldsymbol{k}$ and evaluate

$$\int d\Omega\, e^{i\boldsymbol{k}\cdot\,[\boldsymbol{r}(T+\frac{1}{2}\tau) - \boldsymbol{r}(T-\frac{1}{2}\tau)]} = 4\pi\frac{\sin(ks)}{ks} \tag{10.3.1}$$

where $k = \frac{\omega}{c}$ as always and

$$\begin{aligned} s &= \left|\boldsymbol{r}(T+\tfrac{1}{2}\tau) - \boldsymbol{r}(T-\tfrac{1}{2}\tau)\right| \\ &= \frac{2}{\omega_0}\left|\sin(\tfrac{1}{2}\omega_0\tau)\boldsymbol{v}(T)\right| \\ &= 2R\left|\sin\left(\tfrac{1}{2}\omega_0\tau\right)\right| \end{aligned} \tag{10.3.2}$$

is the chordal distance between the two positions on the circle, so that

$$\int d\Omega\, e^{i\boldsymbol{k}\cdot[\boldsymbol{r}(T+\frac{1}{2}\tau) - \boldsymbol{r}(T-\frac{1}{2}\tau)]} = 4\pi\frac{\sin\left(2\frac{\omega}{c}R\sin\left(\frac{1}{2}\omega_0\tau\right)\right)}{2\frac{\omega}{c}R\sin\left(\frac{1}{2}\omega_0\tau\right)}. \tag{10.3.3}$$

It follows that the total angle-integrated spectral power density is given by

$$\begin{aligned} P(\omega,T) &= \int d\Omega\,\frac{dP(\omega,T)}{d\Omega} \\ &= \frac{\omega}{2\pi}\frac{e^2}{R}\int_{-\infty}^{\infty} d\tau\, e^{-i\omega\tau}\left[\left(\frac{v}{c}\right)^2\cos(\omega_0\tau) - 1\right] \\ &\qquad \times \frac{\sin\left(\frac{2\omega}{c}R\sin\frac{\omega_0\tau}{2}\right)}{\sin\frac{\omega_0\tau}{2}}. \end{aligned} \tag{10.3.4}$$

We note that the factor multiplying $\mathrm{e}^{-\mathrm{i}\omega\tau}$ in the integrand is even in $\tau$, so that we could replace $\mathrm{e}^{-\mathrm{i}\omega\tau}$ by $\cos(\omega\tau)$ without changing the value of the integral — which is just noting that the quantity in question is real and must, therefore, have a vanishing imaginary part. In view of this observation, the $\omega$ integration needed for finding the total radiated power $P(T)$,

$$P(T) = \int_0^\infty \mathrm{d}\omega \int \mathrm{d}\Omega \, \frac{\mathrm{d}P(\omega, T)}{\mathrm{d}\Omega}\,, \tag{10.3.5}$$

is

$$\begin{aligned}
&\int_0^\infty \mathrm{d}\omega\, \omega \cos(\omega\tau) \sin\!\Big(\frac{2\omega R}{c} \sin\frac{\omega_0\tau}{2}\Big) \\
&= \frac{1}{2}\int_{-\infty}^{\infty} \mathrm{d}\omega\, \omega\, \mathrm{e}^{-\mathrm{i}\omega\tau} \frac{1}{2\mathrm{i}}\Big(\mathrm{e}^{\mathrm{i}\omega\tau'} - \mathrm{e}^{-\mathrm{i}\omega\tau'}\Big)\Big|_{\tau' = \frac{2R}{c}\sin\frac{\omega_0\tau}{2}} \\
&= \frac{1}{4}\frac{\partial}{\partial\tau}\int_{-\infty}^{\infty} \mathrm{d}\omega \Big[\mathrm{e}^{-\mathrm{i}\omega(\tau - \tau')} - \mathrm{e}^{-\mathrm{i}\omega(\tau + \tau')}\Big]\Big|_{\tau' = \frac{2R}{c}\sin\frac{\omega_0\tau}{2}} \\
&= \frac{\pi}{2}[\delta'(\tau - \tau') - \delta'(\tau + \tau')]\Big|_{\tau' = \frac{2R}{c}\sin\frac{\omega_0\tau}{2}} \\
&= -\frac{\pi}{2}\Big[\delta'\Big(\tau + \frac{2R}{c}\sin\frac{\omega_0\tau}{2}\Big) - \delta'\Big(\tau - \frac{2R}{c}\sin\frac{\omega_0\tau}{2}\Big)\Big]\,,
\end{aligned} \tag{10.3.6}$$

and so we arrive at

$$\begin{aligned}
P(T) = \frac{e^2}{4R}\int_{-\infty}^{\infty} \mathrm{d}\tau \Big[1 - \Big(\frac{v}{c}\Big)^2 \cos(\omega_0\tau)\Big] \\
\times \frac{\delta'\Big(\tau + \frac{2R}{c}\sin\frac{\omega_0\tau}{2}\Big) - \delta'\Big(\tau - \frac{2R}{c}\sin\frac{\omega_0\tau}{2}\Big)}{\sin\frac{\omega_0\tau}{2}}\,.
\end{aligned} \tag{10.3.7}$$

This becomes more compact after we note that

$$\frac{\delta'\Bigl(\tau+\frac{2R}{c}\sin\frac{\omega_0\tau}{2}\Bigr)-\delta'\Bigl(\tau-\frac{2R}{c}\sin\frac{\omega_0\tau}{2}\Bigr)}{\sin\frac{\omega_0\tau}{2}}$$

$$=\int_{-1}^{1}\mathrm{d}x\;\delta''\Bigl(\tau+\frac{2R}{c}x\sin\frac{\omega_0\tau}{2}\Bigr)\frac{2R}{c}$$

$$=2\frac{v}{c}\omega_0{}^2\int_{-1}^{1}\mathrm{d}x\;\delta''\Bigl(\omega_0\tau+2x\frac{v}{c}\sin\frac{\omega_0\tau}{2}\Bigr), \tag{10.3.8}$$

which uses $\delta''(\omega_0\tau)=\frac{1}{\omega_0{}^3}\delta''(\tau)$ [see at (9.2.4)] and recalls that $v=\omega_0 R$. Upon switching from integration over $\tau$ to integration over $\varphi=\omega_0\tau$, we have

$$P(T)=\frac{1}{2}\frac{e^2}{R}\frac{v}{c}\omega_0\int_{-\infty}^{\infty}\mathrm{d}\varphi\int_{-1}^{1}\mathrm{d}x\;\Bigl[1-\Bigl(\frac{v}{c}\Bigr)^2\cos\varphi\Bigr]\,\delta''\Bigl(\varphi+2x\frac{v}{c}\sin\frac{\varphi}{2}\Bigr). \tag{10.3.9}$$

The argument of the delta function, $y=\varphi+2x\frac{v}{c}\sin\frac{\varphi}{2}$, is a monotonic function of $\varphi$,

$$\mathrm{d}y=\underbrace{\Bigl(1+x\frac{v}{c}\cos\frac{\varphi}{2}\Bigr)}_{>0}\mathrm{d}\varphi, \tag{10.3.10}$$

so that the only contribution to the $\varphi$ integral is at $\varphi=0$ or $y=0$, which we make explicit by switching to integration over $y$,

$$P(T)=\frac{1}{2}\frac{e^2}{R}\frac{v}{c}\omega_0\int_{-1}^{1}\mathrm{d}x\int\mathrm{d}y\;\frac{1-\bigl(\frac{v}{c}\bigr)^2\cos\varphi}{1+x\frac{v}{c}\cos\frac{\varphi}{2}}\,\delta''(y)$$

$$=\frac{1}{2}\frac{e^2}{R}\frac{v}{c}\omega_0\int_{-1}^{1}\mathrm{d}x\Bigl(\frac{\partial}{\partial y}\Bigr)^2\frac{1-\bigl(\frac{v}{c}\bigr)^2\cos\varphi}{1+x\frac{v}{c}\cos\frac{\varphi}{2}}\Bigg|_{y=0}. \tag{10.3.11}$$

Here we need

$$\left(\frac{\partial}{\partial y}\right)^2 \frac{1-\left(\frac{v}{c}\right)^2\cos\varphi}{1+x\frac{v}{c}\cos\frac{\varphi}{2}}\Bigg|_{y=0} = \left(\frac{1}{1+x\frac{v}{c}\cos\frac{\varphi}{2}}\frac{\partial}{\partial\varphi}\right)^2 \frac{1-\left(\frac{v}{c}\right)^2\cos\varphi}{1+x\frac{v}{c}\cos\frac{\varphi}{2}}\Bigg|_{\varphi=0}$$
$$= \frac{\left(\frac{v}{c}\right)^2}{\left(1+x\frac{v}{c}\right)^3} + \frac{\left[1-\left(\frac{v}{c}\right)^2\right]\frac{1}{4}x\frac{v}{c}}{\left(1+x\frac{v}{c}\right)^4} \tag{10.3.12}$$

to first get

$$P(T) = \frac{1}{2}\frac{e^2}{R}\omega_0\frac{v}{c}\int_{-1}^{1} \mathrm{d}x \left[\frac{\left(\frac{v}{c}\right)^2}{\left(1+x\frac{v}{c}\right)^3} + \frac{\left[1-\left(\frac{v}{c}\right)^2\right]\frac{1}{4}x\frac{v}{c}}{\left(1+x\frac{v}{c}\right)^4}\right] \tag{10.3.13}$$

and then

$$P(T) = \frac{2}{3}\frac{e^2}{R}\omega_0\left(\frac{v}{c}\right)^3\frac{1}{\left[1-(v/c)^2\right]^2} \tag{10.3.14}$$

or with $E = \gamma mc^2$, $\gamma = \dfrac{1}{\sqrt{1-(v/c)^2}} = \dfrac{E}{mc^2}$,

$$P(T) = \frac{2}{3}\frac{e^2}{R}\omega_0\left(\frac{v}{c}\right)^3\left(\frac{E}{mc^2}\right)^4 . \tag{10.3.15}$$

In this final expression for the total power of synchrotron radiation, we recognize the nonrelativistic Larmor expression (10.1.15) multiplied by the relativistic energy factor $\left(\dfrac{E}{mc^2}\right)^4$.

To establish the numerical amount of the power of synchrotron radiation, we consider an electron with an energy of $E = 1\,\mathrm{GeV}$ moving on a circle of radius $R = 1\,\mathrm{m}$. Then $(m \to m_{\mathrm{el}},\ e \to -e_0)$

$$\gamma = \frac{E}{m_{\mathrm{el}}c^2} = \frac{10^3\,\mathrm{MeV}}{0.511\,\mathrm{MeV}} = 1.957\times 10^3 ,$$
$$\frac{v}{c} = \left(1-\frac{1}{\gamma^2}\right)^{1/2} = 1-\frac{1}{2\gamma^2} = 1-1.3\times 10^{-7} = 1 ,$$
$$\frac{\hbar c}{R} = \frac{1.9733\times 10^{-11}\,\mathrm{MeV\,cm}}{100\,\mathrm{cm}} = 1.9733\times 10^{-10}\,\mathrm{keV} , \tag{10.3.16}$$

and the total energy radiated during one period of the circular motion is

$$\begin{aligned} P\frac{2\pi}{\omega_0} &= \frac{4\pi}{3}\frac{e_0^2}{\hbar c}\frac{\hbar c}{R}\left(\frac{v}{c}\right)^3\left(\frac{E}{mc^2}\right)^4 \\ &= \frac{4\pi}{3}\frac{1}{137.036}\,1.9733\times 10^{-10}(1.957\times 10^3)^4\,\mathrm{keV} \\ &= 88.5\,\mathrm{keV}\,. \end{aligned} \tag{10.3.17}$$

This says that the energy radiated per cycle, $\Delta E$, is to be calculated as

$$\Delta E[\mathrm{keV}] = 88.5\,\frac{E[\mathrm{GeV}]^4}{R[\mathrm{m}]} \quad \text{for electrons.} \tag{10.3.18}$$

For example, electrons with energy $E = 10\,\mathrm{GeV}$ in a circle of $R = 10\,\mathrm{m}$ will radiate $88.5\times 10^3\,\mathrm{keV} = 88.5\,\mathrm{MeV}$ in each round. This energy needs to be supplied to the orbiting electrons in order to keep the circular motion up, and it makes synchrotrons impractical for high-energy electrons if the objective is to create highly energetic electrons for collision experiments. If, however, you want to have a bright source of synchrotron light, electrons are just fine.

Matters are very different for protons, because the mass ratio

$$\frac{\text{electron mass}}{\text{proton mass}} = \frac{1}{1836}\,, \tag{10.3.19}$$

taken to the fourth power, reduces the radiated power by a factor of $10^{-13}$, so that

$$\Delta E[\mathrm{neV}] = 7.8\,\frac{E[\mathrm{GeV}]^4}{R[\mathrm{m}]} \quad \text{for protons.} \tag{10.3.20}$$

Note that the energy loss is in *nano* eV now, completely irrelevant: There is no substantial synchrotron radiation for protons. This is why one uses proton accelerators in experiments studying the interactions of elementary particles in very-high-energy collisions.

In (10.3.15), earlier in (10.3.4) as well as in equations to come later, such as (10.4.11) and (10.4.15) below, we keep indicating the dependence on $T$ although there is no visible $T$ dependence on the right. But remember that $T$ refers to the macroscopic time, to the epoch of the physical process, and thus accounts for slow parameter changes. Examples are a slowly changing radius $R(T)$ or a slowly varying orbit frequency $\omega_0(T)$, perhaps resulting from a varying strength of the magnetic fields, or from a loss of energy as a consequence of the synchrotron radiation itself.

## 10.4 Power emitted into the $m$th harmonic

We return to the time-dependent spectrum (10.2.11) in order to learn more about the details of synchrotron radiation. First, we remind ourselves that the epoch time $T$ is not well specified on the scale set by the intrinsic frequencies, here: first of all the orbit period $\frac{2\pi}{\omega_0}$. This coarse grain significance of $T$ is taken into account by averaging over one period,

$$\begin{aligned}
\mathrm{e}^{-2\mathrm{i}\frac{\omega}{c}R\sin\theta\sin(\omega_0 T)\sin\left(\frac{1}{2}\omega_0\tau\right)} &\to \frac{\omega_0}{2\pi}\int\limits_{T-\pi/\omega_0}^{T+\pi/\omega_0} \mathrm{d}t\ \mathrm{e}^{-2\mathrm{i}\frac{\omega}{c}R\sin\theta\sin(\omega_0 t)\sin\left(\frac{1}{2}\omega_0\tau\right)} \\
&= \mathrm{J}_0\left(2\frac{\omega}{c}R\sin\theta\ \sin\frac{\omega_0\tau}{2}\right), \qquad (10.4.1)
\end{aligned}$$

where $\mathrm{J}_0(\ )$ is the Bessel* function of 0th order, for which

$$\mathrm{J}_0(z) = \int\limits_{(2\pi)} \frac{\mathrm{d}\varphi}{2\pi}\ \mathrm{e}^{\mathrm{i}z\cos\varphi} \qquad (10.4.2)$$

is one possible definition, with the integration covering any $2\pi$ interval of $\varphi$. Perhaps more useful to memorize is the generating function

$$\mathrm{e}^{\mathrm{i}z\cos\varphi} = \sum_{m=-\infty}^{\infty} \mathrm{i}^m\, \mathrm{e}^{\mathrm{i}m\varphi}\, \mathrm{J}_m(z)\,, \qquad (10.4.3)$$

which states simply that the Bessel functions $\mathrm{J}_m(z)$ are the Fourier coefficients of the periodic function of $\varphi$ on the left. The above integral form of $J_0(z)$ follows when one picks out the $m = 0$ Fourier coefficient.

At this stage we have

$$\begin{aligned}
\frac{\mathrm{d}P(\omega,T)}{\mathrm{d}\Omega} = \frac{\omega^2 e^2}{4\pi^2 c}\int \mathrm{d}\tau\ \mathrm{e}^{-\mathrm{i}\omega\tau}\left[\left(\frac{v}{c}\right)^2\cos(\omega_0\tau) - 1\right] \\
\times\, \mathrm{J}_0\left(2\frac{\omega}{c}R\sin\theta\ \sin\frac{\omega_0\tau}{2}\right) \qquad (10.4.4)
\end{aligned}$$

for the time-dependent power spectrum of synchrotron radiation. The $\tau$ integration is of the form $\mathrm{e}^{-\mathrm{i}\omega\tau}$ times a function that is periodic in $\tau$ with period $\frac{2\pi}{\omega_0}$, which is obvious as soon as one remembers that $\mathrm{J}_0(\ )$ is an even

*Friedrich Wilhelm BESSEL (1784–1846)

function of its argument,

$$\begin{aligned}\mathrm{J}_0(z) &= \sum_{k=0}^{\infty}\left(\frac{1}{k!}\right)^2\left(-\frac{1}{4}z^2\right)^k = \mathrm{J}_0(-z)\\ &= \mathrm{J}_0(|z|) = \mathrm{J}_0\left(\sqrt{z^2}\right),\end{aligned} \tag{10.4.5}$$

and this periodic function has a Fourier series that sums over powers of $\mathrm{e}^{\mathrm{i}\omega_0\tau}$. Therefore, the $\tau$ integration results in a sum over delta functions $\delta(\omega - m\omega_0)$, telling us that synchrotron radiation is composed of contributions with frequencies $\omega_0$, $2\omega_0$, $3\omega_0$, ..., the orbit frequency and its integer multiples, its *higher harmonics*, borrowing a term from acoustics.

We exhibit the harmonics by an application of the addition theorem (derived in Exercise 5)

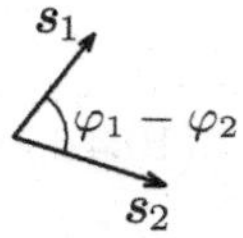


$$\begin{aligned}\mathrm{J}_0(k|\boldsymbol{s}_1 - \boldsymbol{s}_2|) &= \mathrm{J}_0\left(k\sqrt{s_1{}^2 + s_2{}^2 - 2s_1s_2\cos(\varphi_1 - \varphi_2)}\right)\\ &= \sum_{m=-\infty}^{\infty} \mathrm{e}^{\mathrm{i}m\varphi_1}\mathrm{J}_m(ks_1)\,\mathrm{e}^{-\mathrm{i}m\varphi_2}\mathrm{J}_m(ks_2)\end{aligned} \tag{10.4.6}$$

to $k = \frac{\omega}{c}\sin\theta$, $s_1 = s_2 = R$, $\varphi_1 - \varphi_2 = \omega_0\tau$,

$$\begin{aligned}\mathrm{J}_0\left(2\frac{\omega}{c}R\sin\theta\,\sin\frac{\omega_0\tau}{2}\right) &= \mathrm{J}_0\left(\frac{\omega}{c}\sin\theta\,\sqrt{R^2 + R^2 - 2R^2\cos(\omega_0\tau)}\right)\\ &= \sum_{m=-\infty}^{\infty} \mathrm{e}^{\mathrm{i}m\omega_0\tau}\,\mathrm{J}_m\left(\frac{\omega}{c}R\sin\theta\right)^2,\end{aligned} \tag{10.4.7}$$

which yields first

$$\begin{aligned}\frac{\mathrm{d}P(\omega,T)}{\mathrm{d}\Omega} &= \frac{\omega^2e^2}{4\pi^2c}\sum_{m=-\infty}^{\infty}\mathrm{J}_m\left(\frac{\omega}{c}R\sin\theta\right)^2\\ &\quad\times\int\mathrm{d}\tau\;\mathrm{e}^{-\mathrm{i}\omega\tau}\left[\left(\frac{v}{c}\right)^2\frac{\mathrm{e}^{\mathrm{i}\omega_0\tau} + \mathrm{e}^{-\mathrm{i}\omega_0\tau}}{2} - 1\right]\mathrm{e}^{\mathrm{i}m\omega_0\tau}\end{aligned} \tag{10.4.8}$$

and then

$$\frac{\mathrm{d}P(\omega,T)}{\mathrm{d}\Omega} = \frac{\omega^2 e^2}{2\pi c} \sum_{m=-\infty}^{\infty} \mathrm{J}_m\left(\frac{\omega}{c} R\sin\theta\right)^2 \times \left[\frac{1}{2}\left(\frac{v}{c}\right)^2 \left[\delta\big(\omega-(m+1)\omega_0\big)+\delta\big(\omega-(m-1)\omega_0\big)\right] - \delta\big(\omega - m\omega_0\big)\right], \tag{10.4.9}$$

and finally

$$\frac{\mathrm{d}P(\omega,T)}{\mathrm{d}\Omega} = \sum_{m=1}^{\infty} \delta(\omega - m\omega_0)\, \frac{\mathrm{d}P_m(T)}{\mathrm{d}\Omega} \tag{10.4.10}$$

with the power emitted into the $m$th harmonic given by

$$\frac{\mathrm{d}P_m(T)}{\mathrm{d}\Omega} = \frac{m^2 {\omega_0}^2 e^2}{2\pi c} \left[\frac{1}{2}\left(\frac{v}{c}\right)^2 (\mathrm{J}_{m-1}^2 + \mathrm{J}_{m+1}^2) - \mathrm{J}_m^2\right], \tag{10.4.11}$$

where the common argument of the Bessel functions is

$$\frac{m\omega_0}{c} R\sin\theta = m\frac{v}{c}\sin\theta\,. \tag{10.4.12}$$

The summation covers $m = 1, 2, \ldots$, for which $\omega$ is positive, as required by the restriction to positive frequencies adopted in (8.2.2) as a convention.

With the aid of the recurrence relations,

$$\begin{aligned} \mathrm{J}_{m-1}(z) + \mathrm{J}_{m+1}(z) &= \frac{2m}{z}\mathrm{J}_m(z)\,, \\ \mathrm{J}_{m-1}(z) - \mathrm{J}_{m+1}(z) &= 2\frac{\mathrm{d}}{\mathrm{d}z}\mathrm{J}_m(z)\,, \end{aligned} \tag{10.4.13}$$

we can give an alternative form of $\frac{\mathrm{d}P_m}{\mathrm{d}\Omega}$ that only involves the Bessel function of order $m$, see

$$\frac{1}{2}\mathrm{J}_{m-1}(z)^2 + \frac{1}{2}\mathrm{J}_{m+1}(z)^2 = \mathrm{J}_m'(z)^2 + \left(\frac{m}{z}\mathrm{J}_m(z)\right)^2, \tag{10.4.14}$$

giving

$$\frac{\mathrm{d}P_m(T)}{\mathrm{d}\Omega} = m^2 \frac{\omega_0}{2\pi} \frac{e^2}{R} \frac{\omega_0 R}{c} \left(\frac{v}{c}\right)^2 \left[ \mathrm{J}_m'\left(m\frac{v}{c}\sin\theta\right)^2 \right.$$
$$\left. + \left(\left(\frac{1}{\frac{v}{c}\sin\theta}\right)^2 - \frac{1}{(\frac{v}{c})^2}\right) \mathrm{J}_m\left(m\frac{v}{c}\sin\theta\right)^2 \right]$$
$$= m^2 \frac{\omega_0}{2\pi} \frac{e^2}{R} \left(\frac{v}{c}\right)^3 \left[ \mathrm{J}_m'\left(m\frac{v}{c}\sin\theta\right)^2 + \left(\frac{\mathrm{J}_m(m\frac{v}{c}\sin\theta)}{\frac{v}{c}\tan\theta}\right)^2 \right]. \quad (10.4.15)$$

This split of $\frac{\mathrm{d}P_m(T)}{\mathrm{d}\Omega}$ into a sum of two terms has a physical significance that we will recognize below, in Section 10.7.

For the calculation of the total power radiated into the $m$th harmonic we could integrate $\frac{\mathrm{d}P_m}{\mathrm{d}\Omega}$ over the solid angle, but it is easier to return to (10.3.4) and extract the contribution of the $m$th harmonic from this expression for $P(\omega, T)$, which is already integrated over the solid angle. What multiplies $\mathrm{e}^{-\mathrm{i}\omega\tau}$ under the $\tau$ integral is a periodic function of $\omega_0\tau$,

$$\left[\left(\frac{v}{c}\right)^2 \cos(\omega_0\tau) - 1\right] \frac{\sin\left(2\frac{\omega}{c}R\sin\frac{\omega_0\tau}{2}\right)}{\sin\frac{\omega_0\tau}{2}} = \sum_{m=-\infty}^{\infty} \mathrm{e}^{\mathrm{i}m\omega_0\tau} f_m \quad (10.4.16)$$

with the Fourier coefficients

$$f_m = \int_{(2\pi)} \frac{\mathrm{d}\varphi}{2\pi}\, \mathrm{e}^{-\mathrm{i}m\varphi} \left[\left(\frac{v}{c}\right)^2 \cos\varphi - 1\right] \frac{\sin\left(2\frac{\omega}{c}R\sin\frac{\varphi}{2}\right)}{\sin\frac{\varphi}{2}}, \quad (10.4.17)$$

so that

$$P(\omega, T) = \frac{\omega}{2\pi} \frac{e^2}{R} \int_{-\infty}^{\infty} \mathrm{d}\tau\, \mathrm{e}^{-\mathrm{i}\omega\tau} \sum_{m=-\infty}^{\infty} \mathrm{e}^{\mathrm{i}m\omega_0\tau} f_m$$
$$= \sum_{m=1}^{\infty} \delta(\omega - m\omega_0) P_m(T) \quad (10.4.18)$$

with

$$P_m(T) = m\omega_0 \frac{e^2}{R} f_m\,, \quad (10.4.19)$$

where $\frac{\omega}{c}R \to m\frac{\omega_0}{c}R = m\frac{v}{c}$ in the above expression for $f_m$. As we did in the context of (10.4.11), here too we note that only positive $\omega$ values count, and thus the summation is over $m = 1, 2, \ldots$ only, excluding $m = 0, -1, -2, \ldots$.

We expect that it will be possible to express $f_m$ in terms of Bessel functions and, therefore, take a look at

$$J_m(z) = i^{-m} \int\limits_{(2\pi)} \frac{d\varphi}{2\pi}\, e^{-im\varphi}\, e^{iz\cos\varphi}\,. \tag{10.4.20}$$

We substitute $\varphi \to \varphi - \frac{\pi}{2}$ and take the real part explicitly,

$$\begin{aligned} J_m(z) &= \mathrm{Re} \int\limits_{(2\pi)} \frac{d\varphi}{2\pi}\, e^{-im\varphi}\, e^{iz\sin\varphi} \\ &= \int\limits_0^\pi \frac{d\varphi}{\pi} \cos(z\sin\varphi - m\varphi)\,, \end{aligned} \tag{10.4.21}$$

recognizing that the contribution from $-\pi < \varphi < 0$ is exactly the same as from $0 < \varphi < \pi$. Next, we apply the identity

$$\int\limits_0^\pi d\varphi\, f(\varphi) = \int\limits_0^{\pi/2} d\varphi\, [f(\varphi) + f(\pi - \varphi)] \tag{10.4.22}$$

to get

$$\begin{aligned} J_m(z) &= \int\limits_0^{\pi/2} \frac{d\varphi}{\pi} [\cos(z\sin\varphi - m\varphi) + (-1)^m \cos(z\sin\varphi + m\varphi)] \\ &= \int\limits_0^{\pi/2} \frac{d\varphi}{\pi/2} \begin{Bmatrix} \cos(z\sin\varphi)\cos(m\varphi) & \text{for } m \text{ even} \\ \sin(z\sin\varphi)\sin(m\varphi) & \text{for } m \text{ odd} \end{Bmatrix}. \end{aligned} \tag{10.4.23}$$

For the present purpose we need ($\varphi \to \frac{1}{2}\varphi$ substituted)

$$J_{2m}(z) = \int\limits_0^\pi \frac{d\varphi}{\pi} \cos(m\varphi) \cos\left(z \sin\frac{\varphi}{2}\right), \tag{10.4.24}$$

or rather its derivative

$$J'_{2m}(z) = -\int_0^\pi \frac{d\varphi}{\pi} \cos(m\varphi) \sin\left(z \sin\frac{\varphi}{2}\right) \sin\frac{\varphi}{2} \tag{10.4.25}$$

and its antiderivative

$$\int_0^z dz'\, J_{2m}(z') = \int_0^\pi \frac{d\varphi}{\pi} \cos(m\varphi) \frac{\sin\left(z\sin\frac{\varphi}{2}\right)}{\sin\frac{\varphi}{2}}, \tag{10.4.26}$$

in (10.4.17) with $\frac{\omega}{c}R \to m\frac{v}{c}$,

$$\begin{aligned} f_m &= \int_0^\pi \frac{d\varphi}{\pi} \cos(m\varphi) \underbrace{\left[\left(\frac{v}{c}\right)^2 \cos\varphi - 1\right]}_{=(\frac{v}{c})^2 - 1 - 2(\frac{v}{c})^2(\sin\frac{\varphi}{2})^2} \frac{\sin\left(2m\frac{v}{c}\sin\frac{\varphi}{2}\right)}{\sin\frac{\varphi}{2}} \\ &= 2\left(\frac{v}{c}\right)^2 J'_{2m}\left(2m\frac{v}{c}\right) - \left[1 - \left(\frac{v}{c}\right)^2\right] \int_0^{2m\frac{v}{c}} dz\, J_{2m}(z). \end{aligned} \tag{10.4.27}$$

This gives

$$P_m(T) = m\omega_0 \frac{e^2}{R}\left[2\left(\frac{v}{c}\right)^2 J'_{2m}\left(2m\frac{v}{c}\right) - \frac{1}{\gamma^2}\int_0^{2m\frac{v}{c}} dz\, J_{2m}(z)\right] \tag{10.4.28}$$

for the power radiated into the $m$th harmonic. By summing over $m$, we could rederive the total power of (10.3.15),

$$P(T) = \sum_{m=1}^{\infty} P_m(T) = \frac{2}{3}\frac{e^2}{R}\omega_0 \left(\frac{v}{c}\right)^3 \gamma^4, \tag{10.4.29}$$

and all who want to develop their skills in handling Bessel functions should rise to this challenge.

## 10.5 High harmonics

Synchrotrons are operated at parameter values such that the electrons move at relativistic speeds $\frac{v}{c} \lesssim 1$, $\gamma \gg 1$, and under these circumstances one has

$$P_m(T) \cong 2m\omega_0 \frac{e^2}{R} \mathrm{J}'_{2m}(2m) \tag{10.5.1}$$

for the power radiated into the $m$th harmonic and

$$\frac{2\pi}{\omega_0} P_m(T) \cong \frac{e^2}{R} 4\pi\, m\mathrm{J}'_{2m}(2m) \tag{10.5.2}$$

for the energy radiated into the $m$th harmonic per period of the circular motion. What can we say about $\mathrm{J}'_{2m}(2m)$ for large values of $m$?

We return to (10.4.21) and note that it gives

$$\begin{aligned} \mathrm{J}'_m(z) &= -\int_0^\pi \frac{\mathrm{d}\varphi}{\pi} \sin\varphi\ \sin(z\sin\varphi - m\varphi) \\ &= -\mathrm{Im} \int_0^\pi \frac{\mathrm{d}\varphi}{\pi} \sin\varphi\ \mathrm{e}^{\mathrm{i}(z\sin\varphi\, -\, m\varphi)} \end{aligned} \tag{10.5.3}$$

for the derivative of $\mathrm{J}_m(z)$. Now, if $m \gg 1$, the main contribution to

$$\mathrm{J}'_m(m) = -\mathrm{Im} \int_0^\pi \frac{\mathrm{d}\varphi}{\pi} \sin\varphi\ \mathrm{e}^{\mathrm{i}m(\sin\varphi\, -\, \varphi)} \tag{10.5.4}$$

stems from the vicinity of $\varphi = 0$ because that is the only point of stationary phase of the rapidly oscillating exponential factor. We thus approximate the integrand in accordance with

$$\sin\varphi\ \mathrm{e}^{\mathrm{i}m(\sin\varphi\, -\, \varphi)} \cong \varphi\,\mathrm{e}^{-\mathrm{i}m\varphi^3/6} \tag{10.5.5}$$

and extend the range of integration to infinity,

$$\begin{aligned} \mathrm{J}'_m(m) &\cong -\mathrm{Im} \int_0^\infty \frac{\mathrm{d}\varphi}{\pi} \varphi\,\mathrm{e}^{-\mathrm{i}m\varphi^3/6} \\ &\overset{(t = \frac{1}{3}\varphi^3)}{=} -\mathrm{Im} \int_0^\infty \frac{\mathrm{d}t}{\pi} (3t)^{-1/3}\, \mathrm{e}^{-\mathrm{i}\frac{m}{2}t}\,, \end{aligned} \tag{10.5.6}$$

and then make use of Euler's* factorial integral,

$$\int_0^\infty \mathrm{d}t\; t^\nu\,\mathrm{e}^{-xt} = \frac{\nu!}{x^{\nu+1}}\,, \tag{10.5.7}$$

for $x = \mathrm{i}\dfrac{m}{2}$ and $\nu = -\dfrac{1}{3}$ to arrive at

$$\begin{aligned}\mathrm{J}'_m(m) &\cong -\mathrm{Im}\frac{3^{-1/3}}{\pi}\left(\mathrm{i}\frac{m}{2}\right)^{-2/3}\left(-\frac{1}{3}\right)! \\ &= \underbrace{-\mathrm{Im}\left(\mathrm{e}^{\mathrm{i}\frac{\pi}{2}}\right)^{-2/3}}_{=-\mathrm{Im}\,\mathrm{e}^{-\mathrm{i}\frac{\pi}{3}}\,=\,\sin(\pi/3)\,=\,\sqrt{3/4}}\frac{3^{-1/3}\left(-\frac{1}{3}\right)!}{\pi}\left(\frac{m}{2}\right)^{-2/3}\,,\end{aligned} \tag{10.5.8}$$

so that

$$\mathrm{J}'_{2m}(2m) \cong \frac{3^{1/6}\left(-\frac{1}{3}\right)!}{2\pi}\,m^{-2/3} = 0.2588\,m^{-2/3} \tag{10.5.9}$$

for large $m$ values, where the numerical value $\left(-\frac{1}{3}\right)! = 1.35412$ enters.

With this approximation, we have

$$\frac{2\pi}{\omega_0}P_m(T) \cong \frac{e^2}{R}2\cdot 3^{1/6}\left(-\frac{1}{3}\right)!\,m^{1/3} \tag{10.5.10}$$

for the energy emitted into the $m$th harmonic during one period, which cannot be correct, however, for all large $m$ values because the sum

$$\sum_{m=1}^{\infty}\frac{2\pi}{\omega_0}P_m(T) = \frac{e^2}{R}\frac{4\pi}{3}\left(\frac{v}{c}\right)^3\gamma^4 \cong \frac{e^2}{R}\frac{4\pi}{3}\gamma^4 \tag{10.5.11}$$

is finite. We shall, therefore, regard the approximation as reliable only for $m$ values that do not exceed a *critical order* $m_{\mathrm{crit}}$, such that

$$\frac{4\pi}{3}\gamma^4 \cong \sum_{m=1}^{m_{\mathrm{crit}}} 2\cdot 3^{1/6}\left(-\frac{1}{3}\right)!\,m^{1/3} \cong \frac{3}{2}3^{1/6}\left(-\frac{1}{3}\right)!\,m_{\mathrm{crit}}^{4/3}\,. \tag{10.5.12}$$

In view of $\dfrac{8\pi}{3^{13/6}(-\frac{1}{3})!} = 1.7172 = (1.5001)^{4/3}$, this gives us the estimate

$$m_{\mathrm{crit}} \cong \frac{3}{2}\gamma^3 \tag{10.5.13}$$

with no real justification, of course, for taking the $\frac{3}{2}$ factor seriously.

*Leonhard EULER (1707–1783)

The main lesson of these considerations is the insight that there is substantial synchrotron radiation up to the $m_{\text{crit}}$th harmonic, which is $m_{\text{crit}} \cong \gamma^3 \cong 20^3 = 8000$ for a modest electron energy of $E = 10\,\text{MeV}$.

## 10.6 "Roll over"

The approximation we found is clearly breaking down for extremely large $m$ values, and the reason for this break down is that the small deviation of $\frac{v}{c}$ from unity, multiplied by a large $m$ value results in $m\frac{v}{c}$ differing from $m$ by a sizeable amount, in the arguments of the Bessel functions in (10.4.28). We thus reconsider $\mathrm{J}'_{2m}\left(2m\frac{v}{c}\right)$, but now we put

$$2\frac{v}{c} = 2 - 2\left(1-\frac{v}{c}\right) \cong 2 - \left(1+\frac{v}{c}\right)\left(1-\frac{v}{c}\right) = 2 - \frac{1}{\gamma^2}\,, \qquad (10.6.1)$$

rather than $2\frac{v}{c} \cong 2$ as before. Then

$$\begin{aligned}
\mathrm{J}'_{2m}\left(2m\frac{v}{c}\right) &= -\int_0^\pi \frac{\mathrm{d}\varphi}{\pi}\,\sin\varphi\,\sin\left(2m\frac{v}{c}\sin\varphi - 2m\varphi\right) \\
&\cong -\int_0^\pi \frac{\mathrm{d}\varphi}{\pi}\,\varphi\sin\left(\left(2m-\frac{m}{\gamma^2}\right)\left(\varphi-\frac{1}{6}\varphi^3\right)-2m\varphi\right) \\
&= \int_0^\infty \frac{\mathrm{d}\varphi}{\pi}\,\varphi\sin\left(\frac{m}{\gamma^2}\varphi+\frac{m}{3}\varphi^3\right)\,, \qquad (10.6.2)
\end{aligned}$$

where we have consistently taken into account that large $\varphi$ values do not contribute significantly and only the vicinity of $\varphi = 0$ matters. The substitution $\varphi = x/\gamma$, in conjunction with observing that the integrand is even in $\varphi$ (or $x$), then gives

$$\mathrm{J}'_{2m}\left(2m\frac{v}{c}\right) \cong \mathrm{Im}\int_{-\infty}^{\infty} \frac{\mathrm{d}x\,x}{2\pi\gamma^2}\,\mathrm{e}^{\mathrm{i}m(x+\frac{1}{3}x^3)/\gamma^3}\,. \qquad (10.6.3)$$

Here, the rapidly oscillating exponential function has two points of stationary phase, solutions to

$$\frac{\mathrm{d}}{\mathrm{d}x}\left(x+\frac{1}{3}x^3\right) = 1 + x^2 = 0\,, \qquad (10.6.4)$$

namely at $x = \mathrm{i}$ and $x = -\mathrm{i}$, but the solution $x = -\mathrm{i}$ does not count because for $x = -\mathrm{i} + y$ the exponent is increasing rapidly with $y$ rather than decreasing. So, the stationary point at $x = \mathrm{i}$ will give us the value of the integral for $m/\gamma^3 \gg 1$.

We note in passing that the earlier result for $m/\gamma^3 \ll 1$ is contained in (10.6.3) as well. In this case, the main contribution arises from very large $x$ so that $x^3$ dominates over $x$, and we get back to (10.5.6).

Now, for $m \gg \gamma^3$ we write $x = \mathrm{i} + y$, $\mathrm{d}x = \mathrm{d}y$, and

$$x + \frac{1}{3}x^3 = \frac{2\mathrm{i}}{3} + \mathrm{i}y^2, \tag{10.6.5}$$

to get

$$\begin{aligned} \mathrm{J}'_{2m}\left(2m\frac{v}{c}\right) &\cong \mathrm{Im}\int_{-\infty}^{\infty} \frac{\mathrm{d}y\,\mathrm{i}}{2\pi\gamma^2}\,\mathrm{e}^{-\frac{2}{3}m/\gamma^3}\,\mathrm{e}^{-my^2/\gamma^3} \\ &= \frac{1}{2\pi\gamma^2}\,\mathrm{e}^{-\frac{2}{3}m/\gamma^3}\sqrt{\frac{\pi\gamma^3}{m}}\,, \end{aligned} \tag{10.6.6}$$

and the first term in (10.4.28) gives

$$P_m(T) \cong \omega_0\frac{e^2}{R}\sqrt{\frac{m}{\pi\gamma}}\,\mathrm{e}^{-\frac{2}{3}m/\gamma^3}\,. \tag{10.6.7}$$

In summary,

$$\begin{aligned} m < m_{\mathrm{crit}} \cong \gamma^3 &: P_m(T) \propto m^{\frac{1}{3}} = \gamma\left(\frac{m}{\gamma^3}\right)^{\frac{1}{3}}, \\ m > m_{\mathrm{crit}} \cong \gamma^3 &: P_m(T) \propto \left(\frac{m}{\gamma}\right)^{\frac{1}{2}}\mathrm{e}^{-\frac{2}{3}m/\gamma^3} = \gamma\left(\frac{m}{\gamma^3}\right)^{\frac{1}{2}}\mathrm{e}^{-\frac{2}{3}m/\gamma^3}, \end{aligned} \tag{10.6.8}$$

gives this rough picture:

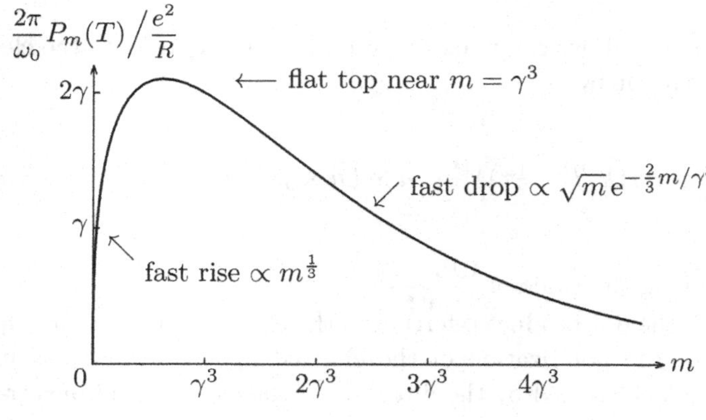


(10.6.9)

for the power emitted into the $m$th harmonic of synchrotron radiation during one period of duration $\frac{2\pi}{\omega_0}$, in units of $\frac{e^2}{R}$. We observe a fast rise $\propto m^{1/3}$ for small $m$ values, a plateau on the approach to $m_{\text{crit}} \sim \gamma^3$, then a fast drop after passing the so-called "roll-over" point.

The figure in (10.6.9) is drawn with an expression for the right-hand side of (10.6.3) that interpolates between the small-$m$ regime and the large-$m$ regime; more about this in Exercise 85. It is worth remembering that the plot gives only a "rough" picture because the second term in (10.4.28) is not taken into account. When one accounts for it, the maximum is slightly below $2\gamma$ rather than slightly above, but the general features of the plot remain the same.

## 10.7 Polarization

For a better understanding of the split of $\frac{\mathrm{d}P_m}{\mathrm{d}\Omega}$ in two in (10.4.15), we return to the general expression for $\frac{\mathrm{d}P(\omega,T)}{\mathrm{d}\Omega}$ in (9.1.6), where the product

$$\left[\boldsymbol{n} \times \boldsymbol{j}\left(\boldsymbol{k}, T + \tfrac{1}{2}\tau\right)\right]^* \cdot \left[\boldsymbol{n} \times \boldsymbol{j}\left(\boldsymbol{k}, T - \tfrac{1}{2}\tau\right)\right] \tag{10.7.1}$$

originated in $\boldsymbol{E} \times \boldsymbol{B} = -(\boldsymbol{n} \times \boldsymbol{B}) \times \boldsymbol{B} = \boldsymbol{n}\, \boldsymbol{B} \cdot \boldsymbol{B}$, but since

$$\boldsymbol{B} \cdot \boldsymbol{B} = (\boldsymbol{n} \times \boldsymbol{B}) \cdot (\boldsymbol{n} \times \boldsymbol{B}) = \boldsymbol{E} \cdot \boldsymbol{E} \tag{10.7.2}$$

in the radiation field, we can make the product (10.7.1) refer to the electric field by replacing it by

$$\left[\boldsymbol{n} \times \left(\boldsymbol{n} \times \boldsymbol{j}\left(\boldsymbol{k}, T + \tfrac{1}{2}\tau\right)\right)\right]^* \cdot \left[\boldsymbol{n} \times \left(\boldsymbol{n} \times \boldsymbol{j}\left(\boldsymbol{k}, T - \tfrac{1}{2}\tau\right)\right)\right] \tag{10.7.3}$$

without changing the value of $\frac{\mathrm{d}P(\omega,T)}{\mathrm{d}\Omega}$.

Now, with the dot product referring to the electric field, we can analyze it in terms of the polarization of the emitted light — where, as usual, polarization is determined by the direction of the electric field, not that of the magnetic field. Recall the geometry defined in (10.1.10),

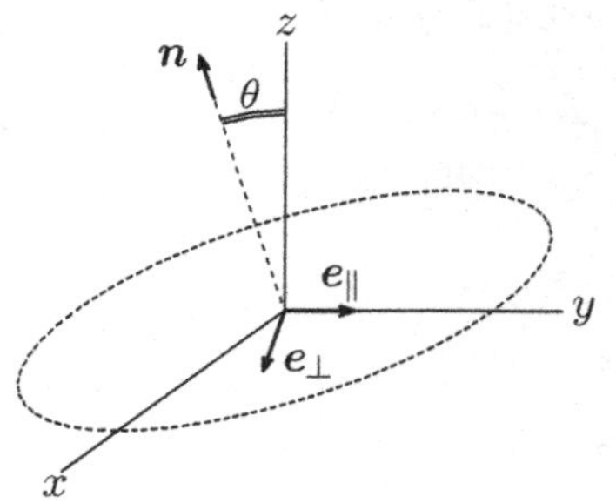


$$e_\perp \times e_\parallel = n \tag{10.7.4}$$

which we supplement by $e_\parallel = \begin{pmatrix} 0 \\ 1 \\ 0 \end{pmatrix}$ for the polarization vector in the $x, y$ plane and $e_\perp = \begin{pmatrix} \cos\theta \\ 0 \\ -\sin\theta \end{pmatrix}$ for the polarization perpendicular to $e_\parallel$. The three vectors — $n$ for propagation and $e_\parallel$, $e_\perp$ for polarization — are, of course, pairwise orthogonal.

Since $n \times (n \times j)$ is surely perpendicular to the direction of $n$, it can be decomposed into components referring to $e_\parallel$ and $e_\perp$,

$$n \times \left(n \times j\left(k, T \pm \tfrac{1}{2}\tau\right)\right) = [\text{`}\pm\text{'}]_\parallel + [\text{`}\pm\text{'}]_\perp \tag{10.7.5}$$

so that

$$\begin{aligned} &\left[n \times \left(n \times j\left(k, T + \tfrac{1}{2}\tau\right)\right)\right]^* \cdot \left[n \times \left(n \times j\left(k, T - \tfrac{1}{2}\tau\right)\right)\right] \\ &= [\text{`}+\text{'}]_\parallel^* \cdot [\text{`}-\text{'}]_\parallel + [\text{`}+\text{'}]_\perp^* \cdot [\text{`}-\text{'}]_\perp \end{aligned} \tag{10.7.6}$$

identifies the contributions of parallel (in plane) and perpendicular (out of plane) polarization.

We turn our attention to the in-plane polarization and note that

$$\begin{aligned} \left[n \times \left(n \times j\left(k, T \pm \tfrac{1}{2}\tau\right)\right)\right]_\parallel &= [n n \cdot j - j]_\parallel \\ &= [n n \cdot j - j] \cdot e_\parallel = -j \cdot e_\parallel \\ &= -j_y\left(k, T \pm \tfrac{1}{2}\tau\right), \end{aligned} \tag{10.7.7}$$

and with

$$\begin{aligned} j_y(\boldsymbol{k},t) &= \int (\mathrm{d}\boldsymbol{r}')\,\mathrm{e}^{-\mathrm{i}\boldsymbol{k}\cdot\boldsymbol{r}'} ev\cos(\omega_0 t)\,\delta(z') \\ &\qquad \times \delta\big(x' - R\cos(\omega_0 t)\big)\,\delta\big(y' - R\sin(\omega_0 t)\big) \\ &= ev\cos(\omega_0 t)\,\mathrm{e}^{-\mathrm{i}kR\sin\theta\,\cos(\omega_0 t)} \end{aligned} \tag{10.7.8}$$

we get $(k = \frac{\omega}{c})$

$$\begin{aligned} \left(\frac{\mathrm{d}P(\omega,T)}{\mathrm{d}\Omega}\right)_{\parallel} &= \frac{\omega^2 e^2 v^2}{4\pi^2 c^3}\int \mathrm{d}\tau\,\mathrm{e}^{-\mathrm{i}\omega\tau} \\ &\qquad \times \mathrm{e}^{\mathrm{i}\frac{\omega}{c}R\sin\theta\left[\cos(\omega_0 T + \frac{1}{2}\omega_0\tau) - \cos(\omega_0 T - \frac{1}{2}\omega_0\tau)\right]} \\ &\qquad \times \cos\big(\omega_0 T + \tfrac{1}{2}\omega_0\tau\big)\cos\big(\omega_0 T - \tfrac{1}{2}\omega_0\tau\big). \end{aligned} \tag{10.7.9}$$

It is time to recall once more that the epoch time $T$ is not defined, nor definable, on the scale set by the period of the circular orbit, and so we average over one period, just like we did earlier in (10.4.1). Here, we carry out this average after first noting that

$$\mathrm{e}^{\pm\mathrm{i}\frac{\omega}{c}R\sin\theta\cos(\omega_0 T \pm \frac{1}{2}\omega_0\tau)} = \sum_{m=-\infty}^{\infty} (\pm\mathrm{i})^m\,\mathrm{e}^{\pm\mathrm{i}m(\omega_0 T \pm \frac{1}{2}\omega_0\tau)}\,\mathrm{J}_m\Big(\frac{\omega}{c}R\sin\theta\Big) \tag{10.7.10}$$

giving

$$\begin{aligned} \left(\frac{\mathrm{d}P(\omega,T)}{\mathrm{d}\Omega}\right)_{\parallel} &= \frac{\omega^2 e^2}{4\pi^2 c}\Big(\frac{v}{c}\Big)^2 \sum_{m,m'} \mathrm{i}^{m-m'}\,\mathrm{J}_m\Big(\frac{\omega}{c}R\sin\theta\Big)\,\mathrm{J}_{m'}\Big(\frac{\omega}{c}R\sin\theta\Big) \\ &\qquad \times \int \mathrm{d}\tau\;\mathrm{e}^{-\mathrm{i}\omega\tau}\,\mathrm{e}^{\mathrm{i}(m+m')\frac{1}{2}\omega_0\tau}\,\mathrm{e}^{\mathrm{i}(m-m')\omega_0 T} \\ &\qquad \times \left[\frac{1}{4}\,\mathrm{e}^{2\mathrm{i}\omega_0 T} + \frac{1}{4}\,\mathrm{e}^{-2\mathrm{i}\omega_0 T} + \frac{1}{2}\cos(\omega_0\tau)\right], \end{aligned} \tag{10.7.11}$$

where

$$\cos\big(\omega_0 T + \tfrac{1}{2}\omega_0\tau\big)\cos\big(\omega_0 T - \tfrac{1}{2}\omega_0\tau\big) = \frac{1}{4}\,\mathrm{e}^{2\mathrm{i}\omega_0 T} + \frac{1}{4}\,\mathrm{e}^{-2\mathrm{i}\omega_0 T} + \frac{1}{2}\cos(\omega_0\tau) \tag{10.7.12}$$

has entered as well. The $T$ average then yields

$$\begin{aligned}
&\mathrm{e}^{\mathrm{i}(m-m')\omega_0 T}\left[\frac{1}{4}\,\mathrm{e}^{2\mathrm{i}\omega_0 T}+\frac{1}{4}\,\mathrm{e}^{-2\mathrm{i}\omega_0 T}+\frac{1}{2}\cos(\omega_0\tau)\right]\\
&\to\int\frac{\mathrm{d}\varphi}{2\pi}\,\mathrm{e}^{\mathrm{i}(m-m')\varphi}\left[\frac{1}{4}\,\mathrm{e}^{2\mathrm{i}\varphi}+\frac{1}{4}\,\mathrm{e}^{-2\mathrm{i}\varphi}+\frac{1}{2}\cos(\omega_0\tau)\right]\\
&=\frac{1}{4}\delta_{m+2,m'}+\frac{1}{4}\delta_{m,m'+2}+\frac{1}{2}\delta_{mm'}\cos(\omega_0\tau)\,,
\end{aligned}\tag{10.7.13}$$

so that ($z\equiv\dfrac{\omega}{c}R\sin\theta$)

$$\begin{aligned}
\left(\frac{\mathrm{d}P(\omega,T)}{\mathrm{d}\Omega}\right)_{\parallel}=\frac{\omega^2e^2}{4\pi^2c}\left(\frac{v}{c}\right)^2\int\mathrm{d}\tau\;\mathrm{e}^{-\mathrm{i}\omega\tau}\sum_m\Bigg[&\frac{1}{4}\mathrm{i}^{-2}\mathrm{J}_m(z)\mathrm{J}_{m+2}(z)\,\mathrm{e}^{\mathrm{i}(m+1)\omega_0\tau}\\
&+\frac{1}{4}\mathrm{i}^{2}\mathrm{J}_m(z)\mathrm{J}_{m-2}(z)\,\mathrm{e}^{\mathrm{i}(m-1)\omega_0\tau}\\
&+\frac{1}{2}\mathrm{J}_m(z)^2\cos(\omega_0\tau)\,\mathrm{e}^{\mathrm{i}m\omega_0\tau}\Bigg]
\end{aligned}\tag{10.7.14}$$

after the averaging over one orbital period. The $\tau$ integration is now elementary, and we arrive at

$$\left(\frac{\mathrm{d}P(\omega,T)}{\mathrm{d}\Omega}\right)_{\parallel}=\sum_{m=1}^{\infty}\delta(\omega-m\omega_0)\left(\frac{\mathrm{d}P_m(T)}{\mathrm{d}\Omega}\right)_{\parallel}\tag{10.7.15}$$

with

$$\left(\frac{\mathrm{d}P_m(T)}{\mathrm{d}\Omega}\right)_{\parallel}=\frac{\omega_0}{2\pi}\frac{e^2}{R}\underbrace{\frac{\omega_0R}{c}}_{=v/c}\left(\frac{v}{c}\right)^2m^2\underbrace{\left[-\frac{1}{2}\mathrm{J}_{m+1}\mathrm{J}_{m-1}+\frac{1}{4}\mathrm{J}_{m+1}^2+\frac{1}{4}\mathrm{J}_{m-1}^2\right]}_{=\mathrm{J}_m'^2},\tag{10.7.16}$$

where the second identity of (10.4.13) is applied and all Bessel functions have the common argument $m\omega_0\dfrac{R}{c}\sin\theta=m\dfrac{v}{c}\sin\theta$.

In summary, then, the power distribution for parallel, in-plane, polarization for the $m$th harmonic is

$$\left(\frac{\mathrm{d}P_m(T)}{\mathrm{d}\Omega}\right)_{\parallel}=\frac{\omega_0}{2\pi}\frac{e^2}{R}\left(\frac{v}{c}\right)^3m^2\,\mathrm{J}_m'\left(m\frac{v}{c}\sin\theta\right)^2,\tag{10.7.17}$$

and since this is the first term on the right-hand side of (10.4.15), we conclude immediately that

$$\left(\frac{\mathrm{d}P_m(T)}{\mathrm{d}\Omega}\right)_\perp = \frac{\omega_0}{2\pi}\frac{e^2}{R}\left(\frac{v}{c}\right)^3 m^2 \left(\frac{\mathrm{J}_m(m\frac{v}{c}\sin\theta)}{\frac{v}{c}\tan\theta}\right)^2 \tag{10.7.18}$$

is the power distribution for perpendicular, out-of-plane, polarization. Accordingly, we have recognized that the two terms in (10.4.15) refer to the two polarizations.

It can be shown, but we do not take the time to actually show it, that the total powers for the two polarizations,

$$\begin{aligned} P_\parallel(T) &= \sum_m \int \mathrm{d}\Omega \left(\frac{\mathrm{d}P_m(T)}{\mathrm{d}\Omega}\right)_\parallel , \\ P_\perp(T) &= \sum_m \int \mathrm{d}\Omega \left(\frac{\mathrm{d}P_m(T)}{\mathrm{d}\Omega}\right)_\perp , \end{aligned} \tag{10.7.19}$$

whose sum is, of course, the total power

$$P = P_\parallel + P_\perp = \frac{2}{3}\frac{e^2}{R}\omega_0\left(\frac{v}{c}\right)^3\left(\frac{E}{mc^2}\right)^4 \tag{10.7.20}$$

of (10.3.15), are simple fractions of $P$, namely

$$P_\parallel = \frac{6+(v/c)^2}{8}P \quad \text{and} \quad P_\perp = \frac{2-(v/c)^2}{8}P , \tag{10.7.21}$$

so that their ratio

$$\frac{P_\parallel}{P_\perp} = \frac{6+(v/c)^2}{2-(v/c)^2} = \begin{cases} 3 & \text{for} \quad v/c \ll 1 \\ 7 & \text{for} \quad v/c \lesssim 1 \end{cases} \tag{10.7.22}$$

testifies to a strong polarization in the ultra-relativistic limit, and sizeable polarization even for nonrelativistic speeds, when the Larmor formula would apply.

## 10.8 Angular distribution

Let us return to Section 10.2 and recognize that, in view of (10.2.8), the exponents in (10.2.7) and (10.2.11) can be written as

$$\begin{aligned}\boldsymbol{k}\cdot\left[\boldsymbol{r}(T+\tfrac{1}{2}\tau)-\boldsymbol{r}(T-\tfrac{1}{2}\tau)\right] &= \boldsymbol{k}\cdot 2\frac{R}{v}\sin\frac{\omega_0\tau}{2}\,\boldsymbol{v}(T)\\ &= 2\frac{\omega}{\omega_0}\sin\frac{\omega_0\tau}{2}\,\boldsymbol{n}\cdot\frac{\boldsymbol{v}(T)}{c}\,, \qquad (10.8.1)\end{aligned}$$

which facilitates the study of the angular distribution of synchrotron radiation [see also (10.3.2)]. The power radiated into solid angle $\mathrm{d}\Omega$ in the direction specified by unit vector $\boldsymbol{n}$, at epoch time $T$, is then given by the frequency integral of $\dfrac{\mathrm{d}P(\omega,T)}{\mathrm{d}\Omega}$ in (10.2.11),

$$\begin{aligned}\frac{\mathrm{d}P(T)}{\mathrm{d}\Omega} &= \frac{e^2}{4\pi^2 c}\int_{-\infty}^{\infty}\mathrm{d}\tau\left[1-\left(\frac{v}{c}\right)^2\cos(\omega_0\tau)\right]\\ &\quad\times\left(-\frac{1}{2}\right)\int_{-\infty}^{\infty}\mathrm{d}\omega\,\omega^2\,\mathrm{e}^{-\mathrm{i}\omega\tau}\,\mathrm{e}^{2\mathrm{i}\frac{\omega}{\omega_0}\sin\frac{\omega_0\tau}{2}\,\boldsymbol{n}\cdot\boldsymbol{v}(T)/c}\\ &= \frac{e^2}{4\pi c}\int_{-\infty}^{\infty}\mathrm{d}\tau\left[1-\left(\frac{v}{c}\right)^2\cos(\omega_0\tau)\right]\delta''\left(\tau-\frac{2}{\omega_0}\boldsymbol{n}\cdot\frac{\boldsymbol{v}(T)}{c}\sin\frac{\omega_0\tau}{2}\right),\end{aligned}$$

$$(10.8.2)$$

where we recognized that the integrand has a real part that is even in $\tau$ and $\omega$, whereas the imaginary part is odd. We have come across an integral of this structure before, in Section 10.3, and just as we did in the transition from (10.3.8) to (10.3.9) we make use of $\delta''(\tau)=\omega_0^3\delta''(\omega_0\tau)$ and switch to $\varphi=\omega_0\tau$ as the new integration variable,

$$\frac{\mathrm{d}P(T)}{\mathrm{d}\Omega} = \frac{e^2}{4\pi c}\omega_0^2\int_{-\infty}^{\infty}\mathrm{d}\varphi\left[1-\left(\frac{v}{c}\right)^2\cos\varphi\right]\delta''\left(\varphi-2\boldsymbol{n}\cdot\frac{\boldsymbol{v}(T)}{c}\sin\frac{\varphi}{2}\right), \qquad (10.8.3)$$

after which we can rely on the evaluation of this integral in the steps from (10.3.9) to (10.3.13), with $x\frac{v}{c} \to -\boldsymbol{n}\cdot\frac{\boldsymbol{v}(T)}{c}$, and thus arrive at

$$\frac{\mathrm{d}P(T)}{\mathrm{d}\Omega} = \frac{\omega_0}{16\pi}\frac{e^2}{R}\frac{v}{c}\left(\frac{1+3\left(\frac{v}{c}\right)^2}{\left(1-\boldsymbol{n}\cdot\frac{\boldsymbol{v}(T)}{c}\right)^3} - \frac{1-\left(\frac{v}{c}\right)^2}{\left(1-\boldsymbol{n}\cdot\frac{\boldsymbol{v}(T)}{c}\right)^4}\right). \quad (10.8.4)$$

With $\phi$ denoting the angle between $\boldsymbol{n}$ and $\boldsymbol{v}(T)$, $\boldsymbol{n}\cdot\frac{\boldsymbol{v}(T)}{c} = \frac{v}{c}\cos\phi$, the angular distribution of synchrotron radiation is then given by

$$\frac{\mathrm{d}P(T)}{\mathrm{d}\Omega} = \frac{\omega_0}{16\pi}\frac{e^2}{R}\frac{v}{c}\left(\frac{1+3\left(\frac{v}{c}\right)^2}{\left(1-\frac{v}{c}\cos\phi\right)^3} - \frac{1-\left(\frac{v}{c}\right)^2}{\left(1-\frac{v}{c}\cos\phi\right)^4}\right), \quad (10.8.5)$$

and we recognize that this differs in two major respects from the expressions we found for impulsive scattering in (8.4.14) or for bremsstrahlung in (8.5.1): There is no factor $(\sin\phi)^2$ in the numerators of (10.8.5), and rather than squares of $1-\frac{v}{c}\cos\phi$ in the denominators, we have third and fourth powers. As a consequence of the absence of the $(\sin\phi)^2$ factor, the maximum of the radiated power is in the direction $\phi = 0$, that is: in the direction of $\boldsymbol{v}(T)$; and the higher powers in the denominator give a faster drop-off to the side of this direction, and thus a stronger concentration of radiated power near $\phi = 0$.

These observations are illustrated by a look at the typical synchrotron situation of $\gamma \gg 1$, when

$$\frac{v}{c} = 1-\frac{1}{2\gamma^2}, \quad 1-\frac{v}{c}\cos\phi = \frac{1}{2\gamma^2}\left(1+\gamma^2\phi^2\right) \quad (10.8.6)$$

apply for the relevant $\phi$ values, $\phi \ll 1$, as in (8.4.15). Then

$$\frac{\mathrm{d}P(T)}{\mathrm{d}\Omega} = \frac{\omega_0}{\pi}\frac{e^2}{R}\gamma^6\frac{1+2\gamma^2\phi^2}{\left(1+\gamma^2\phi^2\right)^4} = \frac{\omega_0}{\pi}\frac{e^2}{R}\begin{cases}\gamma^6 & \text{for} \quad \phi = 0\,, \\ \frac{1}{2}\gamma^6 & \text{for} \quad \phi \cong \frac{3}{5\gamma}\,, \\ \frac{2}{\phi^6} & \text{for} \quad \frac{1}{\gamma} \ll \phi \ll 1\,,\end{cases} \quad (10.8.7)$$

telling us that the maximal intensity $\propto \gamma^6$ is truly large, and that almost all the power is emitted into a small cone of opening angle $\propto \gamma^{-1}$:

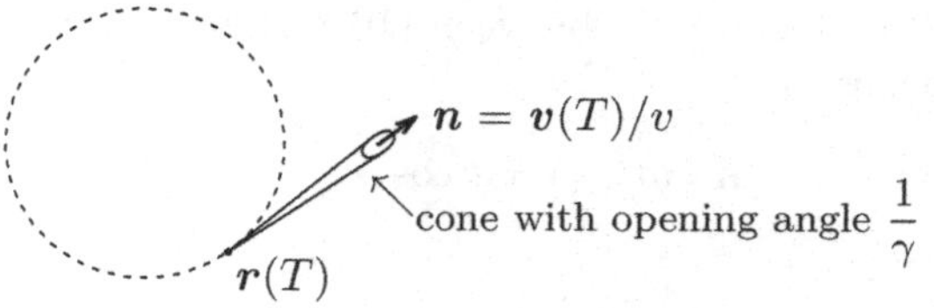


(10.8.8)

A single charge in circular orbit, or a bunch of charges traveling together, thus radiates like a light house: one short and powerful flash of light is observed for each period of the circular motion.

The flash is emitted from an arc of the circular orbit of size $\sim \gamma^{-1}$, which is to say that the epoch time $T$ has an intrinsic uncertainty of the order of $\frac{2\pi}{\gamma\omega_0}$. The instant of emission cannot be specified with any precision better than this, and the average over one orbital period that we applied repeatedly is essentially an average over this short interval during which most of the power is radiated into the cone around $\boldsymbol{n} = \boldsymbol{v}(T)/v$.

## 10.9 Qualitative picture

We close this chapter on synchrotron radiation with a qualitative, semi-quantitative, picture of the situation. For this purpose we return to Section 6.3, where we found

$$\boldsymbol{A}(\boldsymbol{r},t) = \frac{\frac{e}{c}\boldsymbol{v}(t_{\text{ret}})}{\left|\boldsymbol{r} - \boldsymbol{r}(t_{\text{ret}})\right|\left(1 - \frac{\boldsymbol{n}}{c}\cdot\boldsymbol{v}(t_{\text{ret}})\right)} \tag{10.9.1}$$

for the retarded vector potential with

$$\boldsymbol{n} = \frac{\boldsymbol{r} - \boldsymbol{r}(t_{\text{ret}})}{\left|\boldsymbol{r} - \boldsymbol{r}(t_{\text{ret}})\right|}, \tag{10.9.2}$$

the unit vector for the direction of emission that occurs at the retarded time $t_{\text{ret}}$ that is given by

$$t_{\text{ret}} = t - \frac{1}{c}\left|\boldsymbol{r} - \boldsymbol{r}(t_{\text{ret}})\right|. \tag{10.9.3}$$

Here, as we recall, time $t$ is when the radiation is observed at point $\boldsymbol{r}$, and time $t_{\text{ret}}$ is when the radiation is emitted by the charge at $\boldsymbol{r}(t_{\text{ret}})$, a point on the circular orbit.

In a repetition of what we did between (10.8.4) and (10.8.5), we introduce angle $\phi$ by means of

$$\boldsymbol{n} \cdot \boldsymbol{v}(t_{\text{ret}}) = v\cos\phi\,, \tag{10.9.4}$$

and note that, for $v \lesssim c$,

$$\frac{1}{1 - \boldsymbol{n} \cdot \boldsymbol{v}(t_{\text{ret}})/c} \cong \frac{2\gamma^2}{1 + \gamma^2\phi^2}\,, \tag{10.9.5}$$

as in (10.8.6). A relabeled version of figure (10.8.8) is then

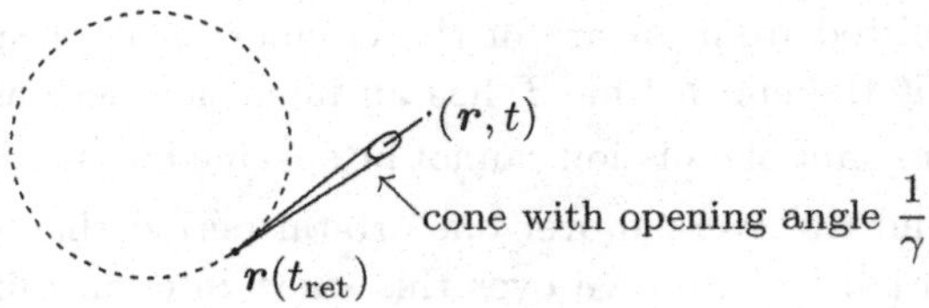


(10.9.6)

As noted at the end of Section 10.8, the light observed at position $\boldsymbol{r}$ at time $t$ is emitted during the short interval $\mathrm{d}t_{\text{ret}} \sim \dfrac{2\pi}{\gamma\omega_0}$, so that typical frequencies are of the order of $\gamma\omega_0$. But this is frequency as measured by *emission time*, whereas *observation time* matters at $\boldsymbol{r}$. Now, since

$$\mathrm{d}t = \mathrm{d}t_{\text{ret}}\left(1 - \frac{1}{c}\boldsymbol{n} \cdot \boldsymbol{v}(t_{\text{ret}})\right) \sim \frac{1}{\gamma^2}\mathrm{d}t_{\text{ret}}\,, \tag{10.9.7}$$

the observed frequency is larger by a factor $\gamma^2$ than the emitted frequency, a manifestation of the Doppler effect under these ultra-relativistic circumstances. In short, the typical observed frequencies are of the order of $\gamma^3\omega_0$, which is reassuringly consistent with our earlier observation, in Section 10.6, about the broad peak in the synchrotron radiation spectrum at $\omega \sim m_{\text{crit}}\omega_0 \sim \gamma^3\omega_0$.

# Chapter 11

# Scattering

We move on to the next subject matter: the scattering of light by charges. A charged particle, exposed to the electric and magnetic fields of incoming radiation, is set in motion and becomes the source of secondary radiation, which appears as scattered light to the observer. We consider two cases: the scattering by a free charge (Thomson* scattering) and the scattering by a bound charge (Rayleigh† scattering).

## 11.1 Thomson scattering

The first scenario that we consider is that of *Thomson scattering*, where a free charge $e$ is exposed to an electric field $\boldsymbol{E}(t)$ (taken at the position of the charge), which gives rise to a force and thus acceleration of the charge, and as a consequence, to radiation emitted by the charge. For the nonrelativistic accelerated motion of the charge, we have Newton's equation of motion

$$m\frac{\mathrm{d}}{\mathrm{d}t}\boldsymbol{v} = e\boldsymbol{E}\,, \tag{11.1.1}$$

which we use in Larmor's formula (7.2.9) to find

$$\begin{aligned}\frac{\mathrm{d}P}{\mathrm{d}\Omega} &= \frac{e^2}{4\pi c^3}\left|\boldsymbol{n}\times\dot{\boldsymbol{v}}\right|^2 = \frac{e^2}{4\pi c^3}\left|\boldsymbol{n}\times\frac{e}{m}\boldsymbol{E}\right|^2\\ &= \frac{e^4}{4\pi m^2c^3}\left|\boldsymbol{n}\times\boldsymbol{E}\right|^2.\end{aligned} \tag{11.1.2}$$

Now, there will clearly be more of the scattered radiation if the electric field is stronger and, therefore, we must normalize the emitted power to

*Sir Joseph John THOMSON (1856–1940)
†John William STRUTT, 3rd Baron Rayleigh (1842–1919)

the incoming energy current density,

$$|\boldsymbol{S}| = \frac{c}{4\pi}|\boldsymbol{E}\times\boldsymbol{B}| = \frac{c}{4\pi}\boldsymbol{E}^2\,, \tag{11.1.3}$$

if we wish, as we do, to characterize the scattering strength of the charge. Accordingly, we write

$$\frac{\mathrm{d}P}{\mathrm{d}\Omega} = \frac{\mathrm{d}\sigma}{\mathrm{d}\Omega}|\boldsymbol{S}| \tag{11.1.4}$$

and so identify the differential scattering cross section $\frac{\mathrm{d}\sigma}{\mathrm{d}\Omega}$, a quantity with the metrical dimension of an area, as the proportionality factor that relates the scattered power $\frac{\mathrm{d}P}{\mathrm{d}\Omega}$ to the incoming energy current density $|\boldsymbol{S}|$.

In the present context of Thomson scattering, this gives the Thomson cross section

$$\frac{\mathrm{d}\sigma_{\mathrm{Th}}}{\mathrm{d}\Omega} = \frac{e^4}{(mc^2)^2}\frac{|\boldsymbol{n}\times\boldsymbol{E}|^2}{|\boldsymbol{E}|^2}\,. \tag{11.1.5}$$

For in-plane scattering, when the unit vector $\boldsymbol{n}$ that specifies the direction into which the light is scattered is coplanar with the wave vector $\boldsymbol{k}$ and the electric field vector $\boldsymbol{E}$ of the incoming plane wave, so that the situation is as illustrated in this figure:

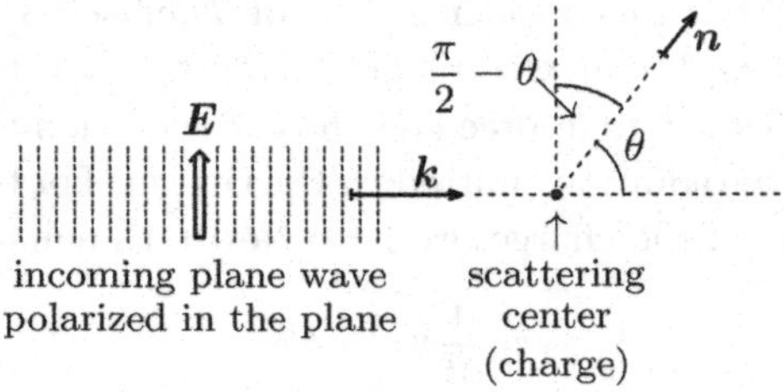


(11.1.6)

we note that $|\boldsymbol{n}\times\boldsymbol{E}|^2 = |\boldsymbol{E}|^2(\cos\theta)^2$ with the scattering angle $\theta$ defined by the projection of $\boldsymbol{k}$ onto $\boldsymbol{n}$, $\boldsymbol{n}\cdot\boldsymbol{k} = k\cos\theta$. If, however, $\boldsymbol{n}$ is perpendicular to the plane defined by $\boldsymbol{E}$ and $\boldsymbol{k}$, we have $|\boldsymbol{n}\times\boldsymbol{E}|^2 = |\boldsymbol{E}|^2$, so that

$$\frac{\mathrm{d}\sigma_{\mathrm{Th}}}{\mathrm{d}\Omega} = \left(\frac{e^2}{mc^2}\right)^2 \begin{cases} (\cos\theta)^2 & \text{for in-plane scattering,} \\ 1 & \text{for out-of-plane scattering,} \end{cases} \tag{11.1.7}$$

and for incoming unpolarized light we get

$$\frac{\mathrm{d}\sigma_{\mathrm{Th}}}{\mathrm{d}\Omega} = \left(\frac{e^2}{mc^2}\right)^2 \frac{1}{2}\left(1+(\cos\theta)^2\right)\,. \tag{11.1.8}$$

For an electron ($e = -e_0$, $m = m_{\rm el}$) as the scattering charge, the ratio $\dfrac{e_0^2}{m_{\rm el}c^2} = r_{\rm cl}$ is the so-called "classical electron radius," a term of historical origin. If you picture — wrongly — the electron as a tiny ball uniformly charged with total charge $-e_0$ and you equate the electrostatic energy of the ball with $m_{\rm el}c^2$, the relativistic rest energy of the electron, then the radius of the ball is of the order of $r_{\rm cl}$.

With this notational convention, then,

$$\frac{{\rm d}\sigma_{\rm Th}}{{\rm d}\Omega} = r_{\rm cl}^2\,\frac{1}{2}\bigl(1+(\cos\theta)^2\bigr) \tag{11.1.9}$$

for unpolarized light. Upon integrating over all directions, we get

$$\sigma_{\rm Th} = \int {\rm d}\Omega\,\frac{{\rm d}\sigma_{\rm Th}}{{\rm d}\Omega} = \frac{8\pi}{3}r_{\rm cl}^2 \tag{11.1.10}$$

for the total Thomson cross section.

Of course, we could have obtained the same result from Larmor's formula (7.2.11) for the total radiated power, see

$$\begin{aligned} P &= \frac{2}{3}\frac{e^2}{c^3}\bigl|\dot{\boldsymbol{v}}\bigr|^2 = \frac{2}{3}\frac{e^2}{c^3}\Bigl|\frac{e}{m}\boldsymbol{E}\Bigr|^2 \\ &= \frac{2}{3}\left(\frac{e^2}{mc^2}\right)^2 c\bigl|\boldsymbol{E}\bigr|^2 \\ &\overset{(e=-e_0,\,m=m_{\rm el})}{=} \frac{2}{3}r_{\rm cl}^2\,4\pi\bigl|\boldsymbol{S}\bigr| = \sigma_{\rm Th}\bigl|\boldsymbol{S}\bigr|\,. \end{aligned} \tag{11.1.11}$$

In this context, it is worth noting the numerical value of $r_{\rm cl}$,

$$\begin{aligned} r_{\rm cl} &= \frac{e_0^2}{m_{\rm el}c^2} = \frac{e_0^2}{\hbar c}\frac{\hbar c}{m_{\rm el}c^2} = \frac{1}{137}\,\frac{1.97\times10^{-11}\,{\rm MeV\,cm}}{0.511\,{\rm MeV}} \\ &= 2.8\times10^{-13}\,{\rm cm} = 2.8\times10^{-5}\,\text{Å}\,, \end{aligned} \tag{11.1.12}$$

a tiny fraction of the size of an atom and roughly the radius of a nucleus.

## 11.2 Rayleigh scattering

If the electron that scatters the radiation is bound, rather than free, we speak of *Rayleigh scattering.* We take the binding potential as an oscillator potential with natural frequency $\omega_0$, and account for a damping force by a simple friction term,

$$m_{\text{el}}\frac{\mathrm{d}^2}{\mathrm{d}t^2}\boldsymbol{r} = -m_{\text{el}}\omega_0^2\boldsymbol{r} - m_{\text{el}}\gamma\frac{\mathrm{d}}{\mathrm{d}t}\boldsymbol{r} - e_0\boldsymbol{E}(t) \tag{11.2.1}$$

with

$$\boldsymbol{E}(t) = \boldsymbol{E}_0\cos(\omega t) = \mathrm{Re}\Big(\boldsymbol{E}_0\,\mathrm{e}^{-\mathrm{i}\omega t}\Big)\,. \tag{11.2.2}$$

Although this is undoubtedly a very simple model dynamics for a bound electron, it is nevertheless a good description provided that the dynamics is dominated by a single intrinsic frequency of the material in question, which is usually the case.

After an initial period, the electron will oscillate with the imposed frequency $\omega$, so that

$$\boldsymbol{r}(t) = \mathrm{Re}\Big(\boldsymbol{r}_0\,\mathrm{e}^{-\mathrm{i}\omega t}\Big)\,, \tag{11.2.3}$$

and we find

$$\boldsymbol{r}_0 = \frac{-\dfrac{e_0}{m_{\text{el}}}\boldsymbol{E}_0}{\omega_0^2 - \omega^2 - \mathrm{i}\gamma\omega} \tag{11.2.4}$$

for the time-independent vector $\boldsymbol{r}_0$. Then we have

$$\boldsymbol{v}(t) = \mathrm{Re}\Big(-\mathrm{i}\omega\boldsymbol{r}_0\,\mathrm{e}^{-\mathrm{i}\omega t}\Big) \tag{11.2.5}$$

for the velocity and

$$\dot{\boldsymbol{v}}(t) = \mathrm{Re}\Big(-\omega^2\boldsymbol{r}_0\,\mathrm{e}^{-\mathrm{i}\omega t}\Big) \tag{11.2.6}$$

for the acceleration. We average $\left|\dot{\boldsymbol{v}}\right|^2$ over one period of the oscillation with the aid of

$$f(t) = \mathrm{Re}\Big(f_0\,\mathrm{e}^{-\mathrm{i}\omega t}\Big) \to \left|f(t)\right|^2 = \frac{1}{4}\big(f_0^2\,\mathrm{e}^{-2\mathrm{i}\omega t} + f_0^{*2}\,\mathrm{e}^{2\mathrm{i}\omega t} + 2f_0^* f_0\big)$$
$$\xrightarrow[\text{one period}]{\text{average over}} \frac{1}{2}\left|f_0\right|^2 \tag{11.2.7}$$

to obtain first

$$\left|\dot{\boldsymbol{v}}\right|^2 \to \frac{1}{2}\left|\frac{-\omega^2\dfrac{e_0}{m_{\text{el}}}\boldsymbol{E}_0}{\omega_0^2 - \omega^2 - \mathrm{i}\gamma\omega}\right|^2 \tag{11.2.8}$$

and then, from the analog of (11.1.4),

$$P = \frac{2}{3}\frac{e_0^2}{c^3}\left|\dot{\boldsymbol{v}}\right|^2 = \sigma_{\text{Ray}} \underbrace{\frac{c}{4\pi}\frac{1}{2}\left|\boldsymbol{E}_0\right|^2}_{\substack{\text{time-average} \\ \text{of } |\boldsymbol{S}|}}, \tag{11.2.9}$$

the Rayleigh cross section

$$\sigma_{\text{Ray}} = \frac{8\pi}{3}\left(\frac{e_0^2}{m_{\text{el}}c^2}\right)^2 \frac{\omega^4}{(\omega^2-\omega_0^2)^2+(\gamma\omega)^2} \tag{11.2.10}$$

or

$$\sigma_{\text{Ray}} = \sigma_{\text{Th}} \frac{\omega^4}{(\omega^2-\omega_0^2)^2+(\gamma\omega)^2}, \tag{11.2.11}$$

upon recognizing that the prefactors compose the Thomson cross section of (11.1.10).

When the imposed frequency is very large compared with the intrinsic frequency and the damping rate, $\omega \gg \omega_0, \gamma$, the internal dynamics cease to matter and the Rayleigh cross section reduces to the Thomson cross section,

$$\sigma_{\text{Ray}} \cong \sigma_{\text{Th}} \quad \text{for} \quad \omega \gg \omega_0, \gamma\,, \tag{11.2.12}$$

because the electron is essentially moving freely on the short time scale associated with the very high frequency $\omega$. The other limiting case is that of a low frequency, or long wavelength of the incoming radiation, $\omega \ll \omega_0, \gamma$, for which

$$\sigma_{\text{Ray}} \cong \sigma_{\text{Th}} \left(\frac{\omega}{\omega_0}\right)^4 \quad \text{for} \quad \omega \ll \omega_0\,. \tag{11.2.13}$$

This is, for example, the situation of scattering visible light off atoms, where the $\omega^4$ dependence implies that blue light is scattered much more efficiently than red light — one of the reason why the sky is blue, and the setting sun is red.

The radiation-reaction force, familiar from Exercises 61 and 62,

$$\boldsymbol{F}_{\text{rad}} = \frac{2}{3}\frac{e_0^2}{c^3}\ddot{\boldsymbol{v}}\,, \tag{11.2.14}$$

becomes

$$\boldsymbol{F}_{\text{rad}} = -\frac{2}{3}\frac{e_0^2}{c^3}\omega^2 \boldsymbol{v} \tag{11.2.15}$$

if the motion is periodic with (angular) frequency $\omega$. When presenting this in the form of a friction force,

$$\boldsymbol{F}_{\mathrm{rad}} = -m_{\mathrm{el}}\gamma_{\mathrm{rad}}\boldsymbol{v}\,, \tag{11.2.16}$$

we identify the radiation-reaction contribution

$$\gamma_{\mathrm{rad}} = \frac{2}{3}\frac{e_0^2}{m_{\mathrm{el}}c^3}\omega^2 = \frac{2}{3}r_{\mathrm{cl}}\frac{\omega^2}{c} \tag{11.2.17}$$

to the total friction constant $\gamma = \gamma_{\mathrm{rad}} + \gamma_{\mathrm{diss}} > \gamma_{\mathrm{rad}}$, where $\gamma_{\mathrm{diss}}$ summarizes the effect of dissipative processes, such as collisions between the light-scattering atoms.

We calculate the mechanical power $\boldsymbol{F}\cdot\boldsymbol{v} = -e_0\boldsymbol{E}\cdot\boldsymbol{v}$ of the energy transferred from the electric field to the electron, time-averaged over a period of the oscillation, from $\boldsymbol{E}(t)$ as in (11.2.2) and the corresponding $\boldsymbol{v}(t)$ given by

$$\boldsymbol{v}(t) = \mathrm{Re}\Bigl(-\frac{e_0}{m_{\mathrm{el}}}\boldsymbol{E}_0\,\mathrm{e}^{-\mathrm{i}\omega t}\frac{-\mathrm{i}\omega}{\omega_0^2-\omega^2-\mathrm{i}\gamma\omega}\Bigr) \tag{11.2.18}$$

with the aid of

$$\begin{aligned}\mathrm{Re}\Bigl(f_0\,\mathrm{e}^{-\mathrm{i}\omega t}\Bigr)\mathrm{Re}\Bigl(g_0\,\mathrm{e}^{-\mathrm{i}\omega t}\Bigr) &= \frac{1}{4}\Bigl(f_0g_0\,\mathrm{e}^{-2\mathrm{i}\omega t} + f_0^*g_0^*\,\mathrm{e}^{2\mathrm{i}\omega t} + f_0^*g_0 + f_0g_0^*\Bigr)\\ &\xrightarrow{\text{average}} \frac{1}{4}(f_0^*g_0 + f_0g_0^*) = \frac{1}{2}\mathrm{Re}(f_0^*g_0)\end{aligned} \tag{11.2.19}$$

and find

$$\overline{-e_0\boldsymbol{E}\cdot\boldsymbol{v}} = \frac{1}{2}\frac{e_0^2}{m_{\mathrm{el}}}|\boldsymbol{E}_0|^2\,\mathrm{Re}\frac{-\mathrm{i}\omega}{\omega_0^2-\omega^2-\mathrm{i}\gamma\omega}\,, \tag{11.2.20}$$

so that the total mechanical power is

$$P_{\mathrm{tot}} = \frac{e_0^2}{m_{\mathrm{el}}}\underbrace{\frac{1}{2}|\boldsymbol{E}_0|^2}_{=\frac{4\pi}{c}|\boldsymbol{S}|\ \text{(time average)}}\frac{\gamma\omega^2}{(\omega^2-\omega_0^2)^2+(\gamma\omega)^2} = \sigma_{\mathrm{tot}}|\boldsymbol{S}| \tag{11.2.21}$$

with

$$\sigma_{\mathrm{tot}} = \frac{4\pi e_0^2}{m_{\mathrm{el}}c}\frac{\gamma\omega^2}{(\omega^2-\omega_0^2)^2+(\gamma\omega)^2}\,. \tag{11.2.22}$$

This $P_{\rm tot}$ accounts for all the work done by the electric field on the electron, so that energy conservation requires that $P_{\rm tot}$ is equal to the sum of the radiated power and the dissipated power.

Now recognizing that $\dfrac{e_0^2}{m_{\rm el}c}\omega^2 = \dfrac{3}{2}c^2\gamma_{\rm rad}$ in accordance with (11.2.17), we get

$$\sigma_{\rm tot} = 6\pi c^2 \frac{\gamma\,\gamma_{\rm rad}}{(\omega^2-\omega_0^2)^2+(\gamma\omega)^2}\,. \tag{11.2.23}$$

At resonance, $\omega = \omega_0$, this is particularly large

$$\begin{aligned}\sigma_{\rm tot} \le \sigma_{\rm tot}\Big|_{\omega=\omega_0} &= 6\pi c^2\frac{\gamma\,\gamma_{\rm rad}}{(\gamma\omega_0)^2} \\ &= 6\pi\left(\frac{c}{\omega_0}\right)^2\frac{\gamma_{\rm rad}}{\gamma} \\ &= 6\pi\lambda\!\!\!^{-}{}_0^2\frac{\gamma_{\rm rad}}{\gamma} \le 6\pi\lambda\!\!\!^{-}{}_0^2\,,\end{aligned} \tag{11.2.24}$$

where $\lambda\!\!\!^{-}{}_0 = \dfrac{c}{\omega_0} = \dfrac{\lambda_0}{2\pi}$ is the reduced wavelength for the resonant frequency, and $\gamma_{\rm rad} \le \gamma$ is used.

We compare with the scattering cross section (11.2.11),

$$\sigma_{\rm Ray} = \sigma_{\rm scat} = 6\pi c^2\frac{\gamma_{\rm rad}^2}{(\omega^2-\omega_0^2)^2+(\gamma\omega)^2}\,, \tag{11.2.25}$$

where we noted that

$$\sigma_{\rm Th}\omega^4 = 6\pi c^2\gamma_{\rm rad}^2\,, \tag{11.2.26}$$

and the energy balance expressed as

$$\sigma_{\rm tot} = \sigma_{\rm scat} + \sigma_{\rm diss} \tag{11.2.27}$$

identifies the cross section for the dissipative processes that turn incoming radiation into heat,

$$\sigma_{\rm diss} = 6\pi c^2\frac{\gamma_{\rm rad}\,\gamma_{\rm diss}}{(\omega^2-\omega_0^2)^2+(\gamma\omega)^2}\,. \tag{11.2.28}$$

Dissipation can be negligible if light is scattered by atoms of a very dilute gas where collisions are rare. Then $\gamma = \gamma_{\rm rad}$, and all that matters are the scattering processes, for which

$$\sigma_{\text{scat}} = \sigma_{\text{Th}} \frac{1}{\left(1-(\frac{\omega_0}{\omega})^2\right)^2 + \left(\frac{2}{3}\frac{\omega}{c} r_{\text{cl}}\right)^2}$$
$$= \frac{\sigma_{\text{Th}}}{\left(1-(\frac{\omega_0}{\omega})^2\right)^2 + \left(\frac{2}{3}\frac{r_{\text{cl}}}{\lambda\!\!\!^{-}}\right)^2} \qquad (11.2.29)$$

is an alternative expression in terms of the reduced wavelength $\lambda\!\!\!^{-} = \frac{c}{\omega}$. This suggests that

$$\sigma_{\text{scat}} \cong \frac{\sigma_{\text{Th}}}{1 + \left(\frac{2}{3}\frac{r_{\text{cl}}}{\lambda\!\!\!^{-}}\right)^2} \quad \text{for} \quad \omega \gg \omega_0 \qquad (11.2.30)$$

and

$$\sigma_{\text{scat}} \cong 6\pi \lambda\!\!\!^{-2} \quad \text{for} \quad \lambda\!\!\!^{-} \ll r_{\text{cl}} \qquad (11.2.31)$$

in addition, but this is taking the expressions too seriously and amounts to applying them in parameter regimes where these classical results are not valid. Specifically, quantum effects become dominating for wavelengths of the order of the Compton* wavelength of the electron,

$$\frac{\hbar}{m_{\text{el}} c} = \frac{r_{\text{cl}}}{\alpha} = 137 r_{\text{cl}} \,, \qquad (11.2.32)$$

and the expressions we derived without any attention to quantum effects do not apply to radiation with wavelengths so short.

*Arthur Holly COMPTON (1882–1962)

# Chapter 12

# Diffraction

## 12.1 Encounter with Huygens's principle

After this brief discussion of scattering, we now turn to diffraction. We shall limit the treatment to the situation depicted here:

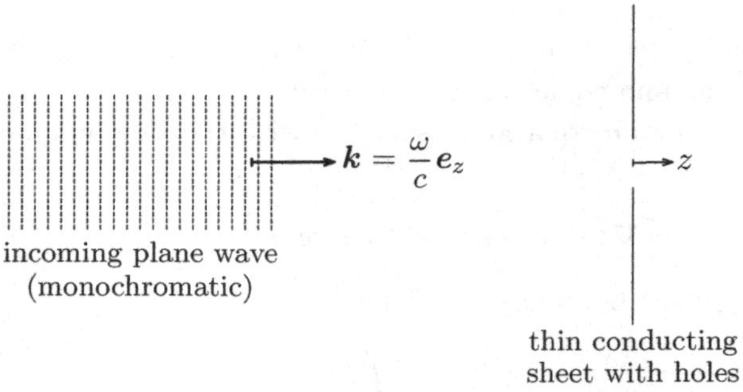


(12.1.1)

where a plane wave propagating in the $z$ direction is incident on a conducting sheet in the $xy$ plane, which has openings, or *apertures*. Electromagnetic radiation reaches the $z > 0$ half-space because it can pass through these apertures.

We assume a fixed frequency, in other words: we consider one Fourier component, and write

$$\begin{aligned} \boldsymbol{E}(\boldsymbol{r},t) &= \mathrm{Re}\Big(\boldsymbol{E}(\boldsymbol{r})\,\mathrm{e}^{-\mathrm{i}\omega t}\Big)\,, \\ \boldsymbol{B}(\boldsymbol{r},t) &= \mathrm{Re}\Big(\boldsymbol{B}(\boldsymbol{r})\,\mathrm{e}^{-\mathrm{i}\omega t}\Big)\,, \end{aligned} \qquad (12.1.2)$$

so that Maxwell's equations (1.1.1) become, after the usual replacement $\frac{\partial}{\partial t} \to -\mathrm{i}\omega$,

$$\begin{aligned} \boldsymbol{\nabla}\cdot\boldsymbol{E}(\boldsymbol{r}) &= 0\,, & \boldsymbol{\nabla}\cdot\boldsymbol{B}(\boldsymbol{r}) &= 0\,,\\ \boldsymbol{\nabla}\times\boldsymbol{B}(\boldsymbol{r}) + \mathrm{i}\frac{\omega}{c}\boldsymbol{E}(\boldsymbol{r}) &= 0\,, & \boldsymbol{\nabla}\times\boldsymbol{E}(\boldsymbol{r}) - \mathrm{i}\frac{\omega}{c}\boldsymbol{B}(\boldsymbol{r}) &= 0\,, \end{aligned} \tag{12.1.3}$$

in the regions to the left and right of the $xy$ plane as well as in the apertures on the $xy$ plane. On the conducting surface, there are surface currents and surface charges, so that these homogeneous Maxwell's equations do not apply there. The implication

$$\begin{aligned} \boldsymbol{\nabla}\times(\boldsymbol{\nabla}\times\boldsymbol{E}) &= -\boldsymbol{\nabla}^2\boldsymbol{E} + \boldsymbol{\nabla}\underbrace{\boldsymbol{\nabla}\cdot\boldsymbol{E}}_{=0}\\ &= \boldsymbol{\nabla}\times\left(\mathrm{i}\frac{\omega}{c}\boldsymbol{B}\right) = \left(\frac{\omega}{c}\right)^2\boldsymbol{E} \end{aligned} \tag{12.1.4}$$

or

$$-\left(\boldsymbol{\nabla}^2 + k^2\right)\boldsymbol{E}(\boldsymbol{r}) = 0\,, \tag{12.1.5}$$

with $k = \frac{\omega}{c}$ as usual, is the *Helmholtz* equation.* Its $k = 0$ version ($\omega = 0$) is the familiar Coulomb equation of electrostatics.

The Green's function equation associated with the Helmholtz equation is

$$-\left(\boldsymbol{\nabla}^2 + k^2\right)G(\boldsymbol{r},\boldsymbol{r}') = 4\pi\delta(\boldsymbol{r}-\boldsymbol{r}')\,, \tag{12.1.6}$$

to which we apply the boundary condition

$$G(\boldsymbol{r},\boldsymbol{r}') = 0 \quad \text{for} \quad \boldsymbol{r} = \begin{pmatrix} x\\ y\\ 0\end{pmatrix} \quad \text{or} \quad z = 0\,. \tag{12.1.7}$$

With

$$\boldsymbol{r}' = \begin{pmatrix} x'\\ y'\\ z'\end{pmatrix} \quad \text{and} \quad \boldsymbol{r}'' = \begin{pmatrix} x'\\ y'\\ -z'\end{pmatrix}, \tag{12.1.8}$$

*Hermann Ludwig Ferdinand von HELMHOLTZ (1821–1894)

so that $\boldsymbol{r}''$ would be the location of the image charge if the charge were at $\boldsymbol{r}'$ and the $xy$ plane were a conducting surface, we have

$$-(\boldsymbol{\nabla}^2+k^2)\frac{\mathrm{e}^{\mathrm{i}k|\boldsymbol{r}-\boldsymbol{r}'|}}{|\boldsymbol{r}-\boldsymbol{r}'|}=4\pi\delta(\boldsymbol{r}-\boldsymbol{r}') \quad \text{for} \quad z,z'>0 \quad \text{or} \quad z,z'<0 \tag{12.1.9}$$

as well as

$$-(\boldsymbol{\nabla}^2+k^2)\frac{\mathrm{e}^{\mathrm{i}k|\boldsymbol{r}-\boldsymbol{r}''|}}{|\boldsymbol{r}-\boldsymbol{r}''|}=0 \quad \text{for} \quad z,z'>0 \quad \text{or} \quad z,z'<0\,. \tag{12.1.10}$$

After putting things together we establish

$$G(\boldsymbol{r},\boldsymbol{r}')=\frac{\mathrm{e}^{\mathrm{i}k|\boldsymbol{r}-\boldsymbol{r}'|}}{|\boldsymbol{r}-\boldsymbol{r}'|}-\frac{\mathrm{e}^{\mathrm{i}k|\boldsymbol{r}-\boldsymbol{r}''|}}{|\boldsymbol{r}-\boldsymbol{r}''|}=G(\boldsymbol{r}',\boldsymbol{r})\,, \tag{12.1.11}$$

which applies when both $\boldsymbol{r}$ and $\boldsymbol{r}'$ are on the same side of the $xy$ plane and correctly vanishes for $z=0$, and also for $z'=0$.

Now, multiplying the Helmholtz equation (12.1.5) by $G(\boldsymbol{r},\boldsymbol{r}')$ and the Green's function equation (12.1.9) by $\boldsymbol{E}(\boldsymbol{r})$, we have

$$G(\boldsymbol{r},\boldsymbol{r}')(\boldsymbol{\nabla}^2+k^2)\boldsymbol{E}(\boldsymbol{r})-\boldsymbol{E}(\boldsymbol{r})(\boldsymbol{\nabla}^2+k^2)G(\boldsymbol{r},\boldsymbol{r}')=4\pi\delta(\boldsymbol{r}-\boldsymbol{r}')\boldsymbol{E}(\boldsymbol{r}) \tag{12.1.12}$$

for the difference of the two equations. On the left-hand side the terms multiplying $k^2$ take care of themselves which leaves us with

$$G(\boldsymbol{r},\boldsymbol{r}')\boldsymbol{\nabla}^2\boldsymbol{E}(\boldsymbol{r})-\boldsymbol{E}(\boldsymbol{r})\boldsymbol{\nabla}^2G(\boldsymbol{r},\boldsymbol{r}')=\boldsymbol{\nabla}\cdot\Big[G(\boldsymbol{r},\boldsymbol{r}')\boldsymbol{\nabla}\boldsymbol{E}(\boldsymbol{r})-\boldsymbol{E}(\boldsymbol{r})\boldsymbol{\nabla}G(\boldsymbol{r},\boldsymbol{r}')\Big] \tag{12.1.13}$$

with the $\cdot$ product between the two gradient operators. Upon integrating over the half-space $z>0$, we have

$$\int_S \mathrm{d}\boldsymbol{S}\cdot\Big[G(\boldsymbol{r},\boldsymbol{r}')\boldsymbol{\nabla}\boldsymbol{E}(\boldsymbol{r})-\boldsymbol{E}(\boldsymbol{r})\boldsymbol{\nabla}G(\boldsymbol{r},\boldsymbol{r}')\Big]=4\pi\boldsymbol{E}(\boldsymbol{r}') \quad \text{for} \quad z'>0\,, \tag{12.1.14}$$

where the surface $S$ of the $z>0$ space consists of an infinitely remote semisphere on which both $G$ and $\boldsymbol{E}$ vanish sufficiently fast with growing $r$, and the $xy$ plane, where $\mathrm{d}\boldsymbol{S}=-\boldsymbol{e}_z\mathrm{d}x\,\mathrm{d}y$. Therefore,

$$\begin{aligned}4\pi\boldsymbol{E}(\boldsymbol{r}') &= -\int \mathrm{d}x\,\mathrm{d}y\,\Big[G(\boldsymbol{r},\boldsymbol{r}')\frac{\partial}{\partial z}\boldsymbol{E}(\boldsymbol{r})-\boldsymbol{E}(\boldsymbol{r})\frac{\partial}{\partial z}G(\boldsymbol{r},\boldsymbol{r}')\Big]\bigg|_{z=0}\\ &= \int \mathrm{d}x\,\mathrm{d}y\,\Big[\boldsymbol{E}(\boldsymbol{r})\frac{\partial}{\partial z}G(\boldsymbol{r},\boldsymbol{r}')\Big]\bigg|_{z=0}\end{aligned} \tag{12.1.15}$$

after recalling that $G(\boldsymbol{r},\boldsymbol{r}')=0$ for $z=0$. The integral over the $xy$ plane consists of the integral over the apertures, where all components of $\boldsymbol{E}$ can be nonzero, and the integral over the surface of the thin metal sheet, where only the normal $z$ component can be nonvanishing, whereas $E_x=0$ and $E_y=0$ on the conducting sheet. So, if we focus on the $x$ component now, we get

$$4\pi E_x(\boldsymbol{r}')=\int\limits_{\text{apertures}}\mathrm{d}x\,\mathrm{d}y\left[E_x(\boldsymbol{r})\frac{\partial}{\partial z}G(\boldsymbol{r},\boldsymbol{r}')\right]\Bigg|_{z=0}, \tag{12.1.16}$$

where the integration is now only over the apertures in the $xy$ plane where we do not have the conducting sheet.

The two terms in $G$ of (12.1.11) involve $z-z'$ or $z+z'$, respectively, so that

$$\frac{\partial}{\partial z}\frac{\mathrm{e}^{\mathrm{i}k|\boldsymbol{r}-\boldsymbol{r}'|}}{|\boldsymbol{r}-\boldsymbol{r}'|}=-\frac{\partial}{\partial z'}\frac{\mathrm{e}^{\mathrm{i}k|\boldsymbol{r}-\boldsymbol{r}'|}}{|\boldsymbol{r}-\boldsymbol{r}'|} \tag{12.1.17}$$

and

$$\frac{\partial}{\partial z}\frac{\mathrm{e}^{\mathrm{i}k|\boldsymbol{r}-\boldsymbol{r}''|}}{|\boldsymbol{r}-\boldsymbol{r}''|}=+\frac{\partial}{\partial z'}\frac{\mathrm{e}^{\mathrm{i}k|\boldsymbol{r}-\boldsymbol{r}''|}}{|\boldsymbol{r}-\boldsymbol{r}''|} \tag{12.1.18}$$

and we note that both terms contribute equal amounts at $z=0$,

$$4\pi E_x(\boldsymbol{r}')=-2\frac{\partial}{\partial z'}\int\limits_{\text{apertures}}\mathrm{d}x\,\mathrm{d}y\,E_x(\boldsymbol{r})\frac{\mathrm{e}^{\mathrm{i}k|\boldsymbol{r}-\boldsymbol{r}'|}}{|\boldsymbol{r}-\boldsymbol{r}'|}\Bigg|_{z=0}. \tag{12.1.19}$$

With $\boldsymbol{r}=\boldsymbol{r}_\perp+z\boldsymbol{e}_z$ and $\mathrm{d}x\,\mathrm{d}y=(\mathrm{d}\boldsymbol{r}_\perp)$ this reads

$$E_x(\boldsymbol{r}')=-\frac{1}{2\pi}\frac{\partial}{\partial z'}\int\limits_{\text{apertures}}(\mathrm{d}\boldsymbol{r}_\perp)\,E_x(\boldsymbol{r}_\perp)\frac{\mathrm{e}^{\mathrm{i}k|\boldsymbol{r}'-\boldsymbol{r}_\perp|}}{|\boldsymbol{r}'-\boldsymbol{r}_\perp|}. \tag{12.1.20}$$

This is an example of a Fresnel*–Kirchhoff† diffraction formula.

The right-hand side of (12.1.20) is clearly reminiscent of Huygens's‡ Principle according to which the points in the apertures act as sources of spherical waves, which then interfere in the $z>0$ half-space on the far side of the plane wave that is incoming from $z<0$. Fine, but there are two problems with this view, problems that Huygens could not have been aware of in the 17th century. First, the physical source of the radiation are the charges on the surface of the conducting sheet, which are set into oscillatory

*Augustin-Jean FRESNEL (1788–1827) †Gustav Robert KIRCHHOFF (1824–1887)
‡Christiaan HUYGENS (1629–1695)

motion by the incident plane wave; there is nothing in the aperture that could be the physical source of radiation. Second, the values of $E_x(\boldsymbol{r}_\perp)$ in the aperture are not known, which limits the usefulness of the integral over the apertures. And to determine the actual field in the apertures, we need to find the induced surface currents flowing in the conducting sheet.

## 12.2 Large apertures

Yet, even without knowing the values of $E_x(\boldsymbol{r}_\perp)$ in the aperture in detail, we can make some progress for large apertures, that is: openings that are characterized by length scales that are very large compared with the wavelength $\lambda = \frac{2\pi}{k}$ of the incident plane wave. Under these circumstances, the edge effects of the boundaries of the apertures are important only in a rather small fraction of the apertures area and we can approximate the field in the aperture by the incoming field

$$E_x(\boldsymbol{r}) \cong E_{\mathrm{inc}}(\boldsymbol{r}) = E_0\, \mathrm{e}^{\mathrm{i}kz} \qquad \text{for } z = 0: \ E_x(\boldsymbol{r})\Big|_{z=0} \cong E_0\,, \tag{12.2.1}$$

thereby recalling the situation depicted in (12.1.1), that is: far to the left we have the plane wave $\boldsymbol{E} = \mathrm{Re}\Big(\boldsymbol{E}_0\, \mathrm{e}^{\mathrm{i}kz - \mathrm{i}\omega t}\Big)$ with $\boldsymbol{E}_0 \cdot \boldsymbol{e}_z = 0$.

The approximation (12.2.1) is, in fact, the way Huygens's principle is usually applied, when the difference between the actual field in the aperture and the incoming field is ignored. This suggests that, perhaps, *Huygens's approximation* is a better choice of terminology than *Huygens's principle.*

With this large-aperture approximation, then, and after interchanging the roles of $\boldsymbol{r}$ and $\boldsymbol{r}'$ in (12.1.19), we have

$$E_x(\boldsymbol{r}) \cong -\frac{E_0}{2\pi}\frac{\partial}{\partial z} \int\limits_{\text{apertures}} (\mathrm{d}\boldsymbol{r}'_\perp)\, \frac{\mathrm{e}^{\mathrm{i}k|\boldsymbol{r}-\boldsymbol{r}'_\perp|}}{|\boldsymbol{r}-\boldsymbol{r}'_\perp|}\,, \tag{12.2.2}$$

where we now apply the usual radiation-field form of the outgoing spherical wave,

$$\frac{\mathrm{e}^{\mathrm{i}k|\boldsymbol{r}-\boldsymbol{r}'_\perp|}}{|\boldsymbol{r}-\boldsymbol{r}'_\perp|} \cong \frac{\mathrm{e}^{\mathrm{i}kr}}{r}\, \mathrm{e}^{-\mathrm{i}k\boldsymbol{n}\cdot\boldsymbol{r}'_\perp} \tag{12.2.3}$$

with $\boldsymbol{n} = \frac{\boldsymbol{r}}{r}$, and

$$\frac{\partial}{\partial z}\frac{\mathrm{e}^{\mathrm{i}k|\boldsymbol{r}-\boldsymbol{r}'_\perp|}}{|\boldsymbol{r}-\boldsymbol{r}'_\perp|} \cong \frac{\mathrm{i}k}{r}\cos\theta\, \mathrm{e}^{\mathrm{i}kr}\, \mathrm{e}^{-\mathrm{i}k\boldsymbol{n}\cdot\boldsymbol{r}'_\perp} \tag{12.2.4}$$

since

$$\frac{\partial r}{\partial z} = \cos\theta \quad \text{for} \quad \boldsymbol{r} = r \begin{pmatrix} \sin\theta\cos\phi \\ \sin\theta\sin\phi \\ \cos\theta \end{pmatrix} = r\boldsymbol{n}\,. \tag{12.2.5}$$

Accordingly,

$$E_x(\boldsymbol{r}) \cong -\frac{E_0}{2\pi}\frac{\mathrm{e}^{ikr}}{r} ik\cos\theta \int\limits_{\text{apertures}} (\mathrm{d}\boldsymbol{r}'_\perp)\, \mathrm{e}^{-ik\boldsymbol{n}\cdot\boldsymbol{r}'_\perp} \tag{12.2.6}$$

approximates the electric field of the diffracted radiation in the situation of apertures that are large on the scale set by the wavelength.

The far-field form (12.2.6) of the general large-aperture expression (12.2.2) is known as the *Fraunhofer* approximation*. It applies in the *Fraunhofer regime*, that is: far way from the diffracting apertures.

## 12.3 Single large circular aperture

### 12.3.1 *Differential cross section*

We apply this to a single circular aperture of radius $a$, so that

$$\int\limits_{\text{apertures}} (\mathrm{d}\boldsymbol{r}'_\perp) \to \int_0^a \mathrm{d}s\, s \int_0^{2\pi} \mathrm{d}\varphi \tag{12.3.1}$$

in (12.2.6) for $(x', y') = s(\cos\varphi, \sin\varphi)$, when

$$\boldsymbol{n}\cdot\boldsymbol{r}'_\perp = s\sin\theta\cos(\phi-\varphi) \tag{12.3.2}$$

and thus

$$\begin{aligned} E_x(\boldsymbol{r}) &\cong -E_0\frac{\mathrm{e}^{ikr}}{r} ik\cos\theta \int_0^a \mathrm{d}s\, s \int \frac{\mathrm{d}\varphi}{2\pi}\, \mathrm{e}^{-iks\sin\theta\cos\varphi} \\ &\overset{(10.4.2)}{=} -E_0\frac{\mathrm{e}^{ikr}}{r} ik\cos\theta \int_0^a \mathrm{d}s\, s\, \mathrm{J}_0(ks\sin\theta) \\ &= -E_0\frac{\mathrm{e}^{ikr}}{r} ik\frac{\cos\theta}{(k\sin\theta)^2} \int_0^{ka\sin\theta} \mathrm{d}t\, t\, \mathrm{J}_0(t)\,. \end{aligned} \tag{12.3.3}$$

---

*Joseph von FRAUNHOFER (1787–1826)

Here we recall the recurrence relations (10.4.13) and recognize that they imply

$$J_{m-1}(t) = \left(\frac{m}{t} + \frac{d}{dt}\right) J_m(t) = t^{-m}\frac{d}{dt}\big(t^m J_m(t)\big) \tag{12.3.4}$$

or

$$t^m J_{m-1}(t) = \frac{d}{dt}\big(t^m J_m(t)\big)\,, \tag{12.3.5}$$

which is $tJ_0(t) = \frac{d}{dt}\big(tJ_1(t)\big)$ for $m = 1$. It follows that

$$\begin{aligned}\frac{E_x(\boldsymbol{r})}{E_0} &\cong -ik\frac{e^{ikr}}{r}\cos\theta\frac{ka\sin\theta}{(k\sin\theta)^2}J_1(ka\sin\theta)\\ &= -ika^2\frac{e^{ikr}}{r}\cos\theta\frac{J_1(ka\sin\theta)}{ka\sin\theta}\,,\end{aligned} \tag{12.3.6}$$

which is a good approximation if $a \gg \lambda$, that is $ka \gg 1$.

Under these circumstances, however, most of the intensity will propagate in the forward direction $\theta \cong 0$, and we can use $\cos\theta \cong 1$, $\sin\theta \cong \theta$, giving

$$\frac{E_x(\boldsymbol{r})}{E_0} \cong -ika^2\frac{e^{ikr}}{r}\frac{J_1(ka\theta)}{ka\theta} \tag{12.3.7}$$

for the amplitude ratio, and

$$\frac{|E_x(\boldsymbol{r})|^2}{|E_0|^2} \cong \frac{1}{r^2}\big(ka^2\big)^2\left(\frac{J_1(ka\theta)}{ka\theta}\right)^2 = \frac{1}{r^2}\frac{d\sigma}{d\Omega} \tag{12.3.8}$$

for the ratio of the intensities or fluxes, thereby identifying the differential diffraction cross section for a large circular aperture,

$$\frac{d\sigma}{d\Omega} = \big(ka^2\big)^2\left(\frac{J_1(ka\theta)}{ka\theta}\right)^2. \tag{12.3.9}$$

### 12.3.2 *Total cross section*

We obtain the total cross section by integrating this differential cross section over all directions,

$$\sigma = \int d\Omega\,\frac{d\sigma}{d\Omega} = 2\pi\big(ka^2\big)^2\int_0^{\pi} d\theta\,\sin\theta\left(\frac{J_1(ka\theta)}{ka\theta}\right)^2 \tag{12.3.10}$$

or

$$\sigma = 2\pi\left(ka^2\right)^2 \int_0^\infty \mathrm{d}\theta\,\theta \left(\frac{\mathrm{J}_1(ka\theta)}{ka\theta}\right)^2 \tag{12.3.11}$$

upon recognizing that only $\theta \ll 1$ matters, and the substitution $ka\theta = t$, $\mathrm{d}\theta\,\theta = \dfrac{\mathrm{d}t\,t}{(ka)^2}$ gives

$$\sigma = 2\pi a^2 \underbrace{\int_0^\infty \mathrm{d}t\,\frac{1}{t}\mathrm{J}_1(t)^2}_{=\frac{1}{2}} = \pi a^2 \,. \tag{12.3.12}$$

As one would expect, the total differential cross section for diffraction by a circular aperture, in the large-aperture approximation, is the area of the opening: $\sigma = \pi a^2$.

To establish the value of the integral in (12.3.12) we recall two implications of the recurrence relations (10.4.13), namely

$$\frac{\mathrm{d}}{\mathrm{d}t}\mathrm{J}_0(t) = -\mathrm{J}_1(t) \quad \text{and} \quad \frac{1}{t}\mathrm{J}_1(t) = \mathrm{J}_0(t) - \frac{\mathrm{d}}{\mathrm{d}t}\mathrm{J}_1(t)\,, \tag{12.3.13}$$

which imply first

$$\begin{aligned} \frac{1}{t}\mathrm{J}_1(t)^2 &= \mathrm{J}_1(t)\Big(\mathrm{J}_0(t) - \frac{\mathrm{d}}{\mathrm{d}t}\mathrm{J}_1(t)\Big) \\ &= -\mathrm{J}_0(t)\frac{\mathrm{d}}{\mathrm{d}t}\mathrm{J}_0(t) - \mathrm{J}_1(t)\frac{\mathrm{d}}{\mathrm{d}t}\mathrm{J}_1(t) \\ &= -\frac{1}{2}\frac{\mathrm{d}}{\mathrm{d}t}\Big(\mathrm{J}_0(t)^2 + \mathrm{J}_1(t)^2\Big) \end{aligned} \tag{12.3.14}$$

and then

$$\int_0^\infty \mathrm{d}t\,\frac{1}{t}\mathrm{J}_1(t)^2 = \frac{1}{2}\Big(\mathrm{J}_0(t)^2 + \mathrm{J}_1(t)^2\Big)\Big|_{t=0} = \frac{1}{2}\,, \tag{12.3.15}$$

indeed, after entering the values $\mathrm{J}_0(0) = 1$ and $\mathrm{J}_1(0) = 0$.

### 12.3.3 *Small diffraction angles*

In the forward direction, $\theta = 0$, we have

$$\left.\frac{\mathrm{d}\sigma}{\mathrm{d}\Omega}\right|_{\theta=0} = \left(ka^2\right)^2 \left.\left(\frac{J_1(ka\theta)}{ka\theta}\right)^2\right|_{\theta=0} = \frac{1}{4}\left(ka^2\right)^2 = \frac{(ka)^2}{4\pi}\sigma \gg \sigma\,, \tag{12.3.16}$$

implying that there is a very strong peak in the $\theta = 0$ direction. This is, of course, consistent with our physical expectations that most of the electromagnetic wave propagates forward if the aperture is large.

We need the asymptotic (that is: large-argument) form of $\mathrm{J}_1(t)$ if we want to understand the diffraction under sizeable angles, for which $ka\theta \gg 1$ while $\theta \ll 1$. We recall the $m = 1$ version of (10.4.21),

$$\mathrm{J}_1(t) = \mathrm{Re}\int_0^\pi \frac{\mathrm{d}\varphi}{\pi}\,\mathrm{e}^{\mathrm{i}(t\sin\varphi\,-\,\varphi)} \tag{12.3.17}$$

and recognize that, for $t \gg 1$, the phase is stationary near $\varphi = \frac{\pi}{2} - \frac{1}{t} = \varphi_0$. Therefore, we write $\varphi = \varphi_0 + \epsilon$ and replace the exponent by

$$t\sin\varphi - \varphi \to t - \frac{1}{2}t\epsilon^2 - \frac{\pi}{2}\,, \tag{12.3.18}$$

discarding terms of order $\frac{1}{t}$ and higher powers thereof. This gives

$$\begin{aligned}\mathrm{J}_1(t) &\cong \mathrm{Re}\int_{-\infty}^{\infty} \frac{\mathrm{d}\epsilon}{\pi}\,\mathrm{e}^{\mathrm{i}(t\,-\,\frac{\pi}{2})}\,\mathrm{e}^{-\frac{\mathrm{i}}{2}t\epsilon^2} \\ &= \mathrm{Re}\frac{1}{\pi}\sqrt{\frac{2\pi}{\mathrm{i}t}}\,\mathrm{e}^{\mathrm{i}(t\,-\,\frac{\pi}{2})} = \sqrt{\frac{2}{\pi t}}\cos\!\left(t - \frac{3\pi}{4}\right),\end{aligned} \tag{12.3.19}$$

valid for $t \gg 1$. Accordingly, we have

$$\begin{aligned}\frac{\mathrm{d}\sigma}{\mathrm{d}\Omega} &\cong \left(ka^2\right)^2\left(\frac{1}{ka\theta}\right)^2 \frac{2}{\pi ka\theta}\cos\!\left(ka\theta - \frac{3\pi}{4}\right)^2 \\ &= \frac{2}{\pi}\left(ka^2\right)^2\left(\frac{1}{ka\theta}\right)^3\cos\!\left(ka\theta - \frac{3\pi}{4}\right)^2,\end{aligned} \tag{12.3.20}$$

which applies when $\theta \ll 1 \ll ka\theta$. The over-all picture is this:

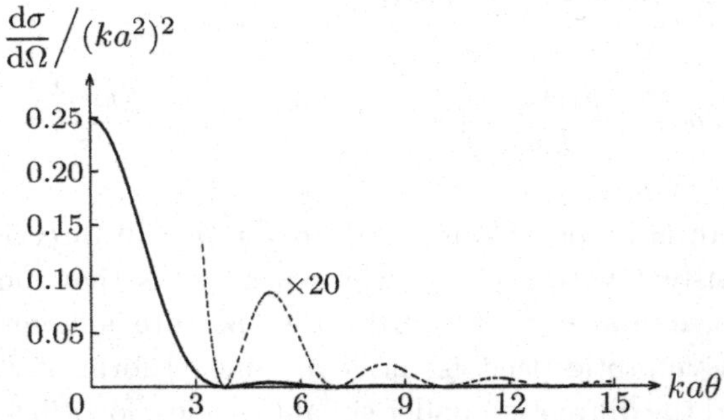


(12.3.21)

with a clearly dominating central maximum and very small side maxima, for which the dashed curve displays ordinate values that are enlarged by a factor of 20. The locations of the minima are given by the zeros of $\mathrm{J}_1(ka\theta)$ and, as told by (12.3.20), these values of $ka\theta$ are approximately given by

$$ka\theta - \frac{3\pi}{4} = \frac{\pi}{2}, \frac{3\pi}{2}, \frac{5\pi}{2}, \ldots \tag{12.3.22}$$

or

$$\begin{aligned} ka\theta &= \frac{5\pi}{4}, \frac{9\pi}{4}, \frac{13\pi}{4}, \ldots \\ &= 3.927, 7.069, 10.210, \ldots \end{aligned} \tag{12.3.23}$$

which are surprisingly good approximations for the exact zeros of $\mathrm{J}_1(t)$ at

$$t = 3.832, 7.016, 10.173, \ldots . \tag{12.3.24}$$

We conclude that a large circular aperture — such as the opening of the objective of a microscope, telescope, or camera — diffracts the incoming light into a central cone with its opening angle given by

$$2\theta = \frac{2 \times 3.832}{ka} \quad \text{or} \quad 2\theta = \frac{3.832}{\pi}\frac{\lambda}{a} = 1.22\frac{\lambda}{a}\,, \tag{12.3.25}$$

a basic statement about the diffraction limit for the resolution of optical instruments.

## 12.4 Induced surface currents

We remarked above, in the context of (12.1.20), that the physical source of the diffracted radiation are the surface currents induced on the conducting sheet in the $xy$ plane in (12.1.1). For simplicity, and without losing any essential physics, we shall assume that the conducting sheet is so thin that we can idealize it as a single surface and that it is perfectly conducting (no ohmic resistance) at the frequency in question. The current density

$$\boldsymbol{j}_{\mathrm{s}}(\boldsymbol{r},t) = \mathrm{Re}\Big(\boldsymbol{j}_{\mathrm{s}}(\boldsymbol{r})\,\mathrm{e}^{-\mathrm{i}\omega t}\Big) \tag{12.4.1}$$

is then that of a pure surface current,

$$\boldsymbol{j}_{\mathrm{s}}(\boldsymbol{r}) = \boldsymbol{K}(\boldsymbol{r}_\perp)\delta(z)\,, \tag{12.4.2}$$

where we recall that $\boldsymbol{r}_\perp$ denotes the two-dimensional tangential part of $\boldsymbol{r} = \boldsymbol{r}_\perp + z\boldsymbol{e}_z$ that is in the $xy$ plane. The surface current density vector $\boldsymbol{K}(\boldsymbol{r}_\perp)$ is a tangential vector as well.

We use the Lorentz gauge so that the vector potential

$$\boldsymbol{A}_{\mathrm{s}}(\boldsymbol{r},t) = \mathrm{Re}\Big(\boldsymbol{A}_{\mathrm{s}}(\boldsymbol{r})\,\mathrm{e}^{-\mathrm{i}\omega t}\Big) \tag{12.4.3}$$

that is associated with this current is given by

$$\boldsymbol{A}_{\mathrm{s}}(\boldsymbol{r}) = \int(\mathrm{d}\boldsymbol{r}')\,\frac{\mathrm{e}^{\mathrm{i}k|\boldsymbol{r}-\boldsymbol{r}'|}}{|\boldsymbol{r}-\boldsymbol{r}'|}\,\frac{1}{c}\,\boldsymbol{j}_{\mathrm{s}}(\boldsymbol{r}')\,, \tag{12.4.4}$$

which we may regard as a special case of (8.1.3), and the scalar potential is

$$\Phi_{\mathrm{s}}(\boldsymbol{r}) = \frac{1}{\mathrm{i}k}\boldsymbol{\nabla}\cdot\boldsymbol{A}_{\mathrm{s}}(\boldsymbol{r}) \tag{12.4.5}$$

as implied by the Lorentz gauge condition (1.3.7) with $\dfrac{\partial}{\partial t} \to -\mathrm{i}\omega = -\mathrm{i}ck$. The total electric field is then

$$\begin{aligned}\boldsymbol{E}(\boldsymbol{r}) &= \boldsymbol{E}_{\mathrm{inc}}(\boldsymbol{r}) + \mathrm{i}k\boldsymbol{A}_{\mathrm{s}}(\boldsymbol{r}) - \boldsymbol{\nabla}\Phi_{\mathrm{s}}(\boldsymbol{r})\\ &= \boldsymbol{E}_{\mathrm{inc}}(\boldsymbol{r}) + \mathrm{i}k\left(\boldsymbol{1} + \frac{\boldsymbol{\nabla}\boldsymbol{\nabla}}{k^2}\right)\cdot\boldsymbol{A}_{\mathrm{s}}(\boldsymbol{r})\,,\end{aligned} \tag{12.4.6}$$

the sum of the incident field and the field radiated by the surface current. We are, of course, reminded of the discussion in Section 7.4 where we met this dyadic differential operator when identifying the vector potential in the radiation gauge.

Upon making the tangential nature of the surface current explicit, we get

$$\boldsymbol{E}(\boldsymbol{r}) = \boldsymbol{E}_{\text{inc}}(\boldsymbol{r}) + \mathrm{i}k\left(\mathbf{1} + \frac{\boldsymbol{\nabla}\boldsymbol{\nabla}}{k^2}\right) \cdot \int (\mathrm{d}\boldsymbol{r}'_\perp)\, \frac{\mathrm{e}^{\mathrm{i}k|\boldsymbol{r}-\boldsymbol{r}'_\perp|}}{|\boldsymbol{r}-\boldsymbol{r}'_\perp|}\, \frac{1}{c}\, \boldsymbol{K}(\boldsymbol{r}'_\perp)\,, \tag{12.4.7}$$

where the $(\mathrm{d}\boldsymbol{r}'_\perp)$ integration covers the whole $x'y'$ plane with

$$\boldsymbol{K}(\boldsymbol{r}_\perp) = 0 \quad \text{if } \boldsymbol{r}_\perp \text{ is in one of the apertures.} \tag{12.4.8}$$

This is one condition imposed on $\boldsymbol{K}(\boldsymbol{r}_\perp)$. It is supplemented by the requirement that the electric field on the perfectly conducting surface is normal and has no tangential components,

$$\boldsymbol{E}_\perp(\boldsymbol{r})\Big|_{z=0} = 0 \quad \text{for} \quad \boldsymbol{r} = \boldsymbol{r}_\perp \quad \text{on the conducting sheet.} \tag{12.4.9}$$

This statement is an integral equation for $\boldsymbol{K}(\boldsymbol{r})$, to be solved under the above constraint (12.4.8) of no current outside the conducting sheet.

The relation between the given incident field and the induced current is linear: if $\boldsymbol{K}^{(1)}(\boldsymbol{r}_\perp)$ and $\boldsymbol{K}^{(2)}(\boldsymbol{r}_\perp)$ are the solutions for $\boldsymbol{E}^{(1)}_{\text{inc}}(\boldsymbol{r})$ and $\boldsymbol{E}^{(2)}_{\text{inc}}(\boldsymbol{r})$ respectively, then $\lambda_1\boldsymbol{K}^{(1)} + \lambda_2\boldsymbol{K}^{(2)}$ is the solution to $\lambda_1\boldsymbol{E}^{(1)}_{\text{inc}} + \lambda_2\boldsymbol{E}^{(2)}_{\text{inc}}$, but despite this linearity the equation pair (12.4.8), (12.4.9) is notoriously difficult to solve. In fact, there is only one known example with an exact solution (first reported by Sommerfeld in 1901), namely the situation of a half-infinite conducting sheet, such as: conductor for $y < 0$, aperture for $y > 0$, diffracting edge at $y = 0$, and $\boldsymbol{E}(\boldsymbol{r}) = E_x(\boldsymbol{r})\boldsymbol{e}_x$ everywhere; more about this in Section 12.7 below. For all other geometries one must rely on approximations, which often require clever educated guesses or sophisticated variational methods.

## 12.5 Large obstacles

No such machinery is necessary if the conducting sheet has no openings at all. Then no radiation gets from the $z < 0$ region of incidence to $z > 0$, which is to say that the incoming plane wave is simply reflected as a whole. In this situation, we have the electric field

$$\boldsymbol{E}(\boldsymbol{r}) = \boldsymbol{E}_0\left(\mathrm{e}^{\mathrm{i}kz} - \mathrm{e}^{-\mathrm{i}kz}\right)\eta(-z) \tag{12.5.1}$$

for the sum of incident and reflected wave, here for normal incidence, and the magnetic field is

$$B(r) = e_z \times E_0\Big(e^{ikz} + e^{-ikz}\Big)\eta(-z)\,, \tag{12.5.2}$$

as follows from Faraday's induction law. The step function $\eta(-z)$ ensures that there is no electromagnetic field in the $z > 0$ region on the far side of the infinite conducting sheet. We apply Ampère's circuital law,

$$\frac{4\pi}{c} j(r) = \nabla \times B(r) + ikE(r)\,, \tag{12.5.3}$$

to determine the electric current and find

$$\frac{4\pi}{c} j(r) = 2E_0\delta(z)\,, \tag{12.5.4}$$

and this tells us that the surface current is given by

$$K(r_\perp) = \frac{c}{2\pi} E_0 \tag{12.5.5}$$

under these circumstances. We note that there is no remaining $r_\perp$ dependence in $K$.

For obstacles — read: conducting sheets — that are large on the scale set by the wavelength $\lambda = \frac{2\pi}{k}$ one expects that edge effects are not dominating because most of the surface is many wavelengths away from the nearest edge. It is then often a good approximation to use

$$K(r_\perp) \cong \begin{cases} \dfrac{c}{2\pi} E_0 & \text{on the conducting sheet} \\ 0 & \text{outside} \end{cases} \tag{12.5.6}$$

as a guess for the induced surface current density (for normal incidence of the incoming plane wave). This approximation is similar in spirit, and of similar quality, as Huygens's approximation in Section 12.3, that is: the replacement of the field in the aperture by the incoming field.

## 12.6 Poisson's spot

### 12.6.1 *Bright center in the shadow*

We apply this approximation to the situation of a conducting circular disk of radius $a$:

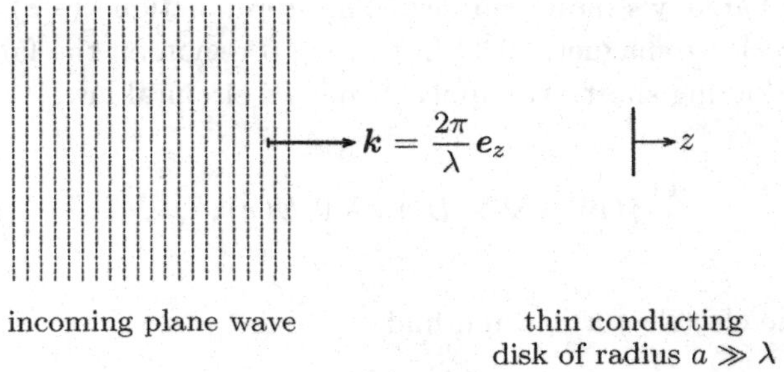


(12.6.1)

With (12.5.6), (12.4.7) turns into

$$E(r) \cong E_0\,\mathrm{e}^{\mathrm{i}kz} + \mathrm{i}kE_0\frac{1}{2\pi}\int_{r'_\perp < a}(\mathrm{d}r'_\perp)\,\frac{\mathrm{e}^{\mathrm{i}k|r - r'_\perp|}}{|r - r'_\perp|}\,, \tag{12.6.2}$$

where, in addition to the current approximation, we dropped the $\nabla\nabla$ term, because it refers to the edge of the disk where the current approximation (12.5.6) cannot be trusted. On the $z$ axis, this gives (with $x' = s\cos\phi$, $y' = s\sin\phi$)

$$\begin{aligned} E(r)\big|_{x=y=0} &\cong E_0\,\mathrm{e}^{\mathrm{i}kz} + \mathrm{i}kE_0\frac{1}{2\pi}\int_0^a \mathrm{d}s\,s\int_{(2\pi)}\mathrm{d}\phi\,\frac{\mathrm{e}^{\mathrm{i}k\sqrt{s^2+z^2}}}{\sqrt{s^2+z^2}} \\ &= E_0\Bigl(\mathrm{e}^{\mathrm{i}kz} + \int_0^a \mathrm{d}s\,\frac{\partial}{\partial s}\,\mathrm{e}^{\mathrm{i}k\sqrt{s^2+z^2}}\Bigr) \\ &= E_0\Bigl(\mathrm{e}^{\mathrm{i}kz} - \mathrm{e}^{\mathrm{i}k|z|} + \mathrm{e}^{\mathrm{i}k\sqrt{a^2+z^2}}\Bigr)\,. \end{aligned} \tag{12.6.3}$$

The $a$ independent terms

$$E_0\Bigl(\mathrm{e}^{\mathrm{i}kz} - \mathrm{e}^{\mathrm{i}k|z|}\Bigr) = \begin{cases} 2\mathrm{i}E_0\sin(kz) & \text{for} \quad z < 0 \\ 0 & \text{for} \quad z > 0 \end{cases} \tag{12.6.4}$$

are just the incoming and reflected plane waves that we would have for an infinite conducting sheet at $z = 0$. The net effect of the finite size of the disk is represented by the $a$ dependent term, which is the only one that matters for $z > 0$, that is: in the geometrical shadow of the disk,

$$\boldsymbol{E}(\boldsymbol{r})\Big|_{x=y=0,\, z>0} \cong \boldsymbol{E}_0\, \mathrm{e}^{\mathrm{i}k\sqrt{a^2+z^2}} . \tag{12.6.5}$$

Most remarkably, we thus find that at the center of the shadow behind the disk the intensity

$$\left|\boldsymbol{E}(\boldsymbol{r})\right|^2 \cong \left|\boldsymbol{E}_0\right|^2 \quad \text{for} \quad x = y = 0,\ z > 0 \tag{12.6.6}$$

is equal to the incoming intensity, so that there is a bright spot at the center of the shadow.

It is known as *Poisson's spot* or as *Arago's* spot* and played a crucial role in convincing people of the wave nature of light. Tersely, the history is this: Poisson predicted that bright center of the shadow, making use of Fresnel's then-new theory of diffraction, and thought that this seemingly absurd implication would demonstrate that Fresnel's theory is fatally flawed. But Arago immediately showed the bright center in an optical experiment, and Fresnel's theory triumphed. This happened in 1818; yet, history has its twists: As Arago noted later, the bright spot had been observed by Maraldi† in 1723, and yet earlier by Delisle‡ in 1715, but at the time this did not attract attention. As a consequence, the phenomenon is not known as *Delisle's spot* nor as *Maraldi's spot.*

What we found in (12.6.5) for the electric field on the axis — a $z$ independent field strength equal to the strength of the incoming plane wave — is incorrect for $0 > z \to 0$ because it gives a nonzero value on the surface of the conducting disk. This observation should not come as a surprise because we did not use the correct form of the induced surface current, which is not known, but the approximation (12.5.6), so that we must not expect that the integral equation (12.4.9) for $\boldsymbol{K}(\boldsymbol{r}_\perp)$ is obeyed. The approximate expression for the electric field cannot be trusted very close to the conducting surface of the disk. But it is reliable for $z \gg \lambda$.

*François Jean Dominique ARAGO (1786–1853)
†Giacomo Filippo MARALDI (1665–1729) ‡Joseph-Nicolas DELISLE (1688–1768)

### 12.6.2 *Size of the bright center*

To determine the size of Poisson's spot, we consider points $\boldsymbol{r}$ close to the $z$ axis in the shadow region behind the disk,

$$\boldsymbol{r} = \begin{pmatrix} s\cos\varphi \\ s\sin\varphi \\ z \end{pmatrix} \quad \text{with} \quad z \gg a \gg s, \lambda\,, \tag{12.6.7}$$

so that

$$\begin{aligned} |\boldsymbol{r} - \boldsymbol{r}'_\perp| &= \sqrt{s^2 - 2ss'\cos(\varphi - \phi) + s'^2 + z^2} \\ &\cong z - \frac{ss'}{z}\cos(\varphi - \phi) + \frac{1}{2}\frac{s'^2}{z} \end{aligned} \tag{12.6.8}$$

applies for

$$\boldsymbol{r}'_\perp = \begin{pmatrix} s'\cos\phi \\ s'\sin\phi \\ 0 \end{pmatrix}, \quad (\mathrm{d}\boldsymbol{r}'_\perp) = \mathrm{d}s'\, s'\, \mathrm{d}\phi \tag{12.6.9}$$

in the exponent in the integrand in (12.6.2), while we use $|\boldsymbol{r} - \boldsymbol{r}'_\perp| \cong z$ in the denominator; these approximations are the usual ones for the radiation field, of course. This brings us to

$$\begin{aligned} \boldsymbol{E} &\cong \boldsymbol{E}_0\, \mathrm{e}^{\mathrm{i}kz} \left( 1 + \frac{\mathrm{i}k}{z} \int_0^a \mathrm{d}s'\, s' \int_{(2\pi)} \frac{\mathrm{d}\phi}{2\pi}\, \mathrm{e}^{-\mathrm{i}(kss'/z)\cos\phi}\, \mathrm{e}^{\frac{1}{2}\mathrm{i}ks'^2/z} \right) \\ &= \boldsymbol{E}_0\, \mathrm{e}^{\mathrm{i}kz} \left( 1 + \frac{\mathrm{i}k}{z} \int_0^a \mathrm{d}s'\, s'\, \mathrm{e}^{\frac{1}{2}\mathrm{i}ks'^2/z}\, \mathrm{J}_0\!\left(k\frac{ss'}{z}\right) \right) \\ &= \boldsymbol{E}_0\, \mathrm{e}^{\mathrm{i}kz} \left( 1 + \int_0^a \mathrm{d}s'\, \mathrm{J}_0\!\left(k\frac{ss'}{z}\right) \frac{\partial}{\partial s'}\, \mathrm{e}^{\frac{1}{2}\mathrm{i}ks'^2/z} \right). \end{aligned} \tag{12.6.10}$$

The last version is an invitation to an integration by parts which gives

$$\boldsymbol{E}(\boldsymbol{r}) \cong \boldsymbol{E}_0\, \mathrm{e}^{\mathrm{i}kz} \left[ \mathrm{J}_0\!\left(\frac{kas}{z}\right) \mathrm{e}^{\frac{1}{2}\mathrm{i}ka^2/z} + \frac{ks}{z} \int_0^a \mathrm{d}s'\, \mathrm{J}_1\!\left(k\frac{ss'}{z}\right) \mathrm{e}^{\frac{1}{2}\mathrm{i}ks'^2/z} \right] \tag{12.6.11}$$

where $\mathrm{J}_0(t=0) = 1$ and $\mathrm{J}_0'(t) = -\mathrm{J}_1(t)$ were used. The second term, the integral with $\mathrm{J}_1(\ )$, is of the order of $\left(\frac{s}{z}\right)^2$ and, therefore, we disregard it, so that the intensity in the shadow region is approximately given by

$$|\boldsymbol{E}(\boldsymbol{r})|^2 \cong |\boldsymbol{E}_0|^2 \mathrm{J}_0\left(k\frac{as}{z}\right)^2 \quad \text{for} \quad z \gg a \gg s, \lambda . \tag{12.6.12}$$

On the axis, $s = 0$, this reproduces (12.6.6): $\left|\boldsymbol{E}(\boldsymbol{r})|_{z=0}\right|^2 \cong \left|\boldsymbol{E}_0\right|^2$, as it should, and off the axis we get a first minimum of the intensity where $\frac{kas}{z} = 2.40$, the first zero of $\mathrm{J}_0(\ )$. The diameter of Poisson's spot is thus approximately given by

$$D = 2 \times 2.40 \frac{z}{ka} = \frac{2.40}{\pi} \frac{\lambda z}{a} = 0.76 \frac{\lambda z}{a} . \tag{12.6.13}$$

As an example, we take a disk of diameter $2a = 1\,\mathrm{cm}$, visible light of wavelength $\lambda = 5 \times 10^{-5}\,\mathrm{cm}$, and look for Poisson's spot at the distance $z = 10\,\mathrm{m} = 10^3\,\mathrm{cm}$, quite far behind the disk. We should find a bright center in the shadow with a diameter of

$$D = 0.76 \frac{5 \times 10^{-5} \times 10^3}{0.5}\,\mathrm{cm} = 0.76\,\mathrm{mm} , \tag{12.6.14}$$

surely visible to the naked eye.

Here is the pictures of an actual diffraction patterns behind a circular obstacle, courtesy of Condylis:*

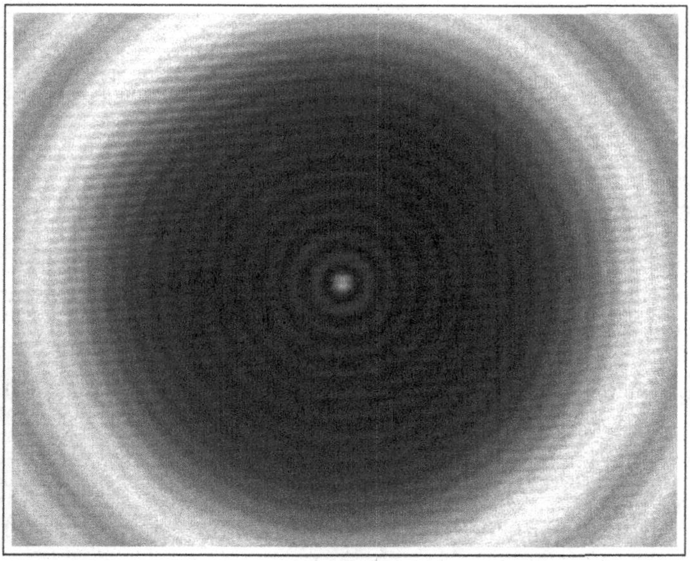

(12.6.15)

*Paul Constantine CONDYLIS (b. 1979)

The pattern was recorded at distance $z = 35.5\,\mathrm{cm}$ behind a bearing ball of $2a = 2\,\mathrm{mm}$ diameter, illuminated by infrared light ($\lambda = 780\,\mathrm{nm}$) from a laser source. The Poisson spot is beautifully visible; it has a diameter of about $0.2\,\mathrm{mm}$. The additional fine fringes are artifacts of an etalon effect in the recording device.

It is not very easy to see Poisson's spot, and in the case of the extremely short wavelengths that characterize diffraction experiments with the de Broglie* waves of electrons, neutrons, or atoms, it may be impossible to resolve the spot if the aperture of the detector is much larger than the spot itself. Let us, therefore, consider the situation of a detector with a circular opening of radius $R$ that is much larger than the radius of Poisson's spot, but still smaller than the radius $a$ of the disk, which is also the radius of the geometrical shadow region: $a > R \gg \dfrac{z}{ka}$.

Far away from the $z$ axis, outside the geometrical shadow, the signal of the detector is proportional to $S_0 = \pi R^2 \bigl|\boldsymbol{E}_0\bigr|^2$, the product of the field intensity and the detector area. This is to be compared with the signal strength when the detector is centered on the $z$ axis and covers all of Poisson's spot,

$$\begin{aligned} S_{\mathrm{spot}} &= 2\pi \int_0^R \mathrm{d}s\, s \,|\boldsymbol{E}_0|^2 \mathrm{J}_0\Bigl(k\frac{as}{z}\Bigr)^2 \\ &= \pi R^2 |\boldsymbol{E}_0|^2\, 2\Bigl(\frac{z}{kaR}\Bigr)^2 \int_0^{kaR/z} \mathrm{d}t\, t\, \mathrm{J}_0(t)^2 . \end{aligned} \tag{12.6.16}$$

With

$$t\,\mathrm{J}_0(t)^2 = \frac{\mathrm{d}}{\mathrm{d}t}\left[\frac{1}{2}t^2\Bigl(\mathrm{J}_0(t)^2 + \mathrm{J}_1(t)^2\Bigr)\right], \tag{12.6.17}$$

which can be verified with the aid of the two relations of (12.3.13), the ratio of the two signal strengths is

$$\frac{S_{\mathrm{spot}}}{S_0} = \Bigl[\mathrm{J}_0(t)^2 + \mathrm{J}_1(t)^2\Bigr]\Big|_{t\,=\,kaR/z\,\gg\,1} = \frac{2z}{\pi kaR} = \frac{\lambda z}{\pi^2 aR}, \tag{12.6.18}$$

where we use the large-$t$ approximation

$$\mathrm{J}_m(t) \cong \sqrt{\frac{2}{\pi t}}\cos\Bigl(t - m\frac{\pi}{2} - \frac{\pi}{4}\Bigr) = \sqrt{\frac{2}{\pi t}}\sin\Bigl(t - m\frac{\pi}{2} + \frac{\pi}{4}\Bigr), \tag{12.6.19}$$

*Prince Louis-Victor DE BROGLIE (1892–1987)

an immediate generalization of the $m = 1$ result in (12.3.19). Upon reading $\frac{15}{2\pi^2}$ for the factor 0.76 in (12.6.13), we have

$$\frac{S_{\text{spot}}}{S_0} = \frac{4}{15}\frac{D}{2R}\,. \tag{12.6.20}$$

So, if a ratio of 3% is considered evidence for observing Poisson's spot, the diameter $2R$ of the detector opening could be 10 times the diameter $D$ of the spot.

### 12.6.3 *Central intensity behind a long strip*

The diffraction pattern of a long strip also shows a bright fringe at the center but, in marked contrast to the Poisson spot of a circular obstacle, the central intensity depends on the geometrical parameters and is not equal to the incident intensity. An example is this diffraction pattern of a strip of width $2a = 1\,\text{mm}$, illuminated by light with wavelength $\lambda = 780\,\text{nm}$, recorded by Condylis at a distance of $z = 15\,\text{cm}$:

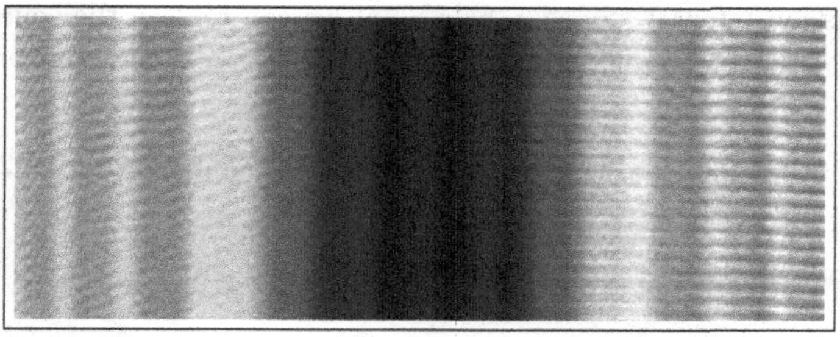

(12.6.21)

Here, too, we have additional fine fringes that are artifacts of the recording device. In this situation of diffraction by a long and wide strip, rather than by a large circular disk, the difference in geometry is quite significant in the context of Poisson's spot.

We return to (12.6.2), but now the integration range is $-a < x' < a$ and $-\infty < y' < \infty$ and the observation point is $\boldsymbol{r} = x\boldsymbol{e}_x + z\boldsymbol{e}_z$ with $z \gg a \gg \lambda$ and $x \ll a$,

$$\frac{E_y(x,z)}{E_0}\,\mathrm{e}^{-\mathrm{i}kz} = 1 + \frac{\mathrm{i}k}{2\pi}\,\mathrm{e}^{-\mathrm{i}kz}\int\limits_{-a}^{a}\mathrm{d}x'\int\limits_{-\infty}^{\infty}\mathrm{d}y'\,\frac{\mathrm{e}^{\mathrm{i}k\sqrt{(x-x')^2+y'^2+z^2}}}{\sqrt{(x-x')^2+y'^2+z^2}}\,, \tag{12.6.22}$$

where we take the incoming and diffracted light to be linearly polarized along the $y$ direction. The substitution

$$y' = \sqrt{(x-x')^2+z^2}\,\sinh\vartheta\,, \qquad \frac{\mathrm{d}y'}{\sqrt{(x-x')^2+y'^2+z^2}} = \mathrm{d}\vartheta \tag{12.6.23}$$

gives

$$\begin{aligned}\frac{E_y(x,z)}{E_0}\,\mathrm{e}^{-\mathrm{i}kz} &= 1+\frac{\mathrm{i}k}{2\pi}\,\mathrm{e}^{-\mathrm{i}kz}\int\limits_{-a}^{a}\mathrm{d}x'\int\limits_{-\infty}^{\infty}\mathrm{d}\vartheta\;\mathrm{e}^{\mathrm{i}k\sqrt{(x-x')^2+z^2}\cosh\vartheta}\\ &= 1+\frac{\mathrm{i}k}{2\pi}\,\mathrm{e}^{-\mathrm{i}kz}\int\limits_{-a}^{a}\mathrm{d}x'\int\limits_{-\infty}^{\infty}\mathrm{d}\vartheta\;\mathrm{e}^{\mathrm{i}k\sqrt{(x-x')^2+z^2}(1+\frac{1}{2}\vartheta^2)}\\ &= 1-\frac{k}{\sqrt{2\mathrm{i}\pi}}\,\mathrm{e}^{-\mathrm{i}kz}\int\limits_{-a}^{a}\mathrm{d}x'\frac{\mathrm{e}^{\mathrm{i}k\sqrt{(x-x')^2+z^2}}}{\left[k\sqrt{(x-x')^2+z^2}\right]^{\frac{1}{2}}}\,,\end{aligned} \tag{12.6.24}$$

after using a stationary-phase approximation near $\vartheta=0$ and evaluating the resulting gaussian integral. Since $z \gg |x'-x|$ we can replace the square root by simply $z$ in the denominator and by $z+\frac{1}{2}(x'-x)^2/z$ in the exponent,

$$\frac{E_y(x,z)}{E_0}\,\mathrm{e}^{-\mathrm{i}kz} = 1-\sqrt{\frac{k}{2\mathrm{i}\pi z}}\int\limits_{-a}^{a}\mathrm{d}x'\,\mathrm{e}^{\frac{1}{2}\mathrm{i}k(x-x')^2/z}\,, \tag{12.6.25}$$

and then substitute $x' = x+\sqrt{\pi z/k}\,t$ to arrive at

$$\frac{E_y(x,z)}{E_0}\,\mathrm{e}^{-\mathrm{i}kz} = 1-\frac{1-\mathrm{i}}{2}\int\limits_{-T_+}^{T_-}\mathrm{d}t\;\mathrm{e}^{\mathrm{i}\frac{\pi}{2}t^2} \tag{12.6.26}$$

with

$$T_\pm = \sqrt{\frac{k}{\pi z}}(a\pm x) = \sqrt{\frac{2}{\lambda z}}(a\pm x)\,. \tag{12.6.27}$$

Expressed in terms of the standard *Fresnel integral*

$$\mathrm{F}(T) = \int\limits_{0}^{T}\mathrm{d}t\;\mathrm{e}^{\mathrm{i}\frac{\pi}{2}t^2}\,, \tag{12.6.28}$$

we thus have

$$\left|\frac{E_y(x,z)}{E_0}\right|^2 = \left|1 - \frac{1-\mathrm{i}}{2}\left[\mathrm{F}(T_-) + \mathrm{F}(T_+)\right]\right|^2 \tag{12.6.29}$$

for the ratio of the diffracted and the incoming intensities. Rather than the 0th Bessel function of (12.6.12), we now have the Fresnel integral. As a consequence of this difference, behind the long strip the central intensity is not equal to the incoming intensity, but is rather given by

$$\left|\frac{E_y(x=0,z)}{E_0}\right|^2 = \left|1 - (1-\mathrm{i})\mathrm{F}\left(\sqrt{2a^2/(\lambda z)}\right)\right|^2 . \tag{12.6.30}$$

For the data of the experiment of (12.6.21), we have $\mathrm{F}(2.07) = 0.089 + \mathrm{i}0.195$ and find 0.044 for this ratio of intensities — indeed, the central intensity is a small fraction of the incident intensity.

We further note that, in contrast to the diffraction pattern (12.6.12) of the circular disk, the right-hand side of (12.6.29) will always be positive as a rule. A vanishing intensity would require $\mathrm{F}(T_-) + \mathrm{F}(T_+) = 1 + i$, which can only happen for exceptional values of $a^2/(\lambda z)$, if at all. But, of course, there will be a succession of intensity maxima and minima on the way from the $x = 0$ plane of symmetry to the borders of the geometrical shadow at $x = \pm a$.

When observing the corresponding fringe pattern of de Broglie waves — of neutrons, say — the detector may not resolve the individual fringes. Then remarks analogous to those at the end of Section 12.6.2 apply.

## 12.7 Diffraction at a straight edge

Finally we observe that the diffraction patterns (12.6.15) and (12.6.21) have in common the feature that the first bright fringe outside the geometrical shadow is particularly bright. In fact, it is brighter than the incoming light by almost 40%. This is typical for diffraction at an edge — and on the scale of the wavelength, the curvature of the circular disks is very small and the edge is essentially straight over many wavelengths.

Here is Condylis's photograph of a straight-edge diffraction pattern:

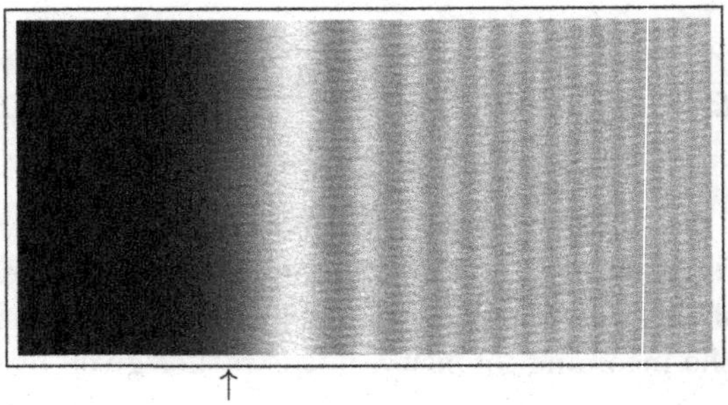

(12.7.1)

The arrow indicates the border of the geometrical shadow, where the intensity is one-fourth of the incoming intensity. A bit further to the right we have the maximum of the first bright fringe, brighter than the incoming intensity that sets the scale on the far right.

### 12.7.1 *Transition region*

We can understand all essential features of the fringe pattern in (12.7.1) by an application of Huygens's approximation, either in the form of the integration over the aperture of (12.2.2) or, equivalently, as the integral (12.4.7) over the induced surface current with the approximate current of (12.5.6). Both procedures give the same result for the fringe pattern in the transition region around the border of the geometrical shadow.

Specifically, we put the straight edge at the $y = z = 0$ line, with the conducting sheet occupying the half-plane with $y < 0$ and $z = 0$:

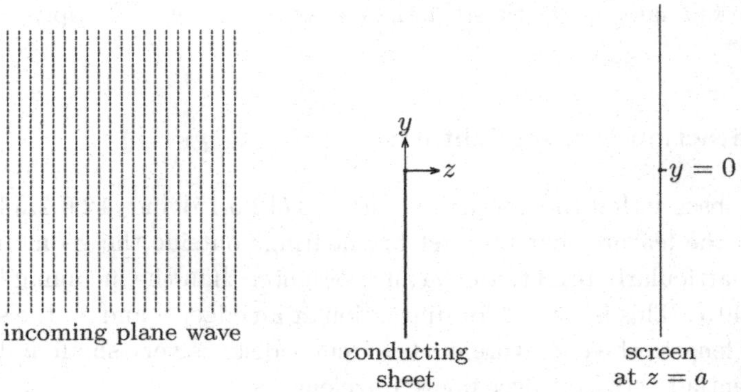


(12.7.2)

The incoming plane wave is polarized in the $x$ direction, that is: parallel to the edge, and we observe the diffraction pattern on a screen at $z = a > 0$ with $ka \gg 1$. The transition region of interest is a strip around $y = 0$ on the screen.

According to (12.2.2), the electric field at the screen is given by

$$E(x, y, z = a) = -\frac{E_0}{2\pi}\frac{\partial}{\partial a}\int_{-\infty}^{\infty} \mathrm{d}x' \int_{0}^{\infty} \mathrm{d}y' \, \frac{\mathrm{e}^{\mathrm{i}k\sqrt{(x-x')^2+(y-y')^2+a^2}}}{\sqrt{(x-x')^2+(y-y')^2+a^2}} \,. \tag{12.7.3}$$

We put

$$x' = x + a\sinh\vartheta\,, \quad \mathrm{d}x' = a\cosh\vartheta\,\mathrm{d}\vartheta \tag{12.7.4}$$

and

$$y' = y + at\,, \quad \mathrm{d}y' = a\,\mathrm{d}t\,, \tag{12.7.5}$$

for which

$$\sqrt{(x-x')^2+(y-y')^2+a^2} = a\sqrt{t^2+(\cosh\vartheta)^2}\,, \tag{12.7.6}$$

and then have

$$E(x, y, a) = -\frac{E_0}{2\pi}\frac{\partial}{\partial a} a \int_{-\infty}^{\infty} \mathrm{d}\vartheta \int_{-y/a}^{\infty} \mathrm{d}t \, \frac{\cosh\vartheta}{\sqrt{t^2+(\cosh\vartheta)^2}} \, \mathrm{e}^{\mathrm{i}ka\sqrt{t^2+(\cosh\vartheta)^2}} \,. \tag{12.7.7}$$

The rapidly oscillating exponential factor has a stationary phase at the point $(\vartheta, t) = (0, 0)$ in the $\vartheta t$ plane of integration. Therefore, only small values of $\vartheta$ and $t$ matter for the integral, and so it is permissible to replace $\sqrt{t^2+(\cosh\vartheta)^2}$ by $1 + \frac{1}{2}t^2 + \frac{1}{2}\vartheta^2$ in the exponent and approximate the ratio of $\cosh\vartheta$ and $\sqrt{t^2+(\cosh\vartheta)^2}$ by unity. Then,

$$\begin{aligned} E(x, y, a) &= -\frac{E_0}{2\pi}\frac{\partial}{\partial a} a\,\mathrm{e}^{\mathrm{i}ka} \int_{-\infty}^{\infty} \mathrm{d}\vartheta\,\mathrm{e}^{\frac{\mathrm{i}}{2}ka\vartheta^2} \int_{-y/a}^{\infty} \mathrm{d}t\;\mathrm{e}^{\frac{\mathrm{i}}{2}kat^2} \\ &= -\frac{E_0}{2\pi}\frac{\partial}{\partial a} a\,\mathrm{e}^{\mathrm{i}ka}\sqrt{\frac{\mathrm{i}2\pi}{ka}}\sqrt{\frac{\pi}{ka}}\left[F(\infty) + F\left(\frac{y}{\sqrt{\pi a/k}}\right)\right]. \end{aligned} \tag{12.7.8}$$

We recognized that the remaining $\vartheta$ integral is of gaussian type and the $t$ integration gives the Fresnel function of (12.6.28) with argument

$$\frac{y}{\sqrt{\pi a/k}} = \frac{y}{\sqrt{\lambda a/2}}\,, \tag{12.7.9}$$

where $\lambda = 2\pi/k$ is the wavelength of the incoming monochromatic plane wave. Only the differentiation of the $\mathrm{e}^{ika}$ factor contributes significantly and, with $F(\infty) = \frac{1}{2}(1+\mathrm{i})$, we arrive at

$$E(x,y,a) = E_0\,\mathrm{e}^{ika}\frac{1}{2}\left[1 + (1-\mathrm{i})F\left(\frac{y}{\sqrt{\lambda a/2}}\right)\right]. \tag{12.7.10}$$

The resulting intensity, normalized to the intensity of the incoming plane wave,

$$\left|\frac{E(x,y,a)}{E_0}\right|^2 = \frac{1}{4}\left|1 + (1-\mathrm{i})F\left(\frac{y}{\sqrt{\lambda a/2}}\right)\right|^2, \tag{12.7.11}$$

reproduces the fringe pattern of (12.7.1) quite convincingly:

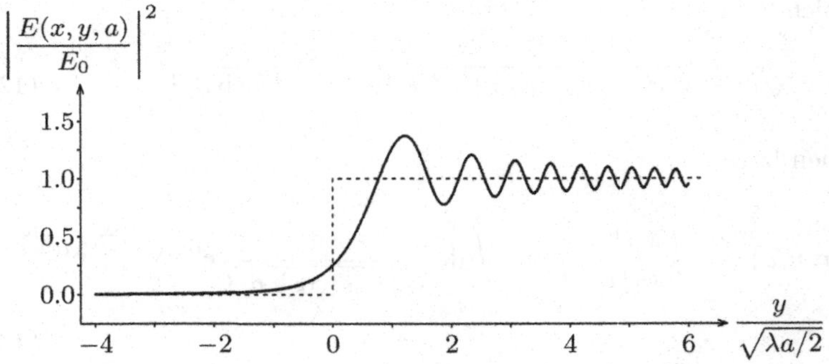


(12.7.12)

The dashed step at $y = 0$ indicates the shadow-light transition of ray optics. Clearly, wave optics is needed to account for the experimental data. In particular, we note that we have one-quarter of the incident intensity at $y = 0$ and the first bright fringe at $y \cong 1.2\sqrt{\lambda a/2}$ has 1.37 times the intensity of the incoming light.

### 12.7.2 *Exact solution*

We return to the situation depicted in (12.7.2), that is: the conducting sheet of (12.1.1) occupies the half-plane $z = 0, y < 0$ and all of the other half-plane with $z = 0, y > 0$ is free of obstacles, and determine the field in the aperture as well as the induced current on the sheet. As mentioned above, at the end of Section 12.4, this diffraction at an infinite straight edge is the problem solved by Sommerfeld in 1901, by a somewhat different method than the one used here. It appears that, more than a century later, it is still the only diffraction problem with a fully known analytical solution.

The incident field

$$\boldsymbol{E}_{\text{inc}}(\boldsymbol{r}) = E_{\text{inc}}(y, z)\,\boldsymbol{e}_x \tag{12.7.13}$$

is linearly polarized in the $x$ direction, which — we recall — is parallel to the edge at $y = 0, z = 0$. As an idealization, we further assume that there is full translational invariance in $x$, as indicated by having no $x$ dependence in $E_{\text{inc}}(y, z)$; remember the lack of $x$ dependence in (12.7.7) and (12.7.10) above. For reasons of symmetry, then, the induced surface current will have only an $x$ component and no $x$ dependence as well,

$$\boldsymbol{K}(\boldsymbol{r}_\perp) = K(y)\,\boldsymbol{e}_x \quad \text{with} \quad K(y) = 0 \text{ for } y > 0\,. \tag{12.7.14}$$

It follows that the surface integral in (12.4.7),

$$\begin{aligned}
&\int (\mathrm{d}\boldsymbol{r}'_\perp)\,\frac{\mathrm{e}^{\mathrm{i}k|\boldsymbol{r}-\boldsymbol{r}'_\perp|}}{|\boldsymbol{r}-\boldsymbol{r}'_\perp|}\,\frac{1}{c}\,\boldsymbol{K}(\boldsymbol{r}'_\perp)\\
&= \int \mathrm{d}x'\,\mathrm{d}y'\,\frac{\mathrm{e}^{\mathrm{i}k\sqrt{(x-x')^2+(y-y')^2+z^2}}}{\sqrt{(x-x')^2+(y-y')^2+z^2}}\,\frac{1}{c}K(y')\,\boldsymbol{e}_x\\
&= \boldsymbol{e}_x \int \mathrm{d}x'\,\mathrm{d}y'\,\frac{\mathrm{e}^{\mathrm{i}k\sqrt{x'^2+(y-y')^2+z^2}}}{\sqrt{x'^2+(y-y')^2+z^2}}\,\frac{1}{c}K(y')\,,
\end{aligned} \tag{12.7.15}$$

does not depend on $x$, so that its divergence vanishes,

$$\boldsymbol{\nabla}\cdot\boldsymbol{e}_x\int \mathrm{d}x'\,\mathrm{d}y'\,[\cdots] = \frac{\partial}{\partial x}\int \mathrm{d}x'\,\mathrm{d}y'\,[\cdots] = 0\,. \tag{12.7.16}$$

As an immediate consequence, we have $\boldsymbol{E}(\boldsymbol{r}) = E(y, z)\,\boldsymbol{e}_x$ in (12.4.7) and arrive at

$$E(y, z) = E_{\text{inc}}(y, z) + \mathrm{i}k\int \mathrm{d}x'\,\mathrm{d}y'\,\frac{\mathrm{e}^{\mathrm{i}k\sqrt{x'^2+(y-y')^2+z^2}}}{\sqrt{x'^2+(y-y')^2+z^2}}\,\frac{1}{c}K(y')\,. \tag{12.7.17}$$

The incident plane wave is $E_{\text{inc}}(y,z) = E_0\,\mathrm{e}^{ikz}$, but it is expedient to allow for a decreasing intensity for large $y$ values,

$$E_{\text{inc}}(y,z) = E_0\,\mathrm{e}^{-\epsilon|y|}\,\mathrm{e}^{ikz} \quad \text{with} \quad 0<\epsilon\to 0 \text{ eventually}, \tag{12.7.18}$$

because we will Fourier transform (12.7.17) in $y$ to exploit the fact that the $y'$ integration is a convolution. The exponential drop-off is useful in this context.

From Section 12.2 we know that the field in the aperture far away from the edge equals the incident field,

$$E(y,0) = E_{\text{inc}}(y,0) = E_0\,\mathrm{e}^{-\epsilon|y|} \quad \text{for} \quad ky \gg 1\,, \tag{12.7.19}$$

and from Section 12.5 we know that the induced surface current far way from the edge is given by the incident field,

$$K(y) = \frac{c}{2\pi}E_{\text{inc}}(y,0) = \frac{c}{2\pi}E_0\,\mathrm{e}^{-\epsilon|y|} \quad \text{for} \quad -ky \gg 1\,, \tag{12.7.20}$$

Together with — these are (12.4.9) and (12.4.8) in the present context —

$$\begin{aligned} E(y,0) &= 0 \quad \text{for} \quad y<0 \quad \text{(perfect conductor)} \\ \text{and} \quad K(y) &= 0 \quad \text{for} \quad y>0 \quad \text{(no conductor there)} \end{aligned} \tag{12.7.21}$$

this tells us that the Fourier transforms $f(t)$ and $g(t)$,

$$\begin{aligned} E(y,0) &= E_0\int\frac{\mathrm{d}t}{2\pi}\,\mathrm{e}^{ity}f(t)\,, \\ K(y) &= \frac{c}{2\pi}E_0\int\frac{\mathrm{d}t}{2\pi}\,\mathrm{e}^{ity}g(t)\,, \end{aligned} \tag{12.7.22}$$

exist and are regular in a bit more than half of the complex $t$ plane,

$$\begin{aligned} &f(t) \text{ is regular for } \mathrm{Im}(t) < \epsilon, \\ &g(t) \text{ is regular for } \mathrm{Im}(t) > -\epsilon, \end{aligned} \tag{12.7.23}$$

respectively.

The integral equation that relates $E(y,0)$ to $K(y)$, this is (12.7.17) for $z=0$,

$$E(y,0) = E_0\,\mathrm{e}^{-\epsilon|y|} + ik\int\mathrm{d}y'\int\mathrm{d}x'\,\frac{\mathrm{e}^{ik\sqrt{x'^2+(y-y')^2}}}{\sqrt{x'^2+(y-y')^2}}\frac{1}{c}K(y')\,, \tag{12.7.24}$$

contains the integral kernel

$$\int \mathrm{d}x \, \frac{\mathrm{e}^{\mathrm{i}k\sqrt{x^2+y^2}}}{\sqrt{x^2+y^2}} \,, \tag{12.7.25}$$

a function of $y$ whose Fourier transform we will need. Actually, this integral defines a so-called Hankel* function which, in a very specific sense, is the analog of $\mathrm{e}^{\mathrm{i}y}$ if the Bessel function $\mathrm{J}_0(y)$ is the analog of $\cos y$, but we do not need to bother with the properties of Hankel functions. Instead, we return to Chapter 6 and recall two versions of the retarded Green's function in (6.1.24) and (6.1.16),

$$\begin{aligned} G_+(\boldsymbol{r},t) &= \int \frac{\mathrm{d}\omega}{2\pi} \, \mathrm{e}^{-\mathrm{i}\omega t} \frac{\mathrm{e}^{\mathrm{i}(\omega/c+\mathrm{i}\epsilon)r}}{r} \\ &= \int \frac{\mathrm{d}\omega}{2\pi} \, \mathrm{e}^{-\mathrm{i}\omega t} \int \frac{(\mathrm{d}\boldsymbol{k})}{(2\pi)^3} \frac{4\pi}{k^2-(\omega/c+\mathrm{i}\epsilon)^2} \, \mathrm{e}^{\mathrm{i}\boldsymbol{k}\cdot\boldsymbol{r}} \end{aligned} \tag{12.7.26}$$

where, as always, $0<\epsilon\to 0$ is understood. For $\boldsymbol{r}$ in the $z=0$ plane, these imply

$$\begin{aligned} \int \mathrm{d}x \, \frac{\mathrm{e}^{\mathrm{i}(\omega/c+\mathrm{i}\epsilon)\sqrt{x^2+y^2}}}{\sqrt{x^2+y^2}} &= \int \frac{(\mathrm{d}\boldsymbol{k})}{(2\pi)^3} \frac{4\pi}{k^2-(\omega/c+\mathrm{i}\epsilon)^2} \, \mathrm{e}^{\mathrm{i}k_y y} 2\pi\delta(k_x) \\ &= \frac{1}{\pi} \int \mathrm{d}k_y \, \mathrm{d}k_z \, \frac{\mathrm{e}^{\mathrm{i}k_y y}}{k_y^2+k_z^2-(\omega/c+\mathrm{i}\epsilon)^2} \\ \overset{(k_y=t,\,k_z=s)}{=} & \int \frac{\mathrm{d}t}{2\pi} \, \mathrm{e}^{\mathrm{i}ty} \underbrace{\int \mathrm{d}s \, \frac{2}{s^2+t^2-(\omega/c+\mathrm{i}\epsilon)^2}}_{=\frac{2\pi\mathrm{i}}{\sqrt{(\omega/c+\mathrm{i}\epsilon)^2-t^2}}} \end{aligned} \tag{12.7.27}$$

or, equivalently,

$$\int \mathrm{d}x \, \frac{\mathrm{e}^{\mathrm{i}(k+\mathrm{i}\epsilon)\sqrt{x^2+y^2}}}{\sqrt{x^2+y^2}} = \int \frac{\mathrm{d}t}{2\pi} \, \mathrm{e}^{\mathrm{i}ty} \frac{2\pi\mathrm{i}}{\sqrt{(k+\mathrm{i}\epsilon)^2-t^2}} \,, \tag{12.7.28}$$

where the reciprocal square root has branch points at $t=\pm(k+\mathrm{i}\epsilon)$, and we choose the cuts such they do not enter the $\epsilon$-strip $-\epsilon<\mathrm{Im}(t)<\epsilon$ around the real $t$ axis. This $\epsilon$-strip is the overlap of the regularity regions for $f(t)$ and $g(t)$ that we identified in (12.7.23) above, which tells us that the various $\epsilon$s are actually one and the same $\epsilon$ parameter.

*Hermann HANKEL (1839–1873)

Upon inserting the Fourier integrals for $E(y,0)$, $K(y)$, and the integral kernel into the integral equation (12.7.24), we get

$$\begin{aligned}\int \frac{\mathrm{d}t}{2\pi}\,\mathrm{e}^{\mathrm{i}yt} f(t) &= \mathrm{e}^{-\epsilon|y|} + \mathrm{i}k \int \mathrm{d}y' \int \frac{\mathrm{d}t}{2\pi}\,\mathrm{e}^{\mathrm{i}t(y-y')} \frac{2\pi\mathrm{i}}{\sqrt{(k+\mathrm{i}\epsilon)^2 - t^2}} \\ &\qquad \times \int \frac{\mathrm{d}t'}{2\pi}\,\mathrm{e}^{\mathrm{i}t'y'} \frac{1}{2\pi} g(t') \\ &= \mathrm{e}^{-\epsilon|y|} - k \int \frac{\mathrm{d}t}{2\pi}\,\mathrm{e}^{\mathrm{i}ty} \frac{g(t)}{\sqrt{(k+\mathrm{i}\epsilon)^2 - t^2}}\,, \end{aligned} \tag{12.7.29}$$

and with

$$\int \mathrm{d}y\ \mathrm{e}^{-\mathrm{i}yt}\,\mathrm{e}^{-\epsilon|y|} = -\frac{\mathrm{i}}{t-\mathrm{i}\epsilon} + \frac{\mathrm{i}}{t+\mathrm{i}\epsilon} \tag{12.7.30}$$

this establishes

$$f(t) = -\frac{\mathrm{i}}{t-\mathrm{i}\epsilon} + \frac{\mathrm{i}}{t+\mathrm{i}\epsilon} - \frac{kg(t)}{\sqrt{(k+\mathrm{i}\epsilon)^2 - t^2}}\,. \tag{12.7.31}$$

In the complex $t$ plane,

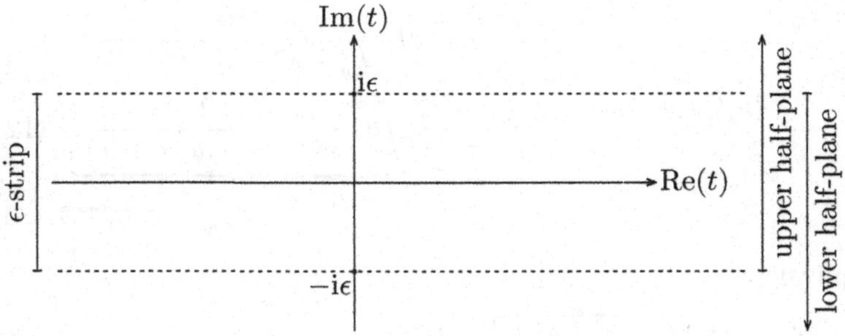


(12.7.32)

we have the $\epsilon$-strip of $-\epsilon < \mathrm{Im}(t) < \epsilon$, the lower half-plane of $\mathrm{Im}(t) < \epsilon$ and the upper half-plane of $\mathrm{Im}(t) > -\epsilon$. We observe that

$$\begin{aligned} &f(t) \text{ and } \frac{1}{t-\mathrm{i}\epsilon} \text{ are regular for } \mathrm{Im}(t) < \epsilon, \\ &g(t) \text{ and } \frac{1}{t+\mathrm{i}\epsilon} \text{ are regular for } \mathrm{Im}(t) > -\epsilon, \\ &\frac{1}{\sqrt{(k+\mathrm{i}\epsilon)^2 - t^2}} \text{ is regular in the } \epsilon \text{ strip.} \end{aligned} \tag{12.7.33}$$

We collect the terms regular for $\mathrm{Im}(t) < \epsilon$ on the left and multiply by $\sqrt{k + \mathrm{i}\epsilon - t}$, which is regular for $\mathrm{Im}(t) < \epsilon$ as well,

$$\underbrace{\sqrt{k+\mathrm{i}\epsilon - t}\left[f(t) + \frac{\mathrm{i}}{t-\mathrm{i}\epsilon}\right]}_{\text{regular for } \mathrm{Im}(t) < \epsilon} = \underbrace{\mathrm{i}\frac{\sqrt{k+\mathrm{i}\epsilon-t}}{t+\mathrm{i}\epsilon}}_{\substack{\text{regular in}\\ \text{the } \epsilon \text{ strip}}} - \underbrace{\frac{kg(t)}{\sqrt{k+\mathrm{i}\epsilon+t}}}_{\substack{\text{regular for}\\ \mathrm{Im}(t) > -\epsilon}} . \tag{12.7.34}$$

The identity

$$\frac{\sqrt{k+\mathrm{i}\epsilon-t}}{t+\mathrm{i}\epsilon} = \underbrace{\frac{\sqrt{k+\mathrm{i}\epsilon-t}-\sqrt{k+2\mathrm{i}\epsilon}}{t+\mathrm{i}\epsilon}}_{\text{regular for } \mathrm{Im}(t) < \epsilon} + \underbrace{\frac{\sqrt{k+2\mathrm{i}\epsilon}}{t+\mathrm{i}\epsilon}}_{\substack{\text{regular for}\\ \mathrm{Im}(t) > -\epsilon}} \tag{12.7.35}$$

decomposes the $\epsilon$-strip-regular term in (12.7.34) into a term that is regular for $\mathrm{Im}(t) < \epsilon$ and a term that is regular for $\mathrm{Im}(t) > -\epsilon$, so that

$$\begin{aligned} &\sqrt{k+\mathrm{i}\epsilon-t}\, f(t) + \mathrm{i}\frac{\sqrt{k+\mathrm{i}\epsilon-t}}{t-\mathrm{i}\epsilon} - \mathrm{i}\frac{\sqrt{k+\mathrm{i}\epsilon-t}-\sqrt{k+2\mathrm{i}\epsilon}}{t+\mathrm{i}\epsilon} \\ &= \mathrm{i}\frac{\sqrt{k+2\mathrm{i}\epsilon}}{t+\mathrm{i}\epsilon} - \frac{k}{\sqrt{k+\mathrm{i}\epsilon+t}}g(t) \end{aligned} \tag{12.7.36}$$

has a function that is regular for $\mathrm{Im}(t) < \epsilon$ on the left-hand side, and a function that is regular for $\mathrm{Im}(t) > -\epsilon$ on the right-hand side. Since either function is the analytical continuation of the other one beyond the common $\epsilon$-strip, the two sides jointly define a function that is regular for all $t$.

Let us consider the behavior of $E(y, 0)$ and $K(y)$ for $k|y| \ll 1$, that is: for $y$ values near the edge. Since there is no intrinsic length scale, we must have

$$\begin{aligned} E(y,0) \propto y^{\alpha} \quad &\text{for} \quad 0 < ky \ll 1 \\ \text{and} \quad K(y) \propto (-y)^{\beta} \quad &\text{for} \quad 0 < -ky \ll 1 \end{aligned} \tag{12.7.37}$$

with $\alpha > 0$ to ensure $E(y \to 0, 0) \to 0$ and $\beta > -1$ for the existence of

$$\frac{c}{2\pi}E_0 g(t) = \int \mathrm{d}y\; \mathrm{e}^{-\mathrm{i}yt} K(y) . \tag{12.7.38}$$

These small-$y$ forms imply that

$$f(t) \propto t^{-(1+\alpha)}, \quad g(t) \propto t^{-(1+\beta)} \quad \text{for} \quad |t| \gg k . \tag{12.7.39}$$

Therefore, the function that is regular for $\mathrm{Im}(t) < \epsilon$ on the left-hand side of (12.7.36) is $\propto t^{-(\frac{1}{2}+\alpha)}$ for large $t$, and the function that is regular for $\mathrm{Im}(t) > -\epsilon$ on the right-hand side is $\propto t^{-(\frac{3}{2}+\beta)}$ for large $t$. It follows that the over-all regular function vanishes for $|t| \to \infty$, which in turn implies that the function vanishes for all $t$ values.

In summary, then, we find

$$\begin{aligned} f(t) &= -\frac{\mathrm{i}}{t-\mathrm{i}\epsilon} + \frac{\mathrm{i}}{t+\mathrm{i}\epsilon} - \mathrm{i}\frac{\sqrt{k+2\mathrm{i}\epsilon}}{\sqrt{k+\mathrm{i}\epsilon-t}\,(t+\mathrm{i}\epsilon)} \\ &= \frac{2\epsilon}{t^2+\epsilon^2} - \mathrm{i}\frac{\sqrt{k}}{\sqrt{k+\mathrm{i}\epsilon-t}\,(t+\mathrm{i}\epsilon)}\,, \end{aligned} \tag{12.7.40}$$

where the first term represents the incident field,

$$\frac{2\epsilon}{t^2+\epsilon^2} \xrightarrow[\epsilon\to 0]{} 2\pi\delta(t) = \int \mathrm{d}y\; \mathrm{e}^{-\mathrm{i}ty}\,, \tag{12.7.41}$$

and the second term represents the diffracted field that is emitted by the induced surface current. For that current we have

$$g(t) = \mathrm{i}\frac{\sqrt{k+2\mathrm{i}\epsilon}}{k}\frac{\sqrt{k+\mathrm{i}\epsilon+t}}{t+\mathrm{i}\epsilon} = \frac{\mathrm{i}}{\sqrt{k}}\frac{\sqrt{k+\mathrm{i}\epsilon+t}}{t+\mathrm{i}\epsilon}\,. \tag{12.7.42}$$

The limit $0 < \epsilon \to 0$ is to be taken when it is convenient and unambiguous.

We shall make no attempt at evaluating the integrals in

$$\begin{aligned} E(y,0) &= E_0 \int \frac{\mathrm{d}t}{2\pi}\,\mathrm{e}^{\mathrm{i}ty}\left[\frac{2\epsilon}{t^2+\epsilon^2} - \frac{\sqrt{k}}{\sqrt{k+\mathrm{i}\epsilon-t}}\frac{\mathrm{i}}{t+\mathrm{i}\epsilon}\right]\Bigg|_{0<\epsilon\to 0} \\ &= E_0 - \mathrm{i}E_0 \int \frac{\mathrm{d}t}{2\pi}\,\frac{\mathrm{e}^{\mathrm{i}ty}}{t+\mathrm{i}\epsilon}\sqrt{\frac{k}{k+\mathrm{i}\epsilon-t}}\,\Bigg|_{0<\epsilon\to 0} \end{aligned} \tag{12.7.43}$$

and

$$K(y) = \frac{c}{2\pi}\mathrm{i}E_0 \int \frac{\mathrm{d}t}{2\pi}\,\frac{\mathrm{e}^{\mathrm{i}ty}}{t+\mathrm{i}\epsilon}\sqrt{\frac{k+\mathrm{i}\epsilon+t}{k}}\,\Bigg|_{0<\epsilon\to 0}\,. \tag{12.7.44}$$

It is possible to get the diffraction cross section directly. The diffracted field is

$$E_{\mathrm{diff}}(\boldsymbol{r}) = \mathrm{i}k \int (\mathrm{d}\boldsymbol{r}'_\perp)\frac{\mathrm{e}^{\mathrm{i}k|\boldsymbol{r}-\boldsymbol{r}'_\perp|}}{|\boldsymbol{r}-\boldsymbol{r}'_\perp|}\frac{1}{c}K(\boldsymbol{r}'_\perp)\,, \tag{12.7.45}$$

where $\boldsymbol{r} \to \begin{pmatrix} 0 \\ y \\ z \end{pmatrix}$ in $E_{\text{diff}}(\boldsymbol{r})$ and $\boldsymbol{r}'_\perp \to \begin{pmatrix} x' \\ y' \\ 0 \end{pmatrix}$ in $K(\boldsymbol{r}'_\perp)$. For the radiation field we have, as usual,

$$|\boldsymbol{r} - \boldsymbol{r}'_\perp| \to r = \sqrt{y^2 + z^2} \tag{12.7.46}$$

in the denominator, and in the phase factor we use

$$\begin{aligned} \boldsymbol{r} &= \begin{pmatrix} 0 \\ s\sin\theta \\ s\cos\theta \end{pmatrix}, \qquad \boldsymbol{r}'_\perp = \begin{pmatrix} x' \\ y' \\ 0 \end{pmatrix}, \\ |\boldsymbol{r} - \boldsymbol{r}'_\perp| &= \sqrt{s^2 - 2sy'\sin\theta + x'^2 + y'^2} \\ &\to s - y'\sin\theta + \frac{1}{2}\frac{x'^2}{s}, \end{aligned} \tag{12.7.47}$$

where we keep the respective leading orders for the integration variables $x'$ and $y'$. This gives

$$\begin{aligned} E_{\text{diff}}(\boldsymbol{r}) &= \mathrm{i}k \int \mathrm{d}x'\,\mathrm{d}y'\, \frac{\mathrm{e}^{\mathrm{i}ks}\,\mathrm{e}^{-\mathrm{i}ky'\sin\theta}\,\mathrm{e}^{\frac{\mathrm{i}}{2}kx'^2/s}}{s}\,\frac{1}{c}K(y') \\ &\overset{(12.7.38)}{=} \mathrm{i}k\sqrt{\frac{\pi}{-\frac{\mathrm{i}}{2}k/s}}\,\frac{\mathrm{e}^{\mathrm{i}ks}}{s}\,\frac{E_0}{2\pi}\,g(k\sin\theta) \\ &= -E_0\sqrt{\frac{k}{\mathrm{i}2\pi s}}\,\mathrm{e}^{\mathrm{i}ks}\,\frac{\mathrm{i}}{\sqrt{k}}\,\frac{\sqrt{k+\mathrm{i}\epsilon+k\sin\theta}}{k\sin\theta+\mathrm{i}\epsilon}\Bigg|_{0<\epsilon\to 0} \\ &= -E_0\sqrt{\frac{\mathrm{i}}{2\pi ks}}\,\mathrm{e}^{\mathrm{i}ks}\,\frac{\sqrt{1+\sin\theta}}{\sin\theta}, \end{aligned} \tag{12.7.48}$$

whereby

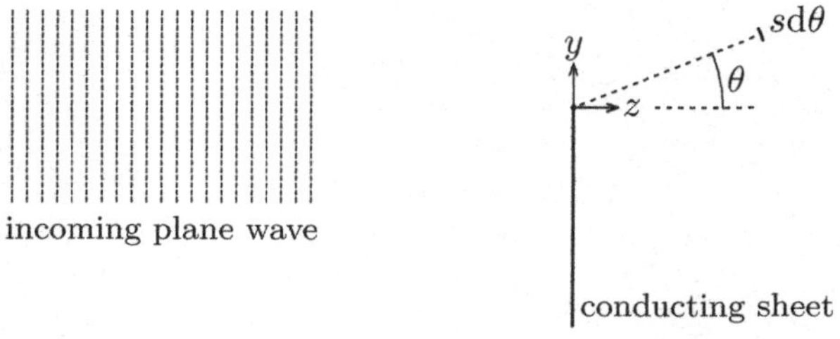


(12.7.49)

states the geometrical significance of the diffraction angle $\theta$.

The comparison between the diffracted intensity and the incident intensity establishes the diffraction cross section,

$$d\sigma = \frac{|E_{\text{diff}}|^2 s d\theta}{|E_0|^2} = \frac{1}{2\pi k} \frac{1 + \sin\theta}{(\sin\theta)^2} d\theta\,, \tag{12.7.50}$$

namely

$$\frac{d\sigma}{d\theta} = \frac{1}{2\pi k} \frac{1 + \sin\theta}{(\sin\theta)^2}\,, \tag{12.7.51}$$

a cross section per unit length (in $x$).

The small-angle cross section,

$$\frac{d\sigma}{d\theta} = \frac{1}{2\pi k} \frac{1}{\theta^2} = \frac{\lambda}{(2\pi\theta)^2} \quad \text{for} \quad |\theta| \ll 1\,, \tag{12.7.52}$$

can also be derived by the standard approximations for larger apertures. Regard this as an exercise, for which Section 12.7.1 could be useful.

Finally, we note that

$$f(t) \propto t^{-\frac{3}{2}}\,, \quad g(t) \propto t^{-\frac{1}{2}} \qquad \text{for large } t, \tag{12.7.53}$$

which identifies the powers $\alpha = \frac{1}{2}$, $\beta = -\frac{1}{2}$ in (12.7.37) and so tells us that

$$E(y, 0) \propto y^{\frac{1}{2}} \quad \text{for} \quad 0 < ky \ll 1 \tag{12.7.54}$$

and

$$K(y) \propto (-y)^{-\frac{1}{2}} \quad \text{for} \quad 0 < -ky \ll 1\,. \tag{12.7.55}$$

These square-root dependences are typical for the near-edge forms of the field in an aperture and the current on the conducting surface.

# Exercises with Hints

## Mathematical tools

**1** Show that

$$\nabla \boldsymbol{A} \cdot \boldsymbol{B} = \boldsymbol{B} \times (\nabla \times \boldsymbol{A}) + \boldsymbol{B} \cdot \nabla \boldsymbol{A} + \boldsymbol{A} \times (\nabla \times \boldsymbol{B}) + \boldsymbol{A} \cdot \nabla \boldsymbol{B},$$

where $\boldsymbol{A}$ and $\boldsymbol{B}$ are vector functions of position vector $\boldsymbol{r}$.

**2** Show that

$$\nabla \cdot (\lambda \boldsymbol{A} \times \boldsymbol{B}) = \lambda \boldsymbol{B} \cdot (\nabla \times \boldsymbol{A}) - \lambda \boldsymbol{A} \cdot (\nabla \times \boldsymbol{B}) + (\boldsymbol{A} \times \boldsymbol{B}) \cdot \nabla \lambda,$$

where $\lambda$ is a scalar function of position vector $\boldsymbol{r}$, and $\boldsymbol{A}$ and $\boldsymbol{B}$ are vector functions of $\boldsymbol{r}$.

**3** Use the familiar Jacobi* identity for double vector products to show that

$$\nabla \times (\boldsymbol{A} \times \boldsymbol{B}) = \boldsymbol{A} \times (\nabla \times \boldsymbol{B}) - \boldsymbol{B} \times (\nabla \times \boldsymbol{A}) - (\boldsymbol{A} \times \nabla) \times \boldsymbol{B} + (\boldsymbol{B} \times \nabla) \times \boldsymbol{A},$$

where $\boldsymbol{A}$ and $\boldsymbol{B}$ are vector functions of $\boldsymbol{r}$. Which alternative version of the right-hand side do you get when expressing all double cross products as differences of two vectors?

**4** Use the generating function (10.4.3) for the Bessel functions $\mathrm{J}_m(t)$ to derive the recurrence relations (10.4.13).

**5** The transverse vectors $\boldsymbol{k}_\perp$, $\boldsymbol{r}_\perp$, $\boldsymbol{r}'_\perp$ are perpendicular to the $z$ direction and have components in the $xy$ plane only. Evaluate the two-dimensional

*Carl Gustav Jacob Jacobi (1804–1851)

integral in

$$\delta(\boldsymbol{r}_\perp - \boldsymbol{r}'_\perp) = \int \frac{(\mathrm{d}\boldsymbol{k}_\perp)}{(2\pi)^2}\, \mathrm{e}^{\mathrm{i}\boldsymbol{k}_\perp \cdot (\boldsymbol{r}_\perp - \boldsymbol{r}'_\perp)}$$

in two different ways to derive the addition theorem for Bessel functions,

$$\mathrm{J}_0\Bigl(k_\perp |\boldsymbol{r}_\perp - \boldsymbol{r}'_\perp|\Bigr) = \sum_{m=-\infty}^{\infty} \mathrm{e}^{im\phi}\, \mathrm{J}_m(k_\perp r_\perp)\, \mathrm{J}_m(k_\perp r'_\perp)\,,$$

where $\phi$ is the angle between $\boldsymbol{r}_\perp$ and $\boldsymbol{r}'_\perp$.

**6** In cylindrical coordinates we parameterize the position vector $\boldsymbol{r}$ by $(x, y, z) = (s\cos\phi, s\sin\phi, z)$ with $s \geq 0$. Show that

$$\frac{1}{r} = \frac{1}{\sqrt{s^2 + z^2}} = \int_0^\infty \mathrm{d}k\, \mathrm{J}_0(ks)\, \mathrm{e}^{-k|z|}\,.$$

**7** Apply the identity of Exercise 6 and the addition theorem of Exercise 5 to express the Coulomb Green's function,

$$\frac{1}{|\boldsymbol{r} - \boldsymbol{r}'|} = \frac{1}{\sqrt{s^2 + s'^2 - 2ss'\cos(\phi - \phi') + (z - z')^2}}\,,$$

as a sum of products in which each factor refers either to $\boldsymbol{r}_\perp$ or to $\boldsymbol{r}'_\perp$.

## Chapter 1

**8** The charge density of an electric point dipole $\boldsymbol{d}$ at rest at $\boldsymbol{r} = 0$ is given by $\rho(\boldsymbol{r}) = -\boldsymbol{d}\cdot\boldsymbol{\nabla}\delta(\boldsymbol{r})$. Verify that $\int (\mathrm{d}\boldsymbol{r})\, \rho(\boldsymbol{r}) = 0$ and $\int (\mathrm{d}\boldsymbol{r})\, \boldsymbol{r}\rho(\boldsymbol{r}) = \boldsymbol{d}$. Then find the electrostatic potential $\Phi(\boldsymbol{r})$ and the electric field $\boldsymbol{E}(\boldsymbol{r})$ of the point dipole. You may find it convenient to make use of the dyadic double gradient of $\dfrac{1}{r}$ that is given by

$$\boldsymbol{\nabla\nabla}\frac{1}{r} = \frac{3\,\boldsymbol{r}\,\boldsymbol{r} - r^2\,\boldsymbol{1}}{r^5} - \frac{4\pi}{3}\boldsymbol{1}\,\delta(\boldsymbol{r})\,.$$

Why do we need to subtract the "contact term" $\dfrac{4\pi}{3}\boldsymbol{1}\,\delta(\boldsymbol{r})$?

**9** The static vector potential $\boldsymbol{A}(\boldsymbol{r}) = \boldsymbol{\nabla} \times \frac{\boldsymbol{\mu}}{r}$, where $\boldsymbol{\mu}$ is a constant magnetic dipole, gives rise to the static magnetic field $\boldsymbol{B}(\boldsymbol{r})$. State $\boldsymbol{B}(\boldsymbol{r})$ and find the current density $\boldsymbol{j}(\boldsymbol{r})$.

**10** A ball of radius $R$ with charge $e$ uniformly distributed over its surface rotates about an axis through its center at an angular frequency $\omega$. What is the electric field $\boldsymbol{E}(\boldsymbol{r})$ and the magnetic field $\boldsymbol{B}(\boldsymbol{r})$, both inside and outside the sphere?

**11** For the situation of Exercise 10, find the total energy and the total angular momentum contained in the electric and magnetic fields. In which direction is the electromagnetic energy current density just outside the surface of the sphere?

**12** Suppose one solves (1.3.12) with the aid of the same Green's function $G(\boldsymbol{r} - \boldsymbol{r}', t - t')$ in

$$\Phi(\boldsymbol{r}, t) = \int (\mathrm{d}\boldsymbol{r}') \int \mathrm{d}t'\, G(\boldsymbol{r} - \boldsymbol{r}', t - t')\, \rho(\boldsymbol{r}', t')$$

and in

$$\boldsymbol{A}(\boldsymbol{r}, t) = \int (\mathrm{d}\boldsymbol{r}') \int \mathrm{d}t'\, G(\boldsymbol{r} - \boldsymbol{r}', t - t')\, \frac{1}{c}\boldsymbol{j}(\boldsymbol{r}', t') \,.$$

Do the resulting potentials obey the Lorentz gauge condition, as they should?

**13** Since (1.3.14) applies in the radiation gauge, the divergence of the left-hand side vanishes and, therefore, consistency requires that the effective current density $\boldsymbol{j}_{\mathrm{eff}} = \boldsymbol{j} - \frac{1}{4\pi}\frac{\partial}{\partial t}\boldsymbol{\nabla}\Phi$ on the right-hand side is divergenceless. Verify this. What is the analogous consistency check for (1.3.12)?

**14** What are the eigenvalues and eigenvectors of the momentum current density dyadic $\mathsf{T}$ of (1.5.9)? Make use of these eigenvalues in finding the determinant of $\mathsf{T}$ and the trace of $\mathsf{T}^2$.

**15** Consider the "electromagnetic rotation"

$$\boldsymbol{E} \to \boldsymbol{E}\cos\phi + \boldsymbol{B}\sin\phi \,, \qquad \boldsymbol{B} \to \boldsymbol{B}\cos\phi - \boldsymbol{E}\sin\phi \,,$$

where $\phi$ is a real angle parameter (see page 40). How do $U$, $\boldsymbol{G}$, and $\mathsf{T}$ appear after the rotation?

**16** Show that one can always choose the potentials $\Phi$ and $\boldsymbol{A}$ such that they obey the gauge condition

$$\boldsymbol{r}\cdot\boldsymbol{A}(\boldsymbol{r},t)=ct\Phi(\boldsymbol{r},t)\,.$$

This gauge is called the "relativistic Poincaré* gauge" by some, the "Fock†–Schwinger gauge" by others, and the "line gauge" by yet others. In case you wonder: The "Poincaré gauge" is specified by $\boldsymbol{r}\cdot\boldsymbol{A}=0$; alternative names include "point gauge."

**17** Verify that

$$\Phi(\boldsymbol{r},t)=-\int_0^1 \mathrm{d}\kappa\,\kappa\boldsymbol{r}\cdot\boldsymbol{E}(\kappa\boldsymbol{r},\kappa t)\,,$$

$$\boldsymbol{A}(\boldsymbol{r},t)=-\int_0^1 \mathrm{d}\kappa\,\Bigl(\kappa ct\boldsymbol{E}(\kappa\boldsymbol{r},\kappa t)+\kappa\boldsymbol{r}\times\boldsymbol{B}(\kappa\boldsymbol{r},\kappa t)\Bigr)$$

are valid potentials in the relativistic Poincaré gauge of Exercise 16.

**18** Consider a collection of charges in nonrelativistic motion, so that

$$U_{\mathrm{ch}}(\boldsymbol{r},t)=\sum_j\delta\bigl(\boldsymbol{r}-\boldsymbol{r}_j(t)\bigr)\,\frac{1}{2}m_j\boldsymbol{v}_j(t)^2\,,$$

$$\boldsymbol{S}_{\mathrm{ch}}(\boldsymbol{r},t)=\sum_j\delta\bigl(\boldsymbol{r}-\boldsymbol{r}_j(t)\bigr)\,\frac{1}{2}m_j\boldsymbol{v}_j(t)^2\boldsymbol{v}_j(t)\,.$$

Show that

$$\frac{\partial}{\partial t}U_{\mathrm{ch}}+\boldsymbol{\nabla}\cdot\boldsymbol{S}_{\mathrm{ch}}=\boldsymbol{j}\cdot\boldsymbol{E}\,.$$

How does the local energy conservation follow from this?

**19** Proceed analogously for

$$\boldsymbol{G}_{\mathrm{ch}}(\boldsymbol{r},t)=\sum_j\delta\bigl(\boldsymbol{r}-\boldsymbol{r}_j(t)\bigr)\,m_j\boldsymbol{v}_j(t)\,,$$

$$\mathsf{T}_{\mathrm{ch}}(\boldsymbol{r},t)=\sum_j\delta\bigl(\boldsymbol{r}-\boldsymbol{r}_j(t)\bigr)\,m_j\boldsymbol{v}_j(t)\boldsymbol{v}_j(t)\,.$$

*Jules Henri POINCARÉ (1854–1912) †Vladimir Alexandrovich FOCK (1898–1974)

**20** And now for angular momentum.

## Chapter 2

**21** Consider an electromagnetic pulse with the fields

$$\begin{aligned} \boldsymbol{E}(\boldsymbol{r},t) &= \int \frac{(\mathrm{d}\boldsymbol{k})}{(2\pi)^3}\, \boldsymbol{E}(\boldsymbol{k}) \sin(\boldsymbol{k}\cdot\boldsymbol{r} - kct)\,, \\ \boldsymbol{B}(\boldsymbol{r},t) &= \int \frac{(\mathrm{d}\boldsymbol{k})}{(2\pi)^3}\, \boldsymbol{B}(\boldsymbol{k}) \sin(\boldsymbol{k}\cdot\boldsymbol{r} - kct)\,, \end{aligned}$$

where $k\boldsymbol{B}(\boldsymbol{k}) = \boldsymbol{k}\times\boldsymbol{E}(\boldsymbol{k})$ and $k\boldsymbol{E}(\boldsymbol{k}) = -\boldsymbol{k}\times\boldsymbol{B}(\boldsymbol{k})$ for all $\boldsymbol{k}$. Express $E$, $P$, and $\boldsymbol{J}$ of (2.1.1)–(2.1.3) as integrals over combinations of $\boldsymbol{E}(\boldsymbol{k})$ and $\boldsymbol{B}(\boldsymbol{k})$ and thus verify that these quantities do not depend on time.

**22** Now find the integral expressions for $\langle\boldsymbol{r}\rangle_E(t)$ and $\langle\boldsymbol{r}\rangle_P(t)$ and verify that (2.2.5) and (2.2.8) are obeyed.

## Chapter 3

**23** For Lorentz boosts in a common direction, parameterized by the rapidity $\theta$ of Section 3.2, show that two successive boosts with $\theta_1$ and $\theta_2$, respectively, amount to a single boost with $\theta = \theta_1 + \theta_2$. What is the corresponding statement about the boost velocities $v_1$, $v_2$, and $v$?

**24** Consider two independent infinitesimal Lorentz transformations as in (3.2.1) carried out successively, whereby $\delta\boldsymbol{v}_1$ and $\delta\boldsymbol{v}_2$ need not be parallel. Does it matter in which order the two transformations are performed? If yes, what is the net difference?

**25** An observer sees a particle move with velocity $\boldsymbol{v}$, which could depend on the observation time $t$. Which *proper time* interval $\mathrm{d}s$, that is: the time interval in the instantaneous rest frame of the particle, corresponds to the observer's time interval $\mathrm{d}t$?

**26** Mr. Ah Beng is eyeing Miss Ah Lian, who sees a particle moving with velocity $\boldsymbol{u}$. If he took his eyes off her and watched the particle as well, which velocity $\boldsymbol{u}'$ would he observe, given that she has velocity $\boldsymbol{v}$ relative to him, with $\boldsymbol{v}$ perpendicular to $\boldsymbol{u}$? Verify that $u' \leq c$ for $v < c$ and $u \leq c$.

**27** Observer A uses unprimed coordinates, observer B uses primed coordinates. They move relative to each other with velocity $\boldsymbol{v}$. By considering both infinitesimal Lorentz transformations and a finite Lorentz transformation, show that both observers see the same total charge, that is

$$\int (\mathrm{d}\boldsymbol{r})\, \rho(\boldsymbol{r},t) = \int (\mathrm{d}\boldsymbol{r}')\, \rho'(\boldsymbol{r}',t')\,,$$

where the integrations cover all of space.

**28** How does $\boldsymbol{E}\cdot\boldsymbol{B}$ transform under Lorentz transformations, like a 4-scalar, like the time-like component of a 4-vector, or like something else? Answer the same question about $\boldsymbol{E}^2 - \boldsymbol{B}^2$.

**29** A uniformly charged ball of radius $a$, which is at rest, carries total charge $e$. Evaluate its electrical energy. If this uniform charge distribution moves with a *small* constant velocity $\boldsymbol{v}$ ($v = |\boldsymbol{v}| \ll c$), what is its magnetic energy? What is its electromagnetic momentum?

**30** We can associate a rest mass $m_1$ with this uniformly charged ball by equating the electrical energy (of the ball at rest) with $m_1c^2$. We can also identify an inertial mass $m_2$ by setting the magnetic energy of the slowly moving ball equal to $\frac{1}{2}m_2v^2$. And we can further identify another inertial mass by equating the electromagnetic momentum with $m_3\boldsymbol{v}$. Show that the two inertial masses are the same and that they differ from the rest mass: $m_1 \neq m_2 = m_3$.

**31** Point charge $e$ moves along the $z$ axis with constant speed $v$. Find the electric field $\boldsymbol{E}(\boldsymbol{r},t)$ by differentiating the potentials in (3.5.7) and also by making use of the Lorentz transformation in (3.5.20)–(3.5.22). Sketch the electric field lines at the instant when the charge is at $\boldsymbol{r} = 0$ and compare with the field lines of a charge at rest.

**32** Electron 1 is moving along the $z$ axis with constant velocity $\boldsymbol{v} = v\,\boldsymbol{e}_z$, so that its trajectory is $\boldsymbol{r}_1(t) = \boldsymbol{v}t$. Electron 2 is co-moving at a fixed distance $\boldsymbol{a}$ from electron 1, so that its trajectory is given by $\boldsymbol{r}_2(t) = \boldsymbol{r}_1(t) + \boldsymbol{a}$. Upon denoting by $\boldsymbol{E}(\boldsymbol{r},t)$ the electric field associated with electron 1, show — by a *very simple* argument — that the corresponding magnetic field $\boldsymbol{B}(\boldsymbol{r},t)$

is given by $\boldsymbol{B} = \frac{\boldsymbol{v}}{c} \times \boldsymbol{E}$. Then determine the force $\boldsymbol{F}$ on electron 2, and express your answer in terms of the parallel and perpendicular components of $\boldsymbol{a} = \boldsymbol{a}_\parallel + \boldsymbol{a}_\perp$ with $\boldsymbol{v} \times \boldsymbol{a}_\parallel = 0$ and $\boldsymbol{v} \cdot \boldsymbol{a}_\perp = 0$. For both $\boldsymbol{a} = \boldsymbol{a}_\perp$ and $\boldsymbol{a} = \boldsymbol{a}_\parallel$, compare $\boldsymbol{F}$ with the force in the common rest frame of the two electrons.

**33** A monochromatic plane wave of frequency $\omega$ propagates with the speed of light at an angle $\vartheta$ with respect to the $z$ axis. Show that this wave will, to an observer moving with relative velocity $v$ along the $z$ axis, have the frequency $\omega'$ and an angle $\vartheta'$ relative to the $z$ axis that are given by

$$\omega' = \frac{\omega}{\sqrt{1-(v/c)^2}}\left(1 - \frac{v}{c}\cos\vartheta\right) \qquad \text{(Doppler effect)}\,,$$

$$\cos\vartheta' = \frac{\cos\vartheta - v/c}{1-(v/c)\cos\vartheta} \qquad \text{(aberration)}\,.$$

**34** A point charge $e$ moves with the constant velocity $\boldsymbol{u}$ in the unprimed coordinate system, so that $\rho(\boldsymbol{r},t) = e\,\delta(\boldsymbol{r}-\boldsymbol{u}t)$, and $\boldsymbol{j}(\boldsymbol{r},t)$ correspondingly. Carry out a finite Lorentz transformation (velocity $\boldsymbol{v}$) to find the charge density and current density in the primed coordinate system. Express $\boldsymbol{u}'$, the velocity of the charge after the transformation, in terms of $\boldsymbol{u}$ and $\boldsymbol{v}$.

**35** The prototype of a monochromatic 4-scalar plane wave is $\lambda(\boldsymbol{r},t) = \mathrm{e}^{\mathrm{i}(\boldsymbol{k}\cdot\boldsymbol{r}-\omega t)}$, where $\boldsymbol{k}$ is the wave vector and $\omega$ is the circular frequency. Show that $\begin{pmatrix}\omega/c\\ \boldsymbol{k}\end{pmatrix}$ is a 4-vector. For a wave that propagates with the speed of light, what is the value of the 4-scalar $(-\frac{\omega}{c}, \boldsymbol{k}^{\mathrm{T}})\begin{pmatrix}\omega/c\\ \boldsymbol{k}\end{pmatrix}$?

## Chapter 4

**36** The tensor $\varepsilon^{\kappa\lambda\mu\nu}$ is totally antisymmetric in its indices, and $\varepsilon^{0123} = +1$. Verify that (4.2.17) can be presented as

$$\partial_\nu {}^*F^{\mu\nu} = 0\,,$$

where the so-called dual field tensor ${}^*F$ is

$${}^*F^{\kappa\lambda} = \frac{1}{2}\varepsilon^{\kappa\lambda\mu\nu}F_{\mu\nu}\,.$$

State the $4 \times 4$ matrix form of ${}^*F^{\mu}{}_{\nu}$. Do you recognize an object that you have met earlier? What is the dual of the dual field tensor?

**37** Verify (4.2.33) and relate the right-hand side to the 4-force density of (3.3.24).

**38** A spherically symmetric distribution of charge $e$ at rest has the potentials $\Phi = ef(r^2)$, $\boldsymbol{A} = 0$, where $f(r^2) \simeq 1/\sqrt{r^2}$ at large distances. Verify that, as observed in uniform relative motion with 4-velocity $u^\mu$, the potentials in the Lorentz gauge are

$$A^\mu(x) = \frac{e}{c} u^\mu f(\xi^\nu \xi_\nu) \qquad \text{with} \quad \xi^\mu = x^\mu + \frac{u^\mu u^\nu x_\nu}{c^2} \,.$$

Evaluate $u^\mu \xi_\mu$ and $\partial_\mu \xi_\nu$, and derive the implied expression for $F^{\mu\nu}$.

**39** For the situation of Exercise 38, find the 4-current density from $\partial_\nu F^{\mu\nu} = \frac{4\pi}{c} j^\mu$, and verify that $\partial_\mu j^\mu = 0$. Then construct $F^{\mu\nu} \frac{1}{c} j_\nu$, and conclude from its 4-vector nature that we can write

$$F^{\mu\nu} \frac{1}{c} j_\nu = -\partial^\mu \, t(\xi^\nu \xi_\nu) \,.$$

What is $t(r^2)$ for $f(r^2) = \dfrac{1}{\sqrt{r^2 + a^2}}$ with $a > 0$?

**40** Next, evaluate the electromagnetic energy-momentum 4-dyadic $T^{\mu\nu}_{\text{field}} = \frac{1}{4\pi}\left(F^{\mu\lambda} F^{\nu}{}_{\lambda} - \frac{1}{4} g^{\mu\nu} F^{\kappa\lambda} F_{\kappa\lambda}\right)$. Which statement in Chapter 4 ensures that

$$T^{\mu\nu} = T^{\mu\nu}_{\text{field}} - g^{\mu\nu} t$$

is a symmetric and divergenceless electromagnetic 4-dyadic, $\partial_\nu T^{\mu\nu} = 0$? This is the basis for a purely electromagnetic relativistic model of mass.

**41** The construction of $T^{\mu\nu}$ in Exercise 40 is not unambiguous because (verify this)

$$\partial_\nu \left( \frac{u^\mu \, u^\nu}{c^2} \, t(\xi^\lambda \xi_\lambda) \right) = 0 \,.$$

Therefore, we could also use

$$T^{\mu\nu} = T^{\mu\nu}_{\text{field}} - \left( g^{\mu\nu} + \frac{u^\mu \, u^\nu}{c^2} \right) t \,,$$

for example. In order to appreciate what speaks in favor of either choice for $T^{\mu\nu}$, show that $T^{0k} = T^{0k}_{\text{field}} = \dfrac{1}{4\pi}(\boldsymbol{E}\times\boldsymbol{B})_k$ for the choice in Exercise 40, whereas one gets $T^{00} = T^{00}_{\text{field}} = \dfrac{1}{8\pi}\boldsymbol{E}^2$ for $\boldsymbol{v} = 0$ for the current choice.

**42** A monochromatic plane light wave of circular frequency $\omega$ is incident on a plane mirror under normal angle $\vartheta$. The mirror moves with constant normal velocity $v$. The reflected wave has frequency $\omega'$ and normal angle $\vartheta'$. For $v = 0$, we have $\omega = \omega'$ and $\vartheta = \vartheta'$. With the convention that the mirror moves *toward* the incoming wave for $v > 0$, express $\omega'$ and $\cos\vartheta'$ in terms of $v$, $\omega$, and $\cos\vartheta$. What do you get for $\omega'$ and $\cos\vartheta'$ in the limit $v \to c$? Why do you expect $\omega' = \omega$ and $\vartheta' = \pi - \vartheta$ when $v = c\cos\vartheta$? Verify that your expressions confirm this expectation. For which (negative) value of $v$ is $\vartheta' = \frac{1}{2}\pi$?

## Chapter 5

**43** The Schwinger-type Lagrange function for a relativistic particle of mass $m$, in force-free motion, is

$$L = \boldsymbol{p}\cdot\left(\frac{\mathrm{d}\boldsymbol{r}}{\mathrm{d}t} - \boldsymbol{v}\right) + mc\left(c - \sqrt{c^2 - v^2}\right).$$

Show that this gives the familiar nonrelativistic expression for $v \ll c$. Then use the implied relation between $\boldsymbol{v}$ and $\boldsymbol{p}$ to eliminate the velocity $\boldsymbol{v}$ and so find the corresponding Hamilton function $H(\boldsymbol{r},\boldsymbol{p})$. What is the physical meaning of the action $W_{12} = \displaystyle\int_{t_2}^{t_1} \mathrm{d}t\, L$ evaluated for an actual trajectory?

**44** In the Lagrange density $\mathcal{L}_{\text{emf}}$ of (5.2.3), that is: $L_{\text{emf}} = \int (\mathrm{d}\boldsymbol{r})\,\mathcal{L}_{\text{emf}}$, we regard $\boldsymbol{E}$, $\boldsymbol{B}$, $\Phi$, and $\boldsymbol{A}$ as independent fields. Use their known transformation laws to establish how $\mathcal{L}_{\text{emf}}$ responds to infinitesimal Lorentz transformations.

**45** In Section 1.3 of the notes, there is the assertion that charge conservation is related to gauge invariance. What can you say about this matter upon considering the response of $L_{\text{int}}$ of (5.1.40) and $L_{\text{emf}}$ of (5.2.3) to an infinitesimal gauge transformation?

## Chapter 6

**46** The retarded time $t_{\text{ret}}$ in (6.3.3) is a function of position $\boldsymbol{r}$ and time $t$. Show that

$$\frac{\partial}{\partial t}t_{\text{ret}} = \left[1 - \frac{\boldsymbol{n}}{c}\cdot\boldsymbol{V}(t_{\text{ret}})\right]^{-1}$$

and

$$\boldsymbol{\nabla}t_{\text{ret}} = -\frac{\boldsymbol{n}}{c}\left[1 - \frac{\boldsymbol{n}}{c}\cdot\boldsymbol{V}(t_{\text{ret}})\right]^{-1}$$

with

$$\boldsymbol{n} = \frac{\boldsymbol{r}-\boldsymbol{R}(t_{\text{ret}})}{|\boldsymbol{r}-\boldsymbol{R}(t_{\text{ret}})|} \quad \text{and} \quad \boldsymbol{V}(t) = \frac{\mathrm{d}}{\mathrm{d}t}\boldsymbol{R}(t)\,.$$

What can you say about $\left(\frac{1}{c}\frac{\partial t_{\text{ret}}}{\partial t}\right)^2 - (\boldsymbol{\nabla}t_{\text{ret}})^2$?

**47** Charge $e$ moves along the trajectory $\boldsymbol{R}(t)$. Use the Liénard–Wiechert potentials to show that the electric field is given by

$$\boldsymbol{E}(\boldsymbol{r},t) = \frac{e}{D^2}\,\boldsymbol{n} + \frac{eD}{c}\frac{\partial}{\partial t}\frac{\boldsymbol{n}}{D^2} + \frac{e}{c^2}\frac{\partial^2}{\partial t^2}\boldsymbol{n}\,,$$

where $D = |\boldsymbol{r}-\boldsymbol{R}(t_{\text{ret}})|$ is the distance from the retarded position of the charge to the observation point, and $\boldsymbol{n} = [\boldsymbol{r}-\boldsymbol{R}(t_{\text{ret}})]/D$ is the unit vector pointing from the retarded position to the observation point.

**48** Next, show that the magnetic field is given by

$$\boldsymbol{B}(\boldsymbol{r},t) = \frac{e}{cD}\,\boldsymbol{n}\times\frac{\partial}{\partial t}\boldsymbol{n} + \frac{e}{c^2}\,\boldsymbol{n}\times\frac{\partial^2}{\partial t^2}\boldsymbol{n}\,,$$

and verify that $\boldsymbol{B} = \boldsymbol{n}\times\boldsymbol{E}$.

**49** A point charge $e$ is moving on a circle of radius $R$ with constant speed $v$, so that $x(t) = R\cos(vt/R)$, $y(t) = R\sin(vt/R)$, $z(t) = 0$ are the cartesian coordinates of the charge as a function of time $t$. Find the retarded potentials for points on the $z$ axis. Which components of $\boldsymbol{E}(\boldsymbol{r},t)$ and $\boldsymbol{B}(\boldsymbol{r},t)$ can you infer from this limited knowledge of the potentials?

## Chapter 7

**50** For $\boldsymbol{E}(\boldsymbol{r},t)$ and $\boldsymbol{B}(\boldsymbol{r},t)$ of Exercises 47 and 48, identify the radiation fields as

$$\boldsymbol{E}_{\mathrm{rad}} = \frac{e}{c^2 r}\,\boldsymbol{n}\times\left(\boldsymbol{n}\times\frac{\partial^2}{\partial t^2}\boldsymbol{R}(t_{\mathrm{ret}})\right), \qquad \boldsymbol{B}_{\mathrm{rad}} = -\frac{e}{c^2 r}\,\boldsymbol{n}\times\frac{\partial^2}{\partial t^2}\boldsymbol{R}(t_{\mathrm{ret}})\,.$$

Verify that $\boldsymbol{E}_{\mathrm{rad}} = -\boldsymbol{n}\times\boldsymbol{B}_{\mathrm{rad}}$ and $\boldsymbol{B}_{\mathrm{rad}} = \boldsymbol{n}\times\boldsymbol{E}_{\mathrm{rad}}$, and find the power radiated per unit emission time.

**51** A time-dependent electric point dipole $\boldsymbol{d}(t)$, located at $\boldsymbol{r}=0$, has the charge density $\rho(\boldsymbol{r},t) = -\boldsymbol{d}(t)\cdot\boldsymbol{\nabla}\delta(\boldsymbol{r})$. Find the corresponding current density $\boldsymbol{j}(\boldsymbol{r},t)$ and verify that the magnetic dipole moment $\boldsymbol{\mu}(t)$ vanishes for this current density. Then check that $\int(\mathrm{d}\boldsymbol{r})\,\boldsymbol{j}(\boldsymbol{r},t)$ has the correct value, and finally find the vector potential in the Lorentz gauge.

**52** The electric dipole of Exercise 51 is now located at position $\boldsymbol{a}$, so that $\rho(\boldsymbol{r},t) = -\boldsymbol{d}(t)\cdot\boldsymbol{\nabla}\delta(\boldsymbol{r}-\boldsymbol{a})$ and similarly for $\boldsymbol{j}(\boldsymbol{r},t)$. Find the magnetic dipole moment $\boldsymbol{\mu}(t)$ and the electric quadrupole moment $\mathbf{Q}(t)$, and determine the total radiated power $P$. Comment on the $\boldsymbol{a}$ dependence of $P$.

**53** A point dipole $\boldsymbol{d}(t)$ moves along the trajectory $\boldsymbol{R}(t)$, so that the charge density is given by

$$\rho(\boldsymbol{r},t) = -\boldsymbol{d}(t)\cdot\boldsymbol{\nabla}\delta\big(\boldsymbol{r}-\boldsymbol{R}(t)\big)\,.$$

Verify that $\int(\mathrm{d}\boldsymbol{r})\,\rho(\boldsymbol{r},t) = 0$ and $\int(\mathrm{d}\boldsymbol{r})\,\boldsymbol{r}\,\rho(\boldsymbol{r},t) = \boldsymbol{d}(t)$. Then find the corresponding current density $\boldsymbol{j}(\boldsymbol{r},t)$ and express the magnetic dipole moment $\boldsymbol{\mu}(t)$ in terms of $\boldsymbol{d}(t)$ and $\boldsymbol{R}(t)$.

**54** An electric point dipole of constant strength $\big|\boldsymbol{d}(t)\big| = d$ is located at $\boldsymbol{r}=0$ and oriented in the $xy$ plane, and rotates around the $z$ axis with constant angular velocity: $\boldsymbol{e}_z\cdot\boldsymbol{d}(t) = 0$ and $\frac{\mathrm{d}}{\mathrm{d}t}\boldsymbol{d}(t) = \boldsymbol{\omega}\times\boldsymbol{d}(t)$ with $\boldsymbol{\omega} = \omega\boldsymbol{e}_z$. Find the angular distribution of the radiated power and the total radiated power, both averaged over one period of the rotation.

**55** If there is only electric quadrupole radiation, what is the total radiated power?

**56** A thin charged ring of radius $R$ is rotating with constant angular velocity $\omega$ about an axis that is perpendicular to the plane of the ring and goes through the center of the ring. For the ring in the $xy$ plane and rotating about the $z$ axis, we have the charge and current densities

$$\rho(\boldsymbol{r},t) = \frac{e}{2\pi R}\,\delta(z)\,\delta(s-R)\,f(\varphi-\omega t)\,, \quad \boldsymbol{j}(\boldsymbol{r},t) = R\omega\rho(\boldsymbol{r},t)\begin{pmatrix} -\sin\varphi \\ \cos\varphi \\ 0 \end{pmatrix},$$

where $(x,y,z) = (s\cos\varphi, s\sin\varphi, z)$ with $s>0$, and $f(\varphi) = f(\varphi+2\pi)$ is a periodic function of the azimuth $\varphi$. Treating this charge distribution as a small system, consider $f(\varphi)=\cos\varphi$ and determine first the angular distribution of the radiated power, averaged over one period of the circular motion, by taking into account electric dipole radiation, magnetic dipole radiation, and electric quadrupole radiation; then find the total radiated power. Repeat for $f(\varphi)=\cos(2\varphi)$.

**57** A charged point particle (mass $m$, charge $e$) is oscillating harmonically, $\boldsymbol{r}(t) = \boldsymbol{r}_0\sin(\omega_0 t)$. Consider dipole radiation and conclude that $\gamma$, the rate of radiative energy loss, is given by

$$\gamma = \frac{2}{3}\frac{e^2\omega_0^2}{mc^3}\,.$$

Verify that $\gamma \ll \omega_0$ if the particle is an electron and $\omega_0$ is the (circular) frequency of visible light.

**58** For $\gamma \ll \omega_0$, the equation of motion of a damped harmonic oscillator,

$$\frac{\mathrm{d}^2}{\mathrm{d}t^2}\boldsymbol{r}(t) = -\omega_0^2\boldsymbol{r}(t) - \gamma\frac{\mathrm{d}}{\mathrm{d}t}\boldsymbol{r}(t)\,,$$

is approximately solved by $\boldsymbol{r}(t) = \boldsymbol{r}_0\sin(\omega_0 t)\,\mathrm{e}^{-\frac{1}{2}\gamma t}$. Confirm this statement. Then find $E(\omega)$, the energy radiated per unit frequency range. Use a consistent approximation to simplify the expression (single Lorentz peak).

**59** Then calculate the total radiated energy $E_{\mathrm{rad}} = \int \mathrm{d}\omega\, E(\omega)$. How does $E_{\mathrm{rad}}$ compare with the initial oscillator energy $E_{\mathrm{init}} = \frac{1}{2}m\omega_0^2{\boldsymbol{r}_0}^2$ if $\gamma$ has the value found in Exercise 57?

**60** An electron moves at speed $v \ll c$ about an infinitely massive proton, in a circular orbit of radius $r$. Compute the rate of radiation — the rate of energy loss — in terms of $v$ or $r$ or both; then express it in terms of the electron energy $E$. Now integrate the resulting differential equation for $E(t)$ to find the time $T$ it takes an electron to fall into the nucleus of this classically-modeled hydrogen atom, if the initial energy is $E_0$. How long (in seconds) is this collapse time $T$ when $E_0$ is the energy of the first Bohr* radius? What do you conclude from your result?

**61** According to Larmor, a charge $e$ in nonrelativistic accelerated motion, emit radiation with a power of $-\frac{\mathrm{d}E}{\mathrm{d}t} = -\frac{2e^2}{3c^3}\boldsymbol{v}(t)\cdot\frac{\mathrm{d}^2}{\mathrm{d}t^2}\boldsymbol{v}(t)$, here presented as a loss of mechanical energy $E$ of the charged particle. Explain why this suggests a radiation-reaction force

$$\boldsymbol{F}_{\mathrm{rad}} = \frac{2e^2}{3c^3}\frac{\mathrm{d}^2}{\mathrm{d}t^2}\boldsymbol{v}(t) = m\tau\frac{\mathrm{d}^3}{\mathrm{d}t^3}\boldsymbol{r}(t)\,.$$

What is the value of the characteristic time $\tau$ for an electron? For a constant external force $\boldsymbol{F}_{\mathrm{ext}} = m\boldsymbol{a}$, what is the general solution of

$$m\frac{\mathrm{d}}{\mathrm{d}t}\boldsymbol{v} = \boldsymbol{F}_{\mathrm{ext}} + \boldsymbol{F}_{\mathrm{rad}}\,,$$

the so-called Abraham*–Lorentz equation, and what initial condition must be imposed to avoid unphysical run-away solutions?

**62** If the external force in the equation of motion of Exercise 61 is that of a harmonic oscillator, $\boldsymbol{F}_{\mathrm{ext}} = -m\omega_0^2\boldsymbol{r}$ with $\omega_0\tau \ll 1$, what initial condition must be imposed to avoid unphysical run-away solutions? State the second-order differential equation for $\boldsymbol{r}(t)$ that is obeyed by the physical solutions thus selected.

**63** Recall that $p^0 = E/c$ is the time-like component of the momentum 4-vector and that the 4-velocity $u^\mu$ has components $u^0 = \gamma_v c$ and $\boldsymbol{u} = \gamma_v\boldsymbol{v}$ with $\gamma_v = 1/\sqrt{1-(\boldsymbol{v}/c)^2}$. Show that the $\mu = 0$ component of

$$-\frac{\mathrm{d}}{\mathrm{d}s}\left(cp^\mu\right)_{\mathrm{rad}} = \frac{2e^2}{3c^3}\frac{u^\mu}{c}\frac{\mathrm{d}u^\nu}{\mathrm{d}s}\frac{\mathrm{d}u_\nu}{\mathrm{d}s}\,,$$

evaluated in the rest frame of the charge $e$, is Larmor's result about the radiative energy loss, whereby $\frac{\mathrm{d}}{\mathrm{d}s} = \gamma_v\frac{\mathrm{d}}{\mathrm{d}t}$ is the Lorentz-invariant differentiation with respect to the proper time $s$ of the charge. — In passing, we note that the 4-velocity is the proper-time derivative of the 4-position,

*Niels Henrik David BOHR (1885–1962) *Max ABRAHAM (1875–1922)

$u^\mu = \frac{\mathrm{d}}{\mathrm{d}s}x^\mu$, and the $\frac{\mathrm{d}}{\mathrm{d}s}u^\mu$ are the contravariant components of the 4-acceleration.

**64** Now, lift the restriction to the rest frame in order to derive

$$-\left.\frac{\mathrm{d}E}{\mathrm{d}t}\right|_{\mathrm{rad}} = \frac{2e^2}{3c^3}\gamma_v^6\left[\left(\frac{\mathrm{d}\boldsymbol{v}}{\mathrm{d}t}\right)^2 - \left(\frac{\boldsymbol{v}}{c}\times\frac{\mathrm{d}\boldsymbol{v}}{\mathrm{d}t}\right)^2\right],$$

which is the relativistic version of Larmor's energy-loss formula.

**65** Consider charge $e$ accelerated from rest by a constant electric field $\boldsymbol{E}$ in accordance with the relativistic equation of motion

$$\frac{\mathrm{d}}{\mathrm{d}t}(\gamma_v m\boldsymbol{v}) = e\boldsymbol{E}\,,$$

and show that the rate of radiative energy loss, as it follows from the equation in Exercise 64, does *not* change as the energy of the charge increases.

**66** In a symmetric tandem accelerator, one first accelerates $H^-$ ions from rest by a constant electric field of strength $E$ toward a thin foil, which is distance $L$ from the $H^-$ source. When passing through the foil, the $H^-$ ion is stripped of both electrons, and the resulting $H^+$ ion is then accelerated further by a constant electric field of the same strength $E$ until it hits the target that is distance $L$ behind the foil. Ignore the small mass difference between $H^-$ and $H^+$ and employ the relativistic version of Larmor's energy-loss formula to determine the total energy that is radiated during the two periods of constant-force acceleration.

**67** An antenna model that is more realistic than that of Section 7.6 has the current

$$\boldsymbol{j}(\boldsymbol{r},t) = \boldsymbol{e}_z I\cos(\omega t)\delta(x)\delta(y)\cos(\pi z/L)\eta(L^2 - 4z^2)\,.$$

Find the angular distribution of the radiated power, averaged over one period of the oscillation, and discuss its properties. Take a particularly careful look at the situation of a so-called "half-wave antenna," specified by $L = \frac{1}{2}\lambda$.

**68** A "full-wave antenna" is modeled by the electric current density

$$\boldsymbol{j}(\boldsymbol{r},t) = \boldsymbol{e}_z I\cos(\omega t)\delta(x)\delta(y)\sin(2\pi z/\lambda)\eta(\lambda^2 - 4z^2)\,,$$

where the current $I$ and the radial frequency $\omega = 2\pi c/\lambda$ are given constants. Find the angular distribution of the radiated power, averaged over one period of the oscillation, and discuss its properties.

**69** Two half-wave antennas (length $L = \frac{1}{2}\lambda = \pi c/\omega$) are parallel to the $z$ axis at a distance $a > 0$, with their centers at $x = \pm\frac{1}{2}a$, so that the electric current density is given by $\boldsymbol{j}(\boldsymbol{r},t) = \boldsymbol{j}_+(\boldsymbol{r},t) + \boldsymbol{j}_-(\boldsymbol{r},t)$ with

$$\boldsymbol{j}_\pm(\boldsymbol{r},t) = \boldsymbol{e}_z I \cos(\omega t \mp \tfrac{1}{2}\beta)\, \delta(x \mp \tfrac{1}{2}a)\, \delta(y)\, \cos(\pi z/L)\eta(L^2 - 4z^2)\,,$$

where $I$ is the current fed into the antennas at frequency $\omega$, and $\beta$ is the relative phase between the currents in the two antennas. Find the angular distribution of the radiated power, averaged over one period of the oscillation, and then determine $a$ and $\beta$ such that the power radiated in the direction $\boldsymbol{n} = +\boldsymbol{e}_x$ is particularly large and the power radiated in direction $\boldsymbol{n} = -\boldsymbol{e}_x$ is particularly small.

**70** Consider this ring antenna: a thin wire in the shape of a circle with radius $a$ in the $xy$ plane, centered at $\boldsymbol{r} = 0$, that carries a periodic current $I\cos(\omega t)$. The electric current density is

$$\boldsymbol{j}(\boldsymbol{r},t) = I\cos(\omega t)\, \delta(r-a)\, \delta(z)\, \boldsymbol{e}_z \times \frac{\boldsymbol{r}}{r}\,,$$

and we are interested in $\dfrac{\mathrm{d}P}{\mathrm{d}\Omega}$, the angular distribution of the radiated power, averaged over one period. Explain why $a\omega \ll c$ means a "small" antenna? What is $\dfrac{\mathrm{d}P}{\mathrm{d}\Omega}$ in the small-antenna limit? What is the corresponding total power $P = \displaystyle\int \mathrm{d}\Omega\, \frac{\mathrm{d}P}{\mathrm{d}\Omega}$?

**71** Now find $\dfrac{\mathrm{d}P}{\mathrm{d}\Omega}$ for the ring antenna *without* any assumption about the value of $a\omega/c$, sketch the pattern of the angular distribution of the radiation emitted by the ring antenna for $a\omega \ll c$ and $a\omega \gg c$, and describe how the patterns differ.

**72** An odd number $N = 2M + 1$ of identical ring antennas are placed along the $z$ axis, so that their centers are at $z = 0, \pm D, \pm 2D, \ldots, \pm MD$ and each antenna ring is parallel to the $xy$ plane. All antennas have the same radius $a$ and carry the same periodic current $I\cos(\omega t)$. Use the known answer for a single ring antenna to find $\dfrac{\mathrm{d}P}{\mathrm{d}\Omega}$, the angular distribution of the radiated power, averaged over one period, for this array of $N$ antennas. How does the many-antenna radiation pattern differ from the single-antenna pattern?

## Chapter 8

**73** At $t = 0$, point charges $e$ and $-e$ are created at $\boldsymbol{r} = 0$, and then move with constant velocities $\boldsymbol{v}$ and $-\boldsymbol{v}$, respectively. Derive the distribution in frequency and angle of the emitted radiation. Why does your result exhibit an unphysical feature, and what is its origin? Sketch the angular distribution for $v \ll c$ and for $v \lesssim c$. Use your favorite plot program to illustrate the angular distribution for $v/c = 0.0001, 0.01, 0.1, 0.5, 0.9, 0.99, 0.9999$.

**74** In a free-electron laser, electrons (mass $m_{\text{el}}$, charge $-e_0$) are injected at ultrahigh speed into a helical magnetic field, which we approximate by

$$\boldsymbol{B}(\boldsymbol{r}) = \begin{pmatrix} B\cos(k_0 z) \\ B\sin(k_0 z) \\ 0 \end{pmatrix} \quad \text{for} \quad 0 < z < L = \frac{2\pi}{k_0} N$$

and $\boldsymbol{B} = 0$ for $z < 0$ and $z > L$. The winding number $N$ is a large integer. We choose $\boldsymbol{r}(t) = \begin{pmatrix} v_\perp t \\ 0 \\ v_\parallel t \end{pmatrix}$ for $t < 0$ as the initial condition for an isolated electron, and take for granted that $v_\perp \ll v_\parallel \lesssim c$, as would be typical for free-electron-laser operation. Under these circumstances, free-electron-laser radiation is predominantly in the forward direction $\boldsymbol{n} = \boldsymbol{e}_z$. State the equations of motion and show that they are solved by

$$\boldsymbol{v}(t) = \begin{pmatrix} v_\perp \cos\big(k_0 z(t)\big) \\ v_\perp \sin\big(k_0 z(t)\big) \\ v_\parallel \end{pmatrix} \quad \text{for} \quad 0 < t < T = L/v_\parallel$$

in conjunction with $z(t) = v_\parallel t$, provided that the various constants $e_0$, $m_{\text{el}}$, $B$, $k_0$, $v_\perp$, $v_\parallel$, $v = \sqrt{v_\parallel^2 + v_\perp^2}$, $\gamma = 1/\sqrt{1-(v/c)^2}$ obey a certain relation.

**75** Then calculate the angular-spectral distribution $\dfrac{\mathrm{d}E(\omega)}{\mathrm{d}\Omega}$ of the free-electron-laser radiation in the forward direction with the aid of (8.2.5), whereby the time integral for $\boldsymbol{j}(\boldsymbol{k}, \omega)$ covers the period of acceleration.

**76** Armed with the result of Exercise 75, determine the frequency $\omega_{\max}$ for which $\frac{\mathrm{d}E(\omega)}{\mathrm{d}\Omega}$ is maximal, and find this maximal value. How do these quantities depend on the winding number $N$? Find also the smallest value of $\Delta\omega$ such that $\frac{\mathrm{d}E(\omega)}{\mathrm{d}\Omega} = 0$ for $\omega = \omega_{\max} \pm \Delta\omega$. How large is the fractional width $\frac{\Delta\omega}{\omega_{\max}}$ of the free-electron-laser frequency peak?

**77** Charge $e$ is moving with constant velocity $\boldsymbol{v}_0$ until it is stopped by a constant acceleration that lasts for duration $T$. Apply the relativistic version of Larmor's energy-loss formula of Exercise 64 to find the total radiated energy $E_{\mathrm{rad}}$ of the bremsstrahlung emitted. Which simplified expression applies when $v_0 \lesssim c$?

## Chapter 9

**78** An electron moves through water in a tank at a speed $v$ so large that Cherenkov radiation of some frequency is emitted. Which relation, between the velocity vector $\boldsymbol{v}$ of the electron and the normal vector $\boldsymbol{e}_z$ of the surface, must be obeyed so that the Cherenkov radiation can be observed above the water?

**79** A charged particle moves with speed $v$ along the axis of a dielectric cylinder with circular cross section of radius $R$. The speed is so large that Cherenkov radiation of wavelength $\lambda \ll R$ is emitted. What fraction of this radiation passes into the surrounding vacuum through the cylindrical surface?

## Chapter 10

**80** Apply the relativistic version of Larmor's energy-loss formula of Exercise 64 to a charge $e$ that moves with constant speed $v \lesssim c$ on a circular orbit of radius $R$, and thus re-derive the total radiated power of synchrotron radiation.

**81** Show that (9.1.6) is equivalent to (10.2.1).

**82** What strength of magnetic field $B$ [kG] is required to keep a relativistic electron of energy $E$ [GeV] moving in a circle of radius $R$ [m] ? — Determine the wavelength $\lambda_{\text{crit}}$ [Å] that is associated with the critical frequency $\omega_{\text{crit}}$ for a relativistic electron of energy $E$ [GeV] moving in a circle of radius $R$ [m] .

**83** An electron is raised to energy $E$ [GeV], while moving in a circle of radius $R$ [m] . Then the power input is cut off. Assuming that the radius $R$ is maintained constant, how long [ms] will it take for the energy to drop to half of its maximum value?

**84** Evaluate the final $x$ integral in (10.3.13) to confirm the result for the total radiated power of synchrotron radiation.

**85** Express the integral in the last line of (10.6.2) in terms of the Airy* function

$$\mathrm{Ai}(x) = \int_{-\infty}^{\infty} \frac{\mathrm{d}y}{2\pi}\, \mathrm{e}^{ixy}\, \mathrm{e}^{iy^3/3} ,$$

and then use this result to write the large-$m$ approximation of the first summand in (10.4.28) in the form

$$P_m(T)\Big|_{\text{1st term}} \cong \frac{\omega_0}{2\pi}\frac{e^2}{R}\gamma f\left(m/\gamma^3\right) .$$

The $m$-independent function $f(\ )$ is what is plotted in (10.6.9).

**86** Derive the statement for $v/c \ll 1$ in (10.7.22) from the expressions in (10.7.17) and (10.7.18), after first establishing the form of $\mathrm{J}_m(z)$ for $|z| \ll 1$.

**87** Proceed from the angular distribution of synchrotron radiation, derived in Section 10.8, and re-derive the total radiated power of (10.3.15).

## Chapter 11

**88** In Section 11.2 we found the total cross section for Rayleigh scattering. What is the differential cross section for unpolarized incoming light?

*Sir George Bidell AIRY (1801–1892)

**89** Assume $\gamma \ll \omega_0$ and determine the value of $\omega$ for which the Rayleigh cross section (11.2.11) is largest. How does your answer change if $\gamma = \gamma_{\text{rad}}$ and the $\omega$ dependence of $\gamma_{\text{rad}}$ is taken into account?

**90** When a dielectric ball of radius $R$ is exposed to an external homogeneous and constant electric field $\boldsymbol{E}$, it acquires an electric dipole moment $\boldsymbol{d} = \dfrac{\varepsilon - 1}{\varepsilon + 2} R^3 \boldsymbol{E}$, where $\varepsilon$ is the static dielectric constant. Assume that this relation also applies to fields with a slow time dependence, and thus find the cross section for Rayleigh scattering.

## Chapter 12

**91** Proceeding from (12.2.6), show that a large aperture has a differential cross section for diffraction that is given by

$$\frac{\mathrm{d}\sigma}{\mathrm{d}\Omega} = \left(\frac{k}{2\pi}\right)^2 \left| \int_{\text{aperture}} (\mathrm{d}\boldsymbol{r}_\perp)\, \mathrm{e}^{-\mathrm{i}\boldsymbol{k}\cdot\boldsymbol{r}_\perp} \right|^2 \quad \text{with } \boldsymbol{k} = k\boldsymbol{n}.$$

Then argue that

$$\mathrm{d}\Omega = \frac{(\mathrm{d}\boldsymbol{k}_\perp)}{k^2}$$

applies here and use this to demonstrate that the total cross section is simply the area of the aperture, irrespective of its shape.

**92** Two circular apertures, each of radius $a \gg \lambda$ are cut into an infinite conducting thin sheet, with their centers the distance $d > 2a$ apart. Find the differential diffraction cross section and compare it with the cross sections of the individual circular apertures.

**93** All of the $xy$ plane is covered by a thin conducting sheet except for an annulus (a ring-shaped opening) whose borders are two concentric circles with radii $a$ and $b$, $0 < a < b$. A plane wave with wavelength $\lambda$ is normally incident from the $z < 0$ side. The wavelength is short in the sense of $\lambda \ll a$ and $\lambda \ll b - a$. Employ the usual approximations and find the differential diffraction cross section.

**94** As a generalization of the discussion around (12.5.1)–(12.5.5), consider an incident plane wave with an angle of incidence $\theta$,

$$\boldsymbol{E}_{\mathrm{inc}}(\boldsymbol{r}) = \boldsymbol{E}_0\, \mathrm{e}^{\mathrm{i}\boldsymbol{k}\cdot\boldsymbol{r}} \qquad \text{with} \quad \boldsymbol{E}_0 \perp \boldsymbol{k} = \boldsymbol{k}_\perp + k_z \boldsymbol{e}_z\,, \quad k_z = k\cos\theta > 0\,.$$

Write the reflected plane wave as

$$\boldsymbol{E}_{\mathrm{refl}}(\boldsymbol{r}) = \boldsymbol{E}_0'\, \mathrm{e}^{\mathrm{i}\boldsymbol{k}'\cdot\boldsymbol{r}} \qquad \text{with} \quad \boldsymbol{k}'\cdot\boldsymbol{E}_0' = 0 \quad \text{and} \quad |\boldsymbol{k}'| = |\boldsymbol{k}| = k\,,$$

but do not assume any other facts about $\boldsymbol{E}_0'$ and $\boldsymbol{k}'$. Use Maxwell's equations to determine how $\boldsymbol{E}_0'$ and $\boldsymbol{k}'$ are related to $\boldsymbol{E}_0$ and $\boldsymbol{k}$ — do you get what you expect? — and find the induced surface current density $\boldsymbol{K}(\boldsymbol{r}_\perp)$.

**95** Look for Poisson's spot with the help of Huygens's approximation (12.2.2).

**96** Use a standard approximation technique to derive the large-$T$ form of the Fresnel integral $\mathrm{F}(T)$ of (12.6.28), and so establish

$$\left|\frac{E(x,y,a)}{E_0}\right|^2 = \begin{cases} 1 + \dfrac{\sqrt{\lambda a}}{\pi y}\cos\left(\dfrac{\pi y^2}{\lambda a} - \dfrac{3\pi}{4}\right) & \text{for} \quad y \gg \sqrt{\lambda a}\,, \\[2ex] \dfrac{\lambda a}{(2\pi y)^2} & \text{for} \quad -y \gg \sqrt{\lambda a}\,, \end{cases}$$

for the $|y| \gg \sqrt{\lambda a}$ form of the knife-edge diffraction pattern of (12.7.12). Use this for a *simple* estimate of the height of the first maximum outside the geometrical shadow.

## Hints

**1–3** By convention, the gradient $\boldsymbol{\nabla}$ differentiates all functions of $\boldsymbol{r}$ on its right. Exploit the product rule and vector identities such as $\boldsymbol{a}\times(\boldsymbol{b}\times\boldsymbol{c}) = \boldsymbol{b}\,\boldsymbol{a}\cdot\boldsymbol{c} - \boldsymbol{c}\,\boldsymbol{a}\cdot\boldsymbol{b}$ and $\boldsymbol{a}\cdot(\boldsymbol{b}\times\boldsymbol{c}) = (\boldsymbol{a}\times\boldsymbol{b})\cdot\boldsymbol{c}$ to ensure that in each term only one function of $\boldsymbol{r}$ is differentiated.

**4** Differentiate with respect to $\varphi$ or $z$ and sort the terms.

**5** Polar coordinates and Poisson's identity

$$\sum_{m=-\infty}^{\infty} \mathrm{e}^{im\varphi} = 2\pi \sum_{l=-\infty}^{\infty} \delta(\varphi - 2\pi l) = 2\pi\delta^{(2\pi)}(\varphi)\,,$$

where $\delta^{(2\pi)}(\varphi) = \delta^{(2\pi)}(\varphi + 2\pi)$ denotes the $2\pi$-periodic version of Dirac's delta function, are useful. If Poisson's identity is not familiar, derive it as an exercise.

**6** One possibility is to combine the two identities

$$\int\limits_{(2\pi)} \frac{\mathrm{d}\varphi}{2\pi}\,\frac{1}{a + \mathrm{i}b\cos\varphi} = \frac{1}{\sqrt{a^2+b^2}} \qquad \text{and} \qquad \int\limits_0^{\infty} \mathrm{d}k\; \mathrm{e}^{-ka} = \frac{1}{a}$$

for $a > 0$ and $b$ real. If the identities are not familiar, derive them as an exercise; which ranges of complex numbers are permissible for $a$ and $b$?

**7** Cylindrical coordinates are understood: $\boldsymbol{r} = \boldsymbol{e}_x s\cos\phi + \boldsymbol{e}_y s\sin\phi + \boldsymbol{e}_z z = \boldsymbol{r}_\perp + \boldsymbol{e}_z z$ and likewise for $\boldsymbol{r}'$.

**8** Remember that

$$\int (\mathrm{d}\boldsymbol{r})\, f(\boldsymbol{r})\boldsymbol{\nabla}\delta(\boldsymbol{r}-\boldsymbol{r}_0) = -\boldsymbol{\nabla}f(\boldsymbol{r})\Big|_{\boldsymbol{r}=\boldsymbol{r}_0} \qquad \text{and} \qquad \boldsymbol{\nabla}^2\frac{1}{r} = -4\pi\delta(\boldsymbol{r})\,.$$

**9** The identity $\boldsymbol{\nabla}\times\big(\boldsymbol{\nabla}\times\boldsymbol{f}(\boldsymbol{r})\big) = \boldsymbol{\nabla}\boldsymbol{\nabla}\cdot\boldsymbol{f}(\boldsymbol{r}) - \boldsymbol{\nabla}^2\boldsymbol{f}(\boldsymbol{r})$ is useful.

**10, 11** This is a static situation with time-independent charge and current densities,

$$\rho(\boldsymbol{r}) = \frac{e}{4\pi R^2}\delta(r - R)\,,$$

$$\boldsymbol{j}(\boldsymbol{r}) = \boldsymbol{\omega}\times\boldsymbol{r}\rho(\boldsymbol{r},t) = \boldsymbol{\nabla}\times\left[\frac{e}{4\pi R}\boldsymbol{\omega}\eta(R-r)\right].$$

Outside the ball, the resulting fields are those of point dipoles; inside the ball, the electric field vanishes and the magnetic field is constant. It is useful to express the magnetic field, and also the vector potential, in terms of the magnetic moment.

**12** Integration by part, as exemplified by

$$\frac{\partial}{\partial x}\int_a^b \mathrm{d}x'\, g(x-x')f(x') = \int_a^b \mathrm{d}x\, g(x-x')\frac{\partial}{\partial x'}f(x') + g(x-a)f(a) - g(x-b)f(b)\,,$$

together with the continuity equation for electric charge will do it.

**13** Use the continuity equation for electric charge and remember the Poisson equation obeyed by $\Phi$ in the radiation gauge.

**14** First note that $\mathsf{T}\cdot\boldsymbol{E}$ and $\mathsf{T}\cdot\boldsymbol{B}$ are linear combinations of $\boldsymbol{E}$ and $\boldsymbol{B}$, then conclude that the eigenvectors of $\mathsf{T}$ are also such linear combinations. Finally, consider $\mathsf{T}\cdot(\boldsymbol{E}\times\boldsymbol{B})$.

**15** It helps to express $U$, $\boldsymbol{G}$, $\mathsf{T}$ in terms of the complex field $\boldsymbol{F}=\boldsymbol{E}+\mathrm{i}\boldsymbol{B}$.

**16** You need to show that you can find a suitable $\lambda(\boldsymbol{r},t)$ for a gauge transformation that converts a permissible choice of $\Phi$ and $\boldsymbol{A}$, perhaps the potentials in the Lorentz gauge, into new potentials that obey the line gauge.

**17** You need to verify (i) that the gauge condition is obeyed, and (ii) that the differentiation of $\Phi$ and $\boldsymbol{A}$ correctly gives $\boldsymbol{E}$ and $\boldsymbol{B}$. Task (ii) requires Maxwell's equations and identities such as $\boldsymbol{\nabla}\boldsymbol{r}\cdot\boldsymbol{F}=\boldsymbol{F}+\boldsymbol{r}\times(\boldsymbol{\nabla}\times\boldsymbol{F})+\boldsymbol{r}\cdot\boldsymbol{\nabla}\boldsymbol{F}$, which is a special case of the identity in Exercise 1.

**18–20** Note that, for example, $\dfrac{\partial}{\partial t}\delta\bigl(\boldsymbol{r}-\boldsymbol{r}_j(t)\bigr) = -\boldsymbol{v}_j(t)\cdot\boldsymbol{\nabla}\delta\bigl(\boldsymbol{r}-\boldsymbol{r}_j(t)\bigr)$ and remember the Lorentz force. Add $U_{\mathrm{ch}}$ and the energy density of the electromagnetic field to obtain the total energy density, and likewise for the energy current density, the momentum density, the momentum current density, the angular momentum density, and the angular momentum current density.

**21, 22** Two important ingredients are

$$\int\frac{(\mathrm{d}\boldsymbol{r})}{(2\pi)^3}\,\sin(\boldsymbol{k}\cdot\boldsymbol{r}-kct)\sin(\boldsymbol{k}'\cdot\boldsymbol{r}-k'ct) = \frac{1}{2}\delta(\boldsymbol{k}-\boldsymbol{k}')-\frac{1}{2}\delta(\boldsymbol{k}+\boldsymbol{k}')\cos(2kct)$$

and $\boldsymbol{B}(\boldsymbol{k}) \cdot \boldsymbol{B}(-\boldsymbol{k}) = -\boldsymbol{E}(\boldsymbol{k}) \cdot \boldsymbol{E}(-\boldsymbol{k})$. Verify both statements before you use them.

**23** Just exploit the familiar addition theorems for the hyperbolic functions.

**24** Keep all terms that are first-order in $\delta\boldsymbol{v}_1$ or $\delta\boldsymbol{v}_2$ or both, that is: do *not* regard products such as $\delta\boldsymbol{v}_1 \times \delta\boldsymbol{v}_2$ as negligibly small. You should find that the order matters, and the difference is a rotation.

**25** The instantaneous rest frame is the frame of reference in which the particle has zero velocity right now. Answer: $\mathrm{d}s = \sqrt{1-(\boldsymbol{v}(t)/c)^2}\,\mathrm{d}t$, where $\boldsymbol{v}(t)$ is the velocity of the particle in the observer's frame.

**26** Recognize two successive finite Lorentz transformations.

**27** The identity

$$f(x-a) = f(x) - \int \mathrm{d}x' \left[\eta(x-x') - \eta(x-a-x')\right] \frac{\mathrm{d}f(x')}{\mathrm{d}x'}$$

could be useful.

**28** Establish the respective response to infinitesimal Lorentz transformations. You should find that both are 4-scalar fields.

**29, 30** Find the electric field of the charge distribution at rest with the help of a statement in Section 3.6. Then note that $\boldsymbol{B} = (\boldsymbol{v}/c) \times \boldsymbol{E}$ to lowest order in the small velocity $\boldsymbol{v}$, whereas the electric field has no first-order modification (other than the coordinate change resulting from the motion). Why is that so?

**31** For the sketch, remember that the field lines are dense where the field is strong and sparse where the field is weak.

**32** The "very simple argument" observes that $\boldsymbol{B}' = (\boldsymbol{v}/c) \times \boldsymbol{E}'$ if $\boldsymbol{B} = 0$ in (3.5.19)–(3.5.21), which applies here because there is no magnetic field in the common rest frame of the two electrons. For the comparison with the force in that rest frame for a particular choice of $\boldsymbol{a}$, determine the corresponding distance between the two electrons in the rest frame.

**33** Describe the wave by a 4-scalar field such as $\lambda(\boldsymbol{r},t) = \lambda_0 \cos\bigl(\omega(z\cos\vartheta + x\sin\vartheta)/c - \omega t\bigr)$ or similar, and perform a Lorentz transformation.

**34** You may want to look at particular cases first, such as $\boldsymbol{u}$ parallel or perpendicular to $\boldsymbol{v}$, thereby re-deriving familiar statements. Then consider the general case and check that your result is correct for the particular cases.

**35** Note that all observers will agree on the number of periods between two events, $(\boldsymbol{r}_1, t_1)$ and $(\boldsymbol{r}_2, t_2)$, say, in the unprimed coordinates, and conclude that $\boldsymbol{k} \cdot \boldsymbol{r} - \omega t$ is a 4-scalar.

**36** Note that, as a consequence of the antisymmetry of $\varepsilon^{\kappa\lambda\mu\nu}$ and $F_{\mu\nu}$, we have ${}^*F^{01} = F_{23}$ and likewise for the other components of ${}^*F^{\mu\nu}$.

**37** Remember (4.2.14).

**38–41** By construction, $A^\mu(x)$ is a 4-vector field, but you need to verify that it has the correct value in the rest frame where $u^0 = c$ and $\boldsymbol{u} = 0$. Then, after establishing that $\partial^\mu \xi^\nu = g^{\mu\nu} + u^\mu u^\nu/c^2$ and $\partial^\mu f(\xi^\lambda \xi_\lambda) = 2f'(\xi^\lambda \xi_\lambda)\xi^\mu$, it follows that $u^\mu \xi_\mu = 0$ and $\partial^\mu \xi_\mu = 3$, and the verification of the Lorentz-gauge condition is straightforward. The 4-field $F^{\mu\nu}$ is a function of $\xi^\lambda \xi_\lambda$ multiplied by $\xi^\mu u^\nu - u^\mu \xi^\nu$; the 4-current $j^\mu$ is a function of $\xi^\lambda \xi_\lambda$ multiplied by $u^\mu$; together these observations imply the stated form of $F^{\mu\nu} j_\nu$. The "statement in Chapter 4" is (4.2.33).

**42** Use the 4-velocity of the mirror and the 4-wave vectors of the incoming and outgoing plane waves to state the fact that "$\omega = \omega'$ and $\vartheta = \vartheta'$ when $v = 0$" in a Lorentz-invariant form.

**43** Note that $1 - v^2/c^2$ and $1 + p^2/(mc)^2$ are reciprocals of each other, and remember Exercise 25.

**44** You should find that $\mathcal{L}_{\text{emf}}$ is a 4-scalar field.

**45** You should find that $L_{\text{emf}}$ is gauge invariant, and that $L_{\text{int}}$ is gauge invariant if local charge conservation is invoked or, equivalently, that requiring the gauge invariance of $L_{\text{int}}$ implies charge conservation. In the case of $L_{\text{int}}$, gauge invariance means that a gauge transformation amounts to adding a total time derivative to $L_{\text{int}}$, thereby resulting in a physically equivalent Lagrange function.

**46** Establish first that the response of $t_{\text{ret}}$ to infinitesimal changes of $\boldsymbol{r}$ and $t$ is such that

$$\delta t_{\text{ret}} + \frac{1}{c}\boldsymbol{n}\cdot\Big(\delta\boldsymbol{r} - \boldsymbol{V}(t_{\text{ret}})\,\delta t_{\text{ret}}\Big) = \delta t\,;$$

then combine this with $\delta t_{\text{ret}} = \Big(\delta\boldsymbol{r}\cdot\boldsymbol{\nabla} + \delta t\frac{\partial}{\partial t}\Big)t_{\text{ret}}$.

**47, 48** Demonstrate the relations

$$\frac{1}{c}\frac{\partial}{\partial t}D = 1 - \frac{\partial}{\partial t}t_{\text{ret}}\,,$$

$$\frac{\partial}{\partial t}\boldsymbol{n} = \frac{1}{D}\boldsymbol{n}\times\big(\boldsymbol{n}\times\boldsymbol{V}(t_{\text{ret}})\big)\frac{\partial t_{\text{ret}}}{\partial t} = \frac{1}{D}\boldsymbol{n}\times\left(\boldsymbol{n}\times\frac{\partial\boldsymbol{R}(t_{\text{ret}})}{\partial t}\right),$$

$$\boldsymbol{n}\times\frac{\partial}{\partial t}\boldsymbol{n} = -\frac{1}{D}\boldsymbol{n}\times\frac{\partial\boldsymbol{R}(t_{\text{ret}})}{\partial t}\,,$$

and then use them when differentiating

$$\Phi(\boldsymbol{r},t) = \frac{e}{D}\frac{\partial t_{\text{ret}}}{\partial t} = \frac{e/c}{t-t_{\text{ret}}}\frac{\partial t_{\text{ret}}}{\partial t}\,,\qquad \boldsymbol{A}(\boldsymbol{r},t) = \frac{1}{c}\boldsymbol{V}(t_{\text{ret}})\Phi(\boldsymbol{r},t)\,.$$

Alternatively, you can differentiate (6.3.2) and the analogous expression for $\boldsymbol{A}(\boldsymbol{r},t)$ before carrying out the $t'$ integration.

**49** Only $E_z(\boldsymbol{r}=z\boldsymbol{e}_z,t)$ is available; all other components require differentiation with respect to $x$ or $y$, for which knowledge of the potentials on the $z$ axis is not enough. But you can use the expressions of Exercises 47 and 48 to determine all components on the $z$ axis from $t_{\text{ret}} = t - \sqrt{R^2+z^2}/c$.

**50** Remember that $|\boldsymbol{E}|, |\boldsymbol{B}| \propto \dfrac{1}{r}$ for the radiation fields and note that $\dfrac{1}{D} = \dfrac{1}{r} + \dfrac{\boldsymbol{r}}{r^3}\cdot\boldsymbol{R}(t_{\text{ret}}) + \cdots$ .

**51** Infer the current density from the continuity equation (1.2.1). The "correct value" is that of (7.3.2).

**52** Physically, there is only dipole radiation. The apparent dependence on $\boldsymbol{a}$ is an artifact of the approximation and would go away if all contributions are added up, not just those the two dipole terms and the electric quadrupole term.

**53** You should find that the current density has two terms: "velocity times charge density" plus a dipole term similar to that of Exercise 51. The magnetic dipole moment is the sum of two corresponding terms.

**54** Just use the Larmor formulas for electric dipole radiation.

**55** The solid-angle averages,

$$\int \frac{\mathrm{d}\Omega}{4\pi}\, \boldsymbol{n}\cdot\mathbf{A}\cdot\boldsymbol{n} = \frac{1}{3}\mathrm{tr}\{\mathbf{A}\}\,,$$

$$\int \frac{\mathrm{d}\Omega}{4\pi}\, \boldsymbol{n}\cdot\mathbf{A}\cdot\boldsymbol{n}\,\boldsymbol{n}\cdot\mathbf{B}\cdot\boldsymbol{n} = \frac{1}{15}\Big(\mathrm{tr}\{\mathbf{A}\}\,\mathrm{tr}\{\mathbf{B}\} + \mathrm{tr}\{\mathbf{A}\cdot\mathbf{B}\} + \mathrm{tr}\{\mathbf{A}^{\mathrm{T}}\cdot\mathbf{B}\}\Big)\,,$$

which are valid for all $\boldsymbol{n}$-independent dyadics $\mathbf{A}$ and $\mathbf{B}$, could be useful; if they are unfamiliar, derive them as an exercise. You should find $P = \dfrac{1}{540c^5}\Big(3\,\mathrm{tr}\Big\{\dddot{\mathbf{Q}}^2\Big\} - \mathrm{tr}\{\dddot{\mathbf{Q}}\}^2\Big)$.

**56** The magnetic moment does not depend on time, irrespective of the choice for $f(\varphi)$; there is only electric dipole radiation for $f(\varphi) = \cos\varphi$; and there is only electric quadrupole radiation for $f(\varphi) = \cos(2\varphi)$.

**57** Extract $\gamma$ from $P = -\dfrac{\mathrm{d}}{\mathrm{d}t}E = \gamma E$, where $P$ is the radiated power averaged over one period of the harmonic oscillation, and $E = \dfrac{1}{2}m\omega_0^2\boldsymbol{r}_0^2$ is the energy of the harmonic oscillator. The amplitude $\boldsymbol{r}_0$ does not change noticeably during one period if $\gamma \ll \omega_0$, which is the case here, so that the slow time dependence of $\boldsymbol{r}_0$ can be ignored.

**58, 59** The approximate solution actually solves the differential equation with $\omega_0^2$ replaced by $\omega_0^2 + \dfrac{1}{4}\gamma^2$, but the difference between $\omega_0^2$ and $\omega_0^2 + \dfrac{1}{4}\gamma^2$ is negligible. The dominating term in $E(\omega)$, which is the only one that matters, has $(\omega - \omega_0)^2 + \dfrac{1}{4}\gamma^2$ in the denominator. The single Lorentz peak at $\omega = \omega_0$ has the very narrow width $\gamma \ll \omega_0$. Keep this in mind when evaluating the $\omega$ integral.

**60** On such a circular orbit, the force has the constant magnitude of $mv^2/r = e^2/r^2$. As a consequence of the virial theorem, we have $2E_{\mathrm{kin}} = -E_{\mathrm{pot}}$ for the period-averaged kinetic and potential energy. Remember also, that the mechanical energy is negative, $E = E_{\mathrm{kin}} + E_{\mathrm{pot}} < 0$. You should find that $-\dfrac{\mathrm{d}}{\mathrm{d}t}E \propto E^4$. The electron "has fallen into the nucleus" when $E = -\infty$ is reached.

**61** Invoke "power = force $\times$ velocity".

**62** The characteristic polynomial $\lambda^2 = -\omega_0^2 + \tau\lambda^3$ has the approximate solutions $\lambda = \pm \mathrm{i}\omega_0 - \frac{1}{2}\omega_0^2\tau$ and $\lambda = 1/\tau$ if terms of relative size $(\omega_0\tau)^2$ are neglected. The second-order differential equation is the one in Exercise 58.

**63** We have $\boldsymbol{v} = 0$, $\gamma_v = 1$, and $\dfrac{\mathrm{d}}{\mathrm{d}s} = \dfrac{\mathrm{d}}{\mathrm{d}t}$ in the rest frame.

**64** Establish that the acceleration 4-vector is

$$\frac{\mathrm{d}u}{\mathrm{d}s} = \gamma_v^4 \begin{pmatrix} \dfrac{\boldsymbol{v}}{c} \cdot \dot{\boldsymbol{v}} \\ \dot{\boldsymbol{v}} + \dfrac{\boldsymbol{v}}{c} \times \left(\dfrac{\boldsymbol{v}}{c} \times \dot{\boldsymbol{v}}\right) \end{pmatrix}$$

and then proceed.

**65** Note that both $\boldsymbol{v}$ and $\dot{\boldsymbol{v}}$ are parallel to $\boldsymbol{E}$; establish $m\gamma_v^3 \boldsymbol{v} \cdot \dot{\boldsymbol{v}} = e\boldsymbol{E} \cdot \boldsymbol{v}$; and infer that $\gamma_v^6 \dot{\boldsymbol{v}}^2 = (e\boldsymbol{E}/m)^2$. Alternatively, you can solve the equation of motion explicitly and then get $\dot{\boldsymbol{v}}(t)$ by differentiation.

**66** This is the situation of Exercise 65 with the force acting for a certain duration.

**67, 68** Repeat the calculation of Section 7.6 with the necessary changes.

**69** There is an interference term in the radiated power.

**70** In the small-antenna limit, the Larmor formula (7.3.16) applies. Only the magnetic dipole momentum is nonzero for this ring antenna.

**71** Use (7.2.4) and recognize that the integral gives a Bessel function $\mathrm{J}_1(\ )$.

**72** You should find that $\dfrac{\mathrm{d}P}{\mathrm{d}\Omega} = \left(\dfrac{\mathrm{d}P}{\mathrm{d}\Omega}\right)_{1\ \mathrm{ring}} \times$ (interference factor). This is quite analogous to Exercise 69, but now the interference term is a sum of many contributions. Evaluate this sum and find a compact expression.

**73** State the current distribution for this situation and apply (8.2.5). You should find features similar to those discussed in Section 8.4.

**74** The equation of motion is $\dfrac{\mathrm{d}}{\mathrm{d}t}(\gamma_v m_{\mathrm{el}} \boldsymbol{v}) = -e_0 \dfrac{\boldsymbol{v}}{c} \times \boldsymbol{B}$ where $|\boldsymbol{v}|$ and $\gamma_v$ do not change in time. That condition is $\gamma_v v_\perp = -e_0 B/(m_{\mathrm{el}} c k_0)$.

**75** The velocity found in Exercise 74 gives the current, and the observation that $1 - \frac{v_\parallel}{c} \cong \frac{1}{2\gamma_v^2}$ leads to a simplification.

**76** Remember that $\omega > 0$ in $\frac{\mathrm{d}E(\omega)}{\mathrm{d}\Omega}$. You should find that $\Delta\omega \propto \omega_{\mathrm{max}}/N$.

**77** It helps to parameterize the velocity by the rapidity, and the approximation $\cosh\vartheta \cong \frac{1}{2}\,\mathrm{e}^{\vartheta}$, valid for $\vartheta \gg 1$, could be useful.

**78, 79** Remember that total internal reflection could prevent the radiation from getting out.

**80** Recall (10.1.6) where $\boldsymbol{\omega}_0 \perp \boldsymbol{v}$.

**81** As stated in Section 10.2, apply (10.2.2) in (9.1.1) and then repeat the steps from (9.1.1) to (9.1.6).

**82** No hint needed.

**83** The reasoning is similar — in spirit, that is, not in detail — to that of Exercise 60.

**84** The antiderivative of the integral is quite simple.

**85** Note that the integrand in (10.6.2) is an even function of $\varphi$; that the Airy function is real; and that the factor $\varphi$ can be replaced by a suitable differentiation.

**86** Use the generating function (10.4.3) or the integral representation (10.4.20) to find $\mathrm{J}_m(z)$ for $|z| \ll 1$.

**87** Angle $\phi$ in (10.8.5) is the polar angle of a suitable system of spherical coordinates.

**88** Remember that Rayleigh scattering differs from Thomson scattering in a particular, simple way.

**89** This does not need a hint.

**90** This is an application of the Larmor formula (7.3.6).

**91** Scattering is predominantly in the forward direction, where $\cos\theta \cong 1$. Use this insight when combining (12.2.6) and (12.3.8), and also to argue that $k^2 \mathrm{d}\Omega = (\mathrm{d}\boldsymbol{k}_\perp)$ under these circumstances.

**92** Pay attention to the interference terms, which are of a similar nature as those in Exercise 69.

**93** Apply the Huygens approximation and modify (12.3.3) and (12.3.8) fittingly.

**94** We expect $\boldsymbol{k}' = \boldsymbol{k}_\perp - k_z \boldsymbol{e}_z$ and $\boldsymbol{E}_0' = -\boldsymbol{E}_{0\perp} + E_{0z}\boldsymbol{e}_z$; this is what you should find.

**95** Consider the field on the symmetry axis only and employ standard approximations when necessary.

**96** Use integration by parts, similar to the procedure in (7.6.19). The "simple estimate" looks for the smallest $y$ value for which the cosine is maximal.

# Electromagnetic Units

In cgs units, distances are measured in centimeters (cm), masses in grams (g), and lapses of time in seconds (s). All other units are derived from these three basic ones: $\mathrm{cm}^2$ for area, $\mathrm{cm}^3$ for volume, $\mathrm{g/cm}^3$ for mass density, cm/s for velocity, $\mathrm{cm/s}^2$ for acceleration, g cm/s for momentum, and so forth.

Some frequently used cgs units have their own names. The cgs unit of force is the dyne (dyn): $1\,\mathrm{dyn} = 1\,\mathrm{g\,cm\,s^{-2}}$; energies are measured in ergons (erg): $1\,\mathrm{erg} = 1\,\mathrm{dyn\,cm} = 1\,\mathrm{g\,cm^2\,s^{-2}}$; and the unit of pressure is called barye (Ba): $1\,\mathrm{Ba} = 1\,\mathrm{dyn\,cm^{-2}} = 1\,\mathrm{g\,cm^{-1}\,s^{-2}}$.

Electric charges are measured in franklins (Fr), with two point charges of 1 Fr each at a distance of 1 cm repelling one another with an electrostatic force of 1 dyn: $1\,\mathrm{Fr} = 1\,\mathrm{dyn}^{1/2}\,\mathrm{cm} = 1\,\mathrm{g}^{1/2}\,\mathrm{cm}^{3/2}\,\mathrm{s}^{-1}$. Accordingly, the unit of the charge density is $\mathrm{Fr/cm}^3$, that of the electric current is Fr/s, and the electric current density is measured in units of $\mathrm{Fr/(cm^2\,s)}$. Alternative names are also in use, where "stat" is put in front of the corresponding SI unit, such as statcoulomb (statC) for franklin or statampere (statA) for Fr/s.

For the unit of the potentials, the standard name is of this kind: the statvolt (statV). An electric field of unit strength exerts a unit force on a unit charge, so that the unit of the electric field is $\mathrm{statV/cm} = \mathrm{dyn/Fr} = \mathrm{dyn}^{1/2}/\mathrm{cm} = \mathrm{g}^{1/2}\,\mathrm{cm}^{-1/2}\,\mathrm{s}^{-1}$. The same cgs unit is used for the magnetic field, and then it is called gauss (G): $1\,\mathrm{G} = 1\,\mathrm{g}^{1/2}\,\mathrm{cm}^{-1/2}\,\mathrm{s}^{-1}$. Both field-strength units are square roots of the unit for the energy density, $1\,\mathrm{statV/cm} = (1\,\mathrm{erg/cm}^3)^{1/2} = 1\,\mathrm{G}$. For the statvolt itself, it follows that $1\,\mathrm{statV} = 1\,\mathrm{dyn}^{1/2} = 1\,\mathrm{Fr/cm}$, as it should.

When converting cgs units and SI units into each other, one needs to keep in mind that the units for charge and current are not related to meter, kilogram, and second in the SI system; rather, the ampere is a separately defined unit for the electric current, and particular conversion constant are introduced to relate the units for electromagnetic quantities to the mechanical units: $\epsilon_0$ and $\mu_0$, the so-called permittivity and permeability of the vacuum. Their values are $\epsilon_0 = (4\pi)^{-1}(3^*)^{-2}10^{-9}\mathrm{C}^2/(\mathrm{J\,m})$ and $\mu_0 = 4\pi \times 10^{-7}\mathrm{N/A}^2$, where $3^* = 2.997\,924\,580 \simeq 3$ originates in the value of the speed of light, $c = 1/\sqrt{\epsilon_0\mu_0} = 3^* \times 10^8\,\mathrm{m/s}$.

This $3^*$ appears also in the conversion factors in the following brief table that summarizes the matter for some of the more important quantities:

| quantity | cgs unit | SI unit | conversion factor |
|---|---|---|---|
| length | cm | m | $10^2$ |
| mass | g | kg | $10^3$ |
| time | s | s | 1 |
| velocity | $\mathrm{cm\,s^{-1}}$ | $\mathrm{m\,s^{-1}}$ | $10^2$ |
| force | $\mathrm{dyn} = \mathrm{g\,cm\,s^{-2}}$ | N | $10^5$ |
| energy | $\mathrm{erg} = \mathrm{g\,cm^2\,s^{-2}}$ | J | $10^7$ |
| power | $\mathrm{erg\,s^{-1}} = \mathrm{g\,cm^2\,s^{-3}}$ | W | $10^7$ |
| charge | $\mathrm{Fr} = \mathrm{g^{1/2}\,cm^{3/2}\,s^{-1}}$ | C | $3^* \times 10^9$ |
| current | $\mathrm{Fr\,s^{-1}} = \mathrm{g^{1/2}\,cm^{3/2}\,s^{-2}}$ | A | $3^* \times 10^9$ |
| potential | $\mathrm{statV} = \mathrm{g^{1/2}\,cm^{1/2}\,s^{-1}}$ | V | $(3^* \times 10^2)^{-1}$ |
| electric field | $\mathrm{statV\,cm^{-1}} = \mathrm{g^{1/2}\,cm^{-1/2}\,s^{-1}}$ | $\mathrm{V\,m^{-1}}$ | $(3^* \times 10^4)^{-1}$ |
| magnetic field | $\mathrm{G} = \mathrm{g^{1/2}\,cm^{-1/2}\,s^{-1}}$ | T | $10^4$ |

For example, a voltage of about 300 V is a voltage of 1 statV, and a body carrying 1 C of charge carries about $3 \times 10^9$ Fr.

We could state the latter equivalence as $1\,\mathrm{C} \mathrel{\widehat{=}} 3^* \times 10^9\,\mathrm{Fr}$, but to express the relation as an equality we need to account for the differences in the definitions of the unit systems and then arrive at $1\,\mathrm{C}/\sqrt{4\pi\epsilon_0} = 3^* \times 10^9\,\mathrm{Fr}$. Analogous statements exist for other units, among them $1\,\mathrm{V}\sqrt{4\pi\epsilon_0} = (3^* \times 10^2)^{-1}\,\mathrm{statV}$ for the voltage units and $1\,\mathrm{T}\sqrt{4\pi/\mu_0} = 10^4\,\mathrm{G}$ for the magnetic-field units.

These relations among cgs and SI units tell us also how to turn the cgs-versions of electromagnetic equations into the SI-versions, namely by the replacements

$$\rho \to \frac{\rho}{\sqrt{4\pi\epsilon_0}}, \qquad j \to \frac{j}{\sqrt{4\pi\epsilon_0}}$$

for the charge and current densities as well as

$$\boldsymbol{E} \to \sqrt{4\pi\epsilon_0}\,\boldsymbol{E}\,, \qquad \boldsymbol{B} \to \sqrt{4\pi/\mu_0}\,\boldsymbol{B}$$

for the electric and magnetic fields; the similarity with (9.3.9) and (9.3.10) is striking and not accidental. When applied to the Maxwell's equations (1.1.1), these replacements give the SI versions,

$$\begin{aligned} \epsilon_0 \boldsymbol{\nabla} \cdot \boldsymbol{E} &= \rho\,, & \boldsymbol{\nabla} \cdot \boldsymbol{B} &= 0\,, \\ \frac{1}{\mu_0}\boldsymbol{\nabla} \times \boldsymbol{B} - \epsilon_0 \frac{\partial}{\partial t}\boldsymbol{E} &= \boldsymbol{j}\,, & \boldsymbol{\nabla} \times \boldsymbol{E} + \frac{\partial}{\partial t}\boldsymbol{B} &= 0\,. \end{aligned}$$

Other examples are the SI-versions of the Lorentz force density (1.5.1),

$$\boldsymbol{f} = \rho\boldsymbol{E} + \boldsymbol{j} \times \boldsymbol{B}\,,$$

and of the energy density (1.4.14),

$$U = \frac{1}{2}\Big(\epsilon_0 \boldsymbol{E}^2 + \frac{1}{\mu_0}\boldsymbol{B}^2\Big)\,.$$

Note that a magnetic field of unit SI strength (1 T) has an energy density that is $(3^* \times 10^8)^2$ times that of an electric field of unit SI strength (1 V/m). By contrast, the energy densities of magnetic and electric fields of unit cgs strength (1 G and 1 statV/cm) are the same.

# Index

Printed in the United States
By Bookmasters